QUESTION EVALUATION METHODS

QUESTION EVALUATION METHODS

Contributing to the Science of Data Quality

Edited by

Jennifer Madans
Kristen Miller
Aaron Maitland
National Center for Health Statistics
Centers for Disease Prevention
Hyattsville, Maryland

Gordon Willis
Applied Research Program
National Cancer Institute
Bethesda, Maryland

A JOHN WILEY & SONS, INC., PUBLICATION

Published by John Wiley & Sons, Inc., Hoboken, New Jersey.
Published simultaneously in Canada.

For general information on our other products and services or for technical support, please contact our Customer Care Department within the United States at (800) 762-2974, outside the United States at (317) 572-3993 or fax (317) 572-4002.

Wiley also publishes its books in a variety of electronic formats. Some content that appears in print may not be available in electronic formats. For more information about Wiley products, visit our web site at www.wiley.com.

Library of Congress Cataloging-in-Publication Data

Question evaluation methods : contributing to the science of data quality / Jennifer Madans ... [et al.].
        p. cm. – (Wiley series in survey methodology)
    Includes bibliographical references and index.
    ISBN 978-0-470-76948-5 (pbk.)
    1. Questionnaires–Design.   2. Social surveys–Methodology.   3. Social surveys–Evaluation.
    4. Sampling (Statistics)–Evaluation.   I. Madans, Jennifer H.
    HM537.Q45 2011
    301.072'3–dc22   2010048273

Printed in the United States of America

oBook ISBN: 9781118037003
ePDF ISBN: 9781118036983

10 9 8 7 6 5 4 3 2 1

CONTENTS

CONTRIBUTORS

DUANE F. ALWIN, Pennsylvania State University, University Park, Pennsylvania

MARCUS BERZOFSKY, RTI International, Research Triangle Park, North Carolina

PAUL P. BIEMER, RTI International, Research Triangle Park, and University of North Carolina at Chapel Hill, Chapel Hill, North Carolina

JOHNNY BLAIR, Abt Associates, Washington, District of Columbia

FREDERICK G. CONRAD, University of Michigan, Ann Arbor, Michigan

JAMES M. DAHLHAMER, National Center for Health Statistics, Hyattsville, Maryland

THERESA DEMAIO, U.S. Bureau of the Census, Washington, District of Columbia

JENNIFER DYKEMA, University of Wisconsin–Madison, Madison, Wisconsin

JENNIFER EDGAR, Bureau of Labor Statistics, Washington, District of Columbia

FLOYD J. FOWLER, JR., University of Massachusetts, Boston, Massachusetts

JANET A. HARKNESS, University of Nebraska–Lincoln, Lincoln, Nebraska

BRIAN A. HARRIS-KOJETIN, U.S. Office of Management and Budget, Washington, District of Columbia

RON D. HAYS, University of California, Los Angeles, California

TIMOTHY P. JOHNSON, University of Illinois at Chicago, Chicago, Illinois

FRAUKE KREUTER, University of Maryland, College Park, Maryland

JON A. KROSNICK, Stanford University, Palo Alto, California

JENNIFER MADANS, National Center for Health Statistics, Hyattsville, Maryland

AARON MAITLAND, National Center for Health Statistics, Hyattsville, Maryland

BRIAN MEEKINS, Bureau of Labor Statistics, Washington, District of Columbia

KRISTEN MILLER, National Center for Health Statistics, Hyattsville, Maryland

PETER PH. MOHLER, Universität Mannheim, Mannheim, Germany

BRYCE B. REEVE, University of North Carolina at Chapel Hill, Chapel Hill, North Carolina

NORA CATE SCHAEFFER, University of Wisconsin–Madison, Madison, Wisconsin

ALISÚ SCHOUA-GLUSBERG, Research Support Services, Chicago, Illinois

CLYDE TUCKER, Bureau of Labor Statistics, Washington, District of Columbia

GORDON WILLIS, National Cancer Institute, Bethesda, Maryland

STEPHANIE WILLSON, National Center for Health Statistics, Hyattsville, Maryland

PREFACE

The goal of this book is to present an interdisciplinary examination of existing methods of question evaluation. Researchers with various backgrounds are faced with the task of designing and evaluating survey questions. While there are several methods available to meet this objective, the field of question evaluation has been hindered by the lack of consensus on the use of terminology and by the technical difficulty of many of the methods available. We have made a concerted effort to start bridging gaps in the knowledge required to utilize the diversity of methods available. Hence, a major challenge for this book was presenting the material in a format that reaches a wide audience.

The idea for the book grew out of discussion among researchers in federal statistical agencies about the need to bring together leading experts to discuss the strengths and weaknesses of different question evaluation methods. During the fall of 2008, Jennifer Madans, Kristen Miller, and Aaron Maitland from the National Center for Health Statistics (NCHS) and Gordon Willis from the National Cancer Institute (NCI) began contacting experts about their willingness to present papers on particular evaluation methods at a workshop. The response was overwhelmingly in favor of such an event, and several key figures in the field elected to participate.

The Workshop on Question Evaluation Methods (http://www.cdc.gov/qbank) was held at NCHS in Hyattsville, Maryland, October 21–23, 2009. Financial support was provided by NCHS and NCI. The workshop agenda was designed to fulfill the objectives of providing a review of the background of each question evaluation method and of identifying and discussing current issues surrounding each method. Each session included three leading experts on a specific method. The initial presenter wrote a primary paper that outlined the background and current issues on the particular method. This was followed by two papers written in response to the primary paper. The timeline for the papers was developed so that the responding authors had a few months to write a paper in response to the primary papers. Time was set aside for discussion between the presenters and audience members following the presentation of all three papers for each method.

Presentations were made on seven question evaluation methods. This organization of the papers at the workshop serves as an outline for the book. This book begins with a brief introduction, and each of the sessions at the workshop is a section in this book. The sections are as follows:

Section I: Behavior Coding
Section II: Cognitive Interviewing
Section III: Item Response Theory
Section IV: Latent Class Analysis
Section V: Split-Sample Experiments
Section VI: Multitrait-Multimethod Experiments
Section VII: Field-Based Data Methods

Following the workshop, the organizers of the workshop constructed a book manuscript out of the papers. Each organizer served as an editor for some of the sections in the manuscript. Jennifer Madans edited Section II; Kristen Miller edited Section VII; Aaron Maitland edited Sections I, III, and IV; and Gordon Willis edited Sections V and VI.

The editors would like to thank all of the workshop participants for their contributions at the workshop and for including their work in this book. Their expertise and willingness to share that expertise made the workshop both exciting and enjoyable. We also thank Ken Waters of the U.S. Department of Agriculture's Animal and Plant Health Inspection Service for performing the thankless role of moderator. He carried out his tasks in a most constructive way. We are most grateful to the National Center for Health Statistics and the National Cancer Institute for providing the funding for the meeting. Finally, we thank Jacqueline Palmieri and Steven Quigley at Wiley for their guidance throughout the production process.

JENNIFER MADANS
KRISTEN MILLER
AARON MAITLAND
GORDON WILLIS

1 Introduction

JENNIFER MADANS, KRISTEN MILLER,
and AARON MAITLAND
National Center for Health Statistics

GORDON WILLIS
National Cancer Institute

If data are to be used to inform the development and evaluation of policies and programs, they must be viewed as credible, unbiased, and reliable. Legislative frameworks that protect the independence of the federal statistical system and codes of conduct that address the ethical aspects of data collection are crucial for maintaining confidence in the resulting information. Equally important, however, is the ability to demonstrate the quality of the data, and this requires that standards and evaluation criteria be accessible to and endorsed by data producers and users. It is also necessary that the results of quality evaluations based on these standards and criteria be made public. Evaluation results not only provide the user with the critical information needed to determine whether a data source is appropriate for a given objective but can also be used to improve collection methods in general and in specific areas. This will only happen if there is agreement in how information on data quality is obtained and presented. In November 2009, a workshop on Question Evaluation Methods (QEM) was held at the National Center for Health Statistics in Hyattsville, Maryland. The objective of the workshop was to advance the development and use of methods to evaluate questions used on surveys and censuses. This book contains the papers presented at that workshop.

To evaluate data quality it is necessary to address the design of the sample, including how that design was carried out, as well as the measurement characteristics of the estimates derived from the data. Quality indicators related to the sample are well developed and accepted. There are also best practices for reporting these indicators. In the case of surveys based on probability samples, the response rate is the most accepted and reported quality indicator. While recent research has questioned the overreliance on the response rate as an indicator of sample bias, the science base for evaluating sample quality is well developed and, for the most part, information on response rates is routinely provided according to agreed-upon methods. The same cannot be said for the quality of the survey content.

Content is generally evaluated according to the reliability and validity of the measures derived from the data. Quality standards for reliability, while generally available, are not often implemented due to the cost of conducting the necessary data collection. While there has been considerable conceptual work regarding the measurement of validity, translating the concepts into measurable standards has been challenging. There is a need for a critical and creative approach to evaluating the quality of the questions used on surveys and censuses. The survey research community has been developing new methodologies to address this need for question evaluation, and the QEM Workshop showcased this work. Since each evaluation method addresses a different aspect of quality, the methods should be used together. Some methods are good at determining that a problem exists while others are better at determining what the problem actually is, and others contribute by addressing what the impact of the problem will be on survey estimates and the interpretation of those estimates. Important synergies can be obtained if evaluations are planned to include more than one method and if each method builds on the strength of the others. To fully evaluate question quality, it will be necessary to incorporate as many of these methods as possible into evaluation plans. Quality standards addressing how the method should be conducted and how the results are to be reported will need to be developed for each method. This will require careful planning, and commitments must be made at the onset of data collection projects with appropriate funding made available. Evaluations cannot be an afterthought but must be an integral part of data collections.

The most direct use of the results of question evaluations is to improve a targeted data collection. The results can and should be included in the documentation for that data collection so that users will have a better understanding of the magnitude and type of measurement error characterizing the resulting data. This information is needed to determine if a data set is fit for an analytic purpose and to inform the interpretation of results of analysis based on the data. A less common but equally if not more important use is to contribute to the body of knowledge about the specific topic that the question deals with as well as more general guidelines for question development. The results of question evaluations are not only the end product of the questionnaire design stage but should also be considered as data which can be analyzed to address generic issues of question design. For this to be the case, the results need to be made available for analysis to the wider research community, and this requires that there be a place where the results can be easily accessed.

A mechanism is being developed to make question test results available to the wider research community. Q-Bank is an online database that houses science-based reports that evaluate survey questions. Question evaluation reports can be accessed by searching for specific questions that have been evaluated. They can also be accessed by searching question topic, key word, or survey title. (For more information, see http://www.cdc.gov/qbank.) Q-Bank was first developed to provide a mechanism for sharing cognitive test results. Historically, cognitive test findings have not been accessible outside of the organization sponsoring the test and sometimes not even shared within the organization. This resulted in lost knowledge and wasted resources as the same questions were tested repeatedly as if no tests had been done. Lack of access to test results also contributed to a lack of transparency and accountability in data quality evaluations. Q-Bank is not a database of good questions but is a database of test results that empowers data users to be able to evaluate the quality of the information for their own uses. Having the results of evaluations in a central repository can also improve the quality of the evaluations themselves, resulting in the development of a true science of question evaluation. The plan is for Q-Bank to expand beyond cognitive test results to include the results of all question evaluation methods addressed in the workshop.

The QEM workshop provided a forum for comparing question evaluation methods, including behavior coding, cognitive interviewing, field-based data studies, item response theory modeling, latent class analysis, and split-sample experiments. The organizers wanted to engage in an interdisciplinary and cross-method discussion of each method, focusing specifically on each method's strengths, weaknesses, and underlying assumptions. A primary paper followed by two response papers outlined key aspects of a method. This was followed by an in-depth discussion among workgroup participants. Because the primary focus for the workgroup was to actively compare methods, each primary author was asked to address the following topics:

- Description of the method
- How it is generally used and in what circumstances it is selected
- The types of data it produces and how these are analyzed
- How findings are documented
- The theoretical or epistemological assumptions underlying use of the method
- The type of knowledge or insight that the method can give regarding questionnaire functioning
- How problems in questions or sources of response error are characterized
- Ways in which the method might be misused or incorrectly conducted
- The capacity of the method for use in comparative studies, such as multicultural or cross-national evaluations

- How other methods best work in tandem with this method or within a mixed-method design
- Recommendations: Standards that should set as criteria for inclusion of results of this method within Q-Bank

Finally, closing remarks, which were presented by Norman Bradburn, Jennifer Madans, and Robert Groves, reflected on common themes across the papers and the ensuing discussions, and the relevance to federal statistics.

One of the goals for the workshop was to support and acknowledge those doing question evaluation and developing evaluation methodology. Encouragement for this work needs to come not only from the survey community but also from data users. Funders, sponsors, and data users should require that information on question quality (or lack thereof) be made public and that question evaluation be incorporated into the design of any data collection. Data producers need to institutionalize question evaluation and adopt and endorse agreed-upon standards. Data producers need to hold themselves and their peers to these standards as is done with standards for sample design and quality evaluation. Workshops like the QEM provide important venues for sharing information and supporting the importance of question evaluation. More opportunities like this are needed. This volume allows the work presented at the Workshop to be shared with a much wider audience—a key requirement if the field is to grow. Other avenues for publishing results of evaluations and of the development of evaluation methods need to be developed and supported.

PART I
Behavior Coding

2 Coding the Behavior of Interviewers and Respondents to Evaluate Survey Questions

FLOYD J. FOWLER, JR.
University of Massachusetts

2.1 INTRODUCTION

Social surveys rely on respondents' answers to questions as measures of constructs. Whether the target construct is an objective fact, such as age or what someone has done, or a subjective state, such as a mood or an opinion, the goal of the survey methodologist is to maximize the relationship between the answers people give and the "true value" of the construct that is to be measured.

When the survey process involves an interviewer and the process goes in the ideal way, the interviewer first asks the respondent a question exactly as written (so that each respondent is answering the same question). Next, the respondent understands the question in the way the researcher intended. Then the respondent searches his or her memory for the information needed to recall or construct an answer to the question. Finally, the respondent provides an answer in the particular form that the question requires.

Of course, the question-and-answer process does not always go so smoothly. The interviewer may not read the question as written, or the respondent may not understand the question as intended. Additionally, the respondent may not have the information needed to answer the question. The respondent may also be unclear about the form in which to put the answer, or may not be able to fit the answer into the form that the question requires.

Question Evaluation Methods: Contributing to the Science of Data Quality, First Edition.
Edited by Jennifer Madans, Kristen Miller, Aaron Maitland, Gordon Willis.
© 2011 John Wiley & Sons, Inc. Published 2011 by John Wiley & Sons, Inc.

In short, the use of behavior coding to evaluate questions rests on three key premises:

1. Deviations from the ideal question-and-answer process pose a threat to how well answers to questions measure target constructs.
2. The way a question is structured or worded can have a direct effect on how closely the question-and-answer process approximates the ideal.
3. The presence of these problems can be observed or inferred by systematically reviewing the behavior of interviewers and respondents.

Coding interviewer and respondent behavior during survey interviews is now a fairly widespread approach to evaluating survey questions. In this chapter, I review the history of behavior coding, describe the way it is done, summarize some of the evidence for its value, and try to describe the place of behavior coding in the context of alternative approaches to evaluating questions.

2.2 A BRIEF HISTORY

Observing and coding behavior has long been part of the social science study of interactions. Early efforts looked at teacher–pupil, therapist–patient, and (perhaps the most developed and widely used) small group interactions (Bales, 1951). However, the first use of the technique to specifically study survey interviews was probably a series of studies led by Charles Cannell (Cannell et al., 1968).

Cannell was studying the sources of error in reporting in the Health Interview Survey, an ongoing survey of health conducted by the National Center for Health Statistics. He had documented that some respondents were consistently worse reporters than others (Cannell and Fowler, 1965). He had also shown that interviewers played a role in the level of motivation exhibited by the respondents (Cannell and Fowler, 1964). He wanted to find out if he could observe which problems the respondents were having and if he could figure out what the successful interviewers were doing to motivate their respondents to be good reporters. There were no real models to follow, so Cannell created a system *de novo* using a combination of ratings and specific behavior codes. Using a strategy of sampling questions as his unit of observation, he had observers code specific behaviors (i.e., was the question read exactly as worded or did the respondent ask for clarification of the question) for some questions. For others, he had observers rate less specific aspects of what was happening, such as whether or not the respondent appeared anxious or bored.

Variations of this scheme were used in a series of studies designed to understand what was happening when respondents did not report accurately. The results of this work are summarized in Cannell et al. (1977). However, elements of the scheme were subsequently put to new purposes.

One obvious application was to use systematic observations to evaluate interviewers. Cannell and Oksenberg (1988) reported on an adaptation of the scheme for use in monitoring telephone interviewers. However, the use of behavior coding to evaluate questions emerged somewhat serendipitously.

Fowler and Mangione (1990) were studying the effects of different protocols for training and supervising interviewers on the quality of data that were produced. As part of their experiments, they used a variation of Cannell's scheme to measure how well interviewers were doing what they were trained to do. Their measure of error for the study was the extent to which interviewers affected the answers they obtained. In the course of their analyses, they discovered that certain questions were always more subject to interviewer-related error, regardless of how much training the interviewers received. Moreover, they were able to determine that questions that required interviewers to do more probing in order to obtain codable answers were particularly likely to have big interviewer effects (Mangione et al., 1992).

This work led to a concerted study of how to use behavior coding to evaluate survey questions. Questions that had been asked in important national health surveys were selected for study. A test survey was conducted in which all interviews were tape-recorded. An adaptation of the previous coding scheme was used to code the behaviors of interviewers and respondents. The questions that stimulated high rates of behaviors that indicated potential problems were revised in an attempt to address the problems. A new survey was conducted to evaluate how the changes affected the behavior during the interview and the resulting data. There was substantial evidence that the changes in the question wording improved both the interactions (as reflected in the behavior coding) and the quality of data (Oksenberg et al., 1991; Fowler, 1992; Fowler and Cannell, 1996).

That work was probably the main foundation on which behavior coding for use in evaluating questions was based. However, there are a few other research streams that deserve mention.

In Europe in the 1970s, there were researchers, particularly Brenner and van der Zouwen, who also were studying behavior in survey interviews. They observed and coded behavior, and in some cases their observations related specifically to how question form or wording affected behavior. Much of their work was compiled in *Response Behavior in Survey Interviews*, edited by Dijkstra and van der Zouwen (1982).

In the United States, those who do conversational analysis (CA) have focused their attention on the survey process. Conversational analysts usually base their studies on detailed analysis of transcripts of tape-recorded conversations. Suchman and Jordan (1990) wrote an early paper using this approach to question how well standardization is realized in surveys and whether or not it is a plausible or effective way to collect survey data. They observed frequent deviations from standardized interviewing. If respondents resist standardization and interviewers have a hard time doing it, then, they argued, the result may be a reduction in respondent willingness to report accurately; perhaps standardization is a bad idea.

While much of the thrust of CA and related ethnographically oriented work with surveys has been aimed at the overall idea of what survey interview protocols should look like, their studies have also produced insights into, and data about, the characteristics of questions that pose special problems for standardized surveys. A lot of the most relevant work appears in Maynard et al.'s (2002) *Standardization and Tacit Knowledge*.

Both of these streams, plus other work that will be cited, produced evidence that there are important relationships between the form of questions and behavior that can be observed. However, most of the systematic behavior coding schemes used to evaluate questions seem to be primarily traceable back to Cannell's work.

2.3   HOW BEHAVIOR CODING IS DONE

The most common application of behavior coding to the question evaluation process is to integrate it into pretests of survey instruments. If a pretest is done that is designed to largely replicate the protocols planned for a survey, it is relatively easy to add behavior coding to the protocol. Whether interviews are being done in person or on the telephone, respondents are asked for permission to tape-record the interview. In practice, those who have agreed to be interviewed almost always will agree to be tape-recorded. It is, of course, possible to use an observer to try to record behaviors during an interview, rather than making a recording. However, experience has shown that the amount of information that can be coded live is limited, and, of course, the coding cannot be check coded for reliability. Therefore, most researchers use tape recordings.

Specially trained coders then listen to the tapes and code the behaviors of interviewers and respondents. The specific codes that are used vary somewhat, but the following are among the most commonly used approaches:

The unit of observation is the question. The codes refer to any and all of the behavior between the time that a specific question is first asked and the following question is first asked.

Among the most common codes:

1. Did the interviewer read the question initially exactly as worded? Coding options often include:
   (a) Exactly as worded
   (b) Minor modifications that did not change the meaning
   (c) Major modifications that had the potential to change the meaning
   (d) Interrupted by respondent so complete question could not be read
   (e) Did not attempt to read question: confirmed information without reading question

(f) Incorrectly skipped the question

(g) Correctly skipped the question

After the initial reading of the question, a number of codes are focused on what the respondent did in the process of providing a codable answer.

2. Did the respondent ask the interviewer to repeat all or part of the question? This would be coded if it occurred at any point before the next question was asked.

3. Did the respondent ask for clarification of some aspect of the question? This would also be coded if it occurred at any point before the next question was asked.

4. Did the respondent give an inadequate answer, one that did not meet the question objectives?

5. Did the respondent give a qualified answer: one that met the objectives, but with the respondent saying something like "I think" or "maybe" or "my guess would be" that indicates less than complete certainty about whether the answer is correct?

6. Did the respondent say he/she "did not know" the answer?

7. Did the respondent say he/she "did not want to answer"?

8. Did the respondent provide a codable answer that met the question objectives?

Some codes focus only on the respondents' behavior, once the way the question was read has been coded. Others include additional codes of what the interviewer did.

9. Did the interviewer provide clarification of some aspect of the question?

10. Did the interviewer repeat the question, in part or in its entirety?

11. Did the interviewer probe in other ways to try to get a better or more complete answer?

The rationale for coding only respondent behavior is that interviewer behaviors are often tightly tied to the behavior of the respondents. For example, requests for clarification usually result in the interviewer providing clarification. Inadequate answers usually result in the interviewer repeating the question or probing in some other way. Thus, one does not get a lot of new information from the coding of the interviewer beyond the way the question was asked. If one was interested in more detail about the interviewers' behavior (e.g., whether the probe was directive or nondirective), that would be a reason for coding the interviewers' behavior more specifically. However, if the main goal is to identify questions that require extra activity in order to produce answers, coding either interviewer or respondent behavior will often produce similar results.

In some ways, the easiest approach is to code whether each of these behaviors happened even once when a particular question was asked. Some of these behaviors, such as asking for clarification or probing, could happen multiple times for the same question. An alternative is to code how many times each of the behaviors occurred.

Depending on which approach is taken, the results can be tabulated either as:

1. The percentage of the times the question was asked that a particular behavior occurred; or
2. The rate per question that a particular behavior occurred (number of behaviors/number of times question was asked).

A reason that the first approach is often preferred is to dampen the effect of one respondent who has a particularly difficult time with a question, and there are numerous exchanges with the interviewer before an answer is given. When that happens, it can result in a lot of behaviors, which in turn gives too much weight to the complicated interviews in the overall tabulation of results if one uses approach 2 rather than approach 1 above.

The output from behavior coding is a table in which the percentage of times the question was asked that a particular behavior occurred is tabulated for each question being tested. It might look something like Table 2.1.

Note that this particular example only displays a subset of the codes listed above. Normally all of the data available would be tabulated and displayed in some form, but the codes in this table are among the most useful. The number of items coded and displayed varies greatly by researcher. Also, van der

TABLE 2.1. Percentage of Times Question Was Asked that Each Behavior Occurred at Least Once, by Question

Question	Major Change in Wording	Question Reading Interrupted	R Asks for Clarification	R Gives Inadequate Answer
1 How many children do you have? Do not count step children.	2%	15%	22%	5%
2 Thinking about all aspects of the way you feel and what you can do and what you have been told by doctors, how would you rate your health—excellent, good, fair, or poor.	25%	1%	4%	3%
3 When did you move to New York?	4%	0%	20%	33%

Zouwen and Smit (2004) who has done a lot of work on behavior coding has a scheme for coding sequences, not just individual behaviors. However, the above table is reasonably representative of the way behavior coding is usually displayed for question evaluation purposes.

The next step is to interpret the results and determine the implications for question wording. The first issue is to determine when an observed pattern of behavior reflects a problem with question wording that should be addressed. The results in the table above are typical in that some deviations from a perfectly smooth interaction occur for all the questions and for most of the behavior coding measures. Behaviors that occur at low rates do not warrant attention, but there is an issue in deciding when a behavior suggests a question wording issue is worthy of attention. A guideline that is often used is that when any of the problem behaviors occurs in 15% of the interviews, it at least warrants some effort to figure out why it is occurring; requests for clarification, which occur less often than some of the other behaviors, in excess of 10% may warrant attention. However, there is obviously a degree of arbitrariness in any such guideline.

The tabulations above display a number of behaviors that occurred at high enough rates to warrant further study; the patterns and likely problems are different for the three questions. However, behavior coding by itself does not tell us what the problem is or how to fix it.

There are several ways to gain insights into what aspects of questions may be causing problems:

1. If the questions have had some kind of cognitive testing before the behavior coding, there may be insights from that testing that are reflected in the problems.
2. Debriefing interviewers who carried out the interviews for their thoughts about what is causing the problems.
3. Those who coded the behaviors can be debriefed.
4. If the interviews were recorded in a digital form, it may be possible to play back the interactions around certain questions for all interviews or for those where the behaviors of interest occurred.
5. There is a growing body of principles about how question form affects these behaviors that can be drawn on.

For the three questions above, I will draw on some common observations about question issues and behavior that constitute likely hypotheses about what is happening.

Question 1. The high rate of interruptions is typical of questions that include explanatory material after the question itself has been asked. The reason for the requests for clarification is less obvious, but may have to do with which children should be included (e.g., include only minor children or children of all ages; only children living at home or all children?)

Question 2. The lengthy introduction which does not have much to do with answering the question seems to be a likely reason to explain why interviewers are not reading the question as worded.

Question 3. Requests for clarification could easily be spurred by ambiguity about what is being asked about: New York State, somewhere in the New York City area, or New York City itself. The inadequate answers no doubt are caused by the fact that the question does not tell respondents how to answer the question: is an exact year wanted, a number of years ago? Since the kind of answer is not specified, it is not surprising that there are a lot of answers that are not what the researcher was looking for.

Although the evidence regarding interruptions for question 1 is well established and is almost certainly correct, as is the analysis of the source of the inadequate answers in question 3, most researchers would want further support from interviewers, coders, or listening to the tapes to confirm the source of the other behavior coding results.

A final point about behavior coding schemes: Researchers have a trade-off to make between detail and efficiency. The protocol outlined above focuses on a fairly small number of easily coded behaviors and simply codes whether or not they occurred. It does not require transcription of the interviews. It requires another step to gather details about the problems that are identified. It can be done quickly and at low cost, which is what most survey organizations would value for routine use.

More elaborate protocols can provide more information. Naturally, the more information coded, the greater the time and effort required. Researchers have to decide how to balance the amount of detail coders record and the value of the added effort.

2.4   EVIDENCE FOR THE SIGNIFICANCE OF BEHAVIOR CODING RESULTS

The first question to be asked is whether behavior coding results are reliably linked to the characteristics of questions. If that was not the case, then behavior coding would not be a reliable way to identify question problems. Evidence for the reliability of behavior coding results come from a study in which the same interview schedule was administered by two survey organizations in parallel, and the interviews were behavior coded. The researchers then correlated the rates of various behaviors observed in the two organizations for each question. The rates at which questions were misread, question reading was interrupted, clarification was requested, and inadequate answers were given correlated from 0.6 to 0.8. If a question was frequently misread by Staff A, the interviewers in Staff B were very likely to misread that same question at a similar rate. Thus, we can conclude that the design of questions has a consistent, predictable effect on the behaviors of interviewers and respondents when it is asked in an interview (Fowler and Cannell, 1996).

There are three kinds of studies of behavior coding and questions. The most common simply links observed behaviors to the characteristics of questions. The second links observed behaviors to interviewer-related error. The third links observed behaviors to the "validity" of estimates from surveys.

Behavior-wording links are among the best documented. A few examples:

1. Providing definitions or other explanatory material after a question has been asked leads to a high rate of interrupted question reading (Fowler and Mangione, 1990; Oksenberg et al., 1991; Houtcoop-Steenstra, 2000, 2002).

2. Questions worded so that it is not clear that a set of response alternatives is going to be offered, from which the respondent must choose, also creates interruptions (e.g., van der Zouwen and Dijkstra, 1995).

3. Inadequate answers, and the resulting need for interviewers to probe, is most commonly associated with questions that do not clearly specify how to answer the question; that is, what kind of answer will meet the question's objectives (Fowler and Mangione, 1990).

4. Questions that require interviewers to code answers given in narrative form into fixed categories (field coding) are likely to require probing, particularly directive probing (Houtcoop-Steenstra, 2000).

5. Lengthier questions, particularly with awkward sentence structures, are likely to be misread (Oksenberg et al., 1991).

6. It seems tautological to say that questions that contain poorly defined terms or concepts are likely to produce requests for clarification, but it has been shown that defining the apparently unclear terms leads to a decrease in such requests (Fowler, 1992).

Researchers have also measured response latency, the time from when the reading of the question is complete to the time the respondent proffers an answer. That time seems to be related to the cognitive complexity of the question (Bassili, 1996).

While these are typical of the kinds of findings that have been reported, they are by no means exhaustive. However, the main point is that it is well documented that characteristics of questions can have predictable, observable effects on the behaviors of interviewers and respondents.

Interviewer-related error is one of the ways that the effect of question asking and answering has been evaluated. The "true value" for an answer should not be associated with who asks the question. To the extent that answers are related to the interviewer, it is obvious evidence of error.

The clearest finding on this topic is that questions that require interviewers to probe in order to obtain an adequate answer produce significantly more interviewer-related error (Fowler and Mangione, 1990; Mangione et al., 1992). Since behavior coding is a highly reliable way to identify questions that require interviewer probing, the potential to reduce interviewer-related error through reducing the need for interviewers to probe is one of the most straightforward and well-documented ways that behavior coding can help reduce total survey error. Not telling respondents enough about how they are supposed to answer the question is the most common characteristic of questions that require a lot of probing.

In contrast, studies have failed to find a relationship between interviewer-related error and misreading the questions (e.g., Groves and Magilavy, 1980; Mangione et al., 1992).

A few direct studies of validity have been done by comparing answers to survey questions with some independent external data. Once again, relating how well questions are read to validity of answers does not yield evidence supporting the hypothesis that misreading leads to invalidity (Dykema et al., 1997). However, there is evidence that qualified answers and response latency are linked to the accuracy of answers (Dykema et al., 1997; Mathiowetz, 1998; Draisma and Dijkstra, 2004). Dykema et al. (1997) also found that a composite measure reflecting whether respondents exhibited any of several problem behaviors was indicative of answers that were not accurate.

Validity has also been inferred from the effects on the resulting data of changing questions to address problems uncovered by behavior coding. For example, Fowler (1992) presents several examples from split-ballot studies in which possible question problems were identified via behavior coding. Questions were revised to address the problems. The revised questions not only reduced problematic behaviors, such as requests for clarification and inadequate answers, but they also produced changes in the resulting data that were consistent with hypotheses that they were more valid. Similar results were also reported in Fowler (2004).

2.5   STRENGTHS AND LIMITATIONS OF BEHAVIOR CODING

Behavior coding is a low-cost add-on to a pretest and/or ongoing interviewer-administered survey that provides useful information about characteristics of the questions that may affect the quality of survey data. The existence of stable relationships between certain features of questions and how the question-and-answer process is carried out is clearly established. The information about behavior can be indicative of potential error in surveys in two different ways:

1. It appears that questions that routinely require interviewers to probe in order to obtain adequate answers may be distinctively associated with interviewer-related error. Because the question interferes with the ideal standardized administration of the interview, the interviewer behaviors that result themselves cause error in the data.

2. Behavior coding can also provide suggestions that a question is problematic for respondents to answer. When questions frequently require clarification, for example, or cause respondents to either take a long time to answer them or provide qualified answers, it is a likely sign that respondents are having trouble either understanding what is called for or providing the answer that they think is required. In such cases, identifying why respondents are having trouble and improving the question is likely to improve the validity of the resulting data.

An additional attractive feature of behavior coding is that it provides objective, quantitative data. In contrast, cognitive interviewing, perhaps the most commonly used approach to evaluating a question, depends heavily on the judgments of the interviewers and often involves relatively small numbers of respondents. Results from debriefing pretest interviewers are usually even more subjective and less systematic.

Finally, behavior coding provides evidence of how questions perform under realistic conditions, generally with representative samples of respondents and interviewers—a contrast with some of the other question evaluation techniques.

The most frequently cited limitation of behavior coding is that the results themselves do not tell us what the problem is. While some of the generalizations, such as those presented earlier in this chapter, provide researchers with good ideas about the likely causes of noteworthy behavior coding issues, there is still an imperfect diagnostic process that is necessary.

Second, behavior coding does not identify all problems with questions. In particular, many respondents answer questions that include ambiguous or confusing concepts without showing any evidence that they do not really understand the question as the researcher intended. A favorite example: "Did you eat breakfast this morning?" Testing shows that people have widely varying ideas about what constitutes breakfast, but this question is routinely asked and answered with no indication in the behavior of interviewers or respondents that there is a comprehension problem.

Third, some of the "problems" identified in behavior coding do not have much or any effect on the resulting data. The example noted above is that questions that interviewers misread have not been linked with increased risk of response error. Of course, there is an extensive literature showing that the details of the way questions are worded affects responses (e.g., Schuman and Presser, 1981). Knowing how questions are worded is fundamental to our confidence in being able to interpret the data and in our ability to replicate our studies. On those grounds alone, it would seem worth fixing questions that interviewers find difficult to read. Moreover, we do not have a wealth of good studies with validating data that we can use to critically evaluate the importance of some of the problems identified with behavior coding. Nonetheless, we have to say that the uncertain relationship between behavior coding findings and survey error in some cases constitutes a limitation of the method.

2.6   THE ROLE OF BEHAVIOR CODING IN QUESTION EVALUATION PROTOCOLS

Before starting to use a survey instrument under realistic data collection conditions, expert review, question appraisal protocols (e.g., Lessler and Forsyth, 1996; Fowler and Cosenza, 2008), focus groups, and cognitive interviewing (e.g., Willis, 2005) may be used to help learn how to word questions and to provide a preliminary assessment of proposed questions. The expert reviews and appraisals can apply generalizations from existing literature to flag issues that have been shown to potentially affect the usability of a question or the resulting data. Focus groups are excellent ways to examine vocabulary issues (how people understand words and what words they use) and what people know and can report. Cognitive interviews are the best way to find out how people understand questions and the extent to which they can provide answers that meet question objectives. Behavior coding of pretest interviews is no substitute for any of these activities.

However, behavior coding can substantially add to the value of a field pretest designed to learn how a near-final survey instrument works in the real world. Debriefing interviewers has long been a standard part of pretests, but interviewers cannot replicate what behavior coding provides. They can report what they find to be problematic, particularly about usability, but it turns out they cannot even report reliably about whether or not they can read the questions as worded. Furthermore, the quantitative nature of the coding tends to make the results of behavior coding more reliable and meaningful to users than interviewers' recollections and qualitative opinions.

The fact that behavior coding does not depend on human judgment makes it especially appealing as an approach to evaluating survey instruments across languages for cross-cultural studies. Its quantitative output also permits comparison of how well questions are working in the various languages in which they are being administered.

The fact that behavior coding results are quantitative and reliable also makes them a strong candidate for routine use in a data bank, such as Q-Bank. An issue would be exactly which results one would want routinely reported. At this point, however, one could probably pick four or five results (e.g., % interrupted, % read exactly, % requests of clarification, and % had 1+ inadequate answers) that would provide a meaningful and reliable profile of how well a question works according to a standardized measure under realistic survey conditions.

In terms of how behavior coding compares and contrasts with the results of other techniques, it can be best thought of as complementary. Studies comparing the problems found through behavior coding with those identified by expert ratings or cognitive testing show some overlap, but each provides some unique results as well (Presser and Blair, 1994; Forsyth et al., 2004). Mainly, none of the other methods provides the same kind of evidence about how questions perform under realistic conditions and, in particular, on the rates at which important interviewer and respondent behaviors are affected by the characteristics of the questions. Two of the most important unique contributions of behavior coding to question evaluation are that it uniquely provides information on how often questions are read as written and can be answered without extra interviewer probing.

2.7 CONCLUSION

An ideal question evaluation protocol should probably include both cognitive testing and behavior coding. The former provides information about how questions are dealt with cognitively, which behavior coding cannot do, while behavior coding provides information about how the question-and-answer process proceeds under realistic conditions, which cannot be addressed by cognitive testing. If a pretest is going to be done, it makes little sense not to collect systematic, reliable data to help identify those questions that interviewers find hard to ask or respondents find hard to answer.

REFERENCES

Bales RF (1951). Interaction Process Analysis. Cambridge, MA: Addison-Wesley.

Bassili JN (1996). The how and why of response latency measurement in telephone surveys. In: Schwarz NA, Sudman S, editors. Answering Questions. San Francisco, CA: Jossey-Bass; pp. 319–346.

Cannell CF, Fowler FJ (1964). A note on interviewer effects on self-enumerative procedures. American Sociological Review; 29:269.

Cannell CF, Fowler FJ (1965). Comparison of Hospitalization Reporting in Three Survey Procedures. Washington, DC: U.S. Department of Health, Education and Welfare, Public Health Service.

Cannell C, Oksenberg L (1988). Observation of behaviour in telephone interviewers. In: Groves RM, Biemer PN, Lyberg LE, Massey JT, Nichols WL II, Waksberg J, editors. Telephone Survey Methodology. New York: John Wiley; pp. 475–495.

Cannell CF, Fowler FJ, Marquis K (1968). The Influence of Interviewer and Respondent Psychological and Behavioral Variables on the Reporting in Household Interviews. Washington, DC: U.S. Department of Health, Education and Welfare, Public Health Service.

Cannell C, Marquis K, Laurent A (1977). A summary of studies. In: Vital & Health Statistics, Series 2. Washington, DC: Government Printing Office; p. 69.

Dijkstra W, van der Zouwen J, editors (1982). Response Behavior in the Survey Interview. London: Academic Press.

Draisma S, Dijkstra W (2004). Response latency and (para) linguistic expressions as indicators of response error. In: Presser S, Rothgeb JM, Couper MP, Lessler JT, Martin E, Martin J, Singer E, editors. Methods for Testing and Evaluating Survey Questionnaires. New York: Wiley; pp. 131–148.

Dykema J, Lepkowski J, Blixt S (1997). The effect of interviewer and respondent behavior on data quality. In: Lyberg LE, Biemer P, Collins M, de Leeuw ED, Dippos C, Schwarz N, Trewin D, editors. Survey Measurement and Process Quality. New York: Wiley; pp. 287–310.

Forsyth BH, Rothgeb J, Willis G (2004). Does pretesting make a difference? Presser S, Rothgeb JM, Couper MP, Lessler JT, Martin E, Martin J, Singer E, editors. Questionnaire Development Evaluation and Testing Methods. New York: John Wiley; pp. 525–546.

Fowler FJ (1992). How unclear terms affect survey data. Public Opinion Quarterly; 56:218–231.

Fowler FJ (2004). The case for more split-ballot experiments in developing survey instruments. In: Presser S, Rothgeb JM, Couper MP, Lessler JT, Martin E, Martin J, Singer E, editors. Methods for Testing and Evaluating Survey Questionnaires. New York: Wiley; pp. 173–188.

Fowler FJ, Cannell CF (1996). Using behavioral coding to identify cognitive problems with survey questions. In: Schwarz NA, Sudman S, editors. Answering Questions. San Francisco, CA: Jossey-Bass; pp. 15–36.

Fowler FJ, Cosenza C (2008). Writing effective survey questions. In: Hox J, de Leeuw E, Dillman D, editors. The International Handbook of Survey Methodology. New York: Erlbaum; pp. 136–260.

Fowler FJ, Mangione TW (1990). Standardized Survey Interviewing: Minimizing Interviewer Related Error. Newbury Park, CA: Sage.

Groves RM, Magilavy LJ (1980). Estimates of interviewer variance in telephone surveys. In: American Statistical Association 1980 Proceedings of the Section on Survey Research Methods. Washington, DC: American Statistical Association; pp. 622–627.

Houtcoop-Steenstra H (2000). Interaction and the Standardized Survey Interview: The Living Questionnaire. Cambridge: Cambridge University Press.

Houtcoop-Steenstra H (2002). Question turn format and turn-taking problems in standardized interviewing. In: Maynard DW, Houtcoop-Steenstra H, Schaeffer NC, van der Zouwen J, editors. Standardization and Tacit Knowledge. New York: Wiley; pp. 243–260.

Lessler JT, Forsyth BH (1996). A coding system for appraising questionnaires. In: Schwartz NA, Sudman S, editors. Answering Questions. San Francisco, CA: Jossey-Bass; pp. 259–292.

Mangione TW, Fowler FJ, Louis TA (1992). Question characteristics and interviewer effects. Journal of Official Statistics; 8(3):293–307.

Mathiowetz NA (1998). Respondent expressions of uncertainty: data source for imputation. Public Opinion Quarterly; 62:47–56.

Maynard DW, Houtcoop-Steenstra H, Schaeffer NC, van der Zouwen J (2002). Standardization and Tacit Knowledge. New York: Wiley.

Oksenberg L, Cannell C, Kalton G (1991). New strategies of pretesting survey questions. Journal of Official Statistics; 7(3):349–366.

Presser S, Blair J (1994). Survey pretesting: do different methods produce different results? Marsden P, editors. Sociological Methodology. San Francisco, CA: Jossey-Bass; pp. 73–104.

Schuman H, Presser S (1981). Question and Answers in Attitude Surveys. New York: Academic Press.

Suchman L, Jordan B (1990). Interactional troubles in face-to-face survey interviews. Journal of the American Statistical Association; 85:232–241.

van der Zouwen J, Dijkstra W (1995). Trivial and non-trivial question answer sequences: types, determinants and effects on data quality. In: Proceedings, International Conference on Survey Measurement and Process Quality in Bristol, UK. Alexandria, VA: American Statistical Association; pp. 81–86.

van der Zouwen J, Smit JH (2004). Evaluating survey questions by analyzing patterns of behavior codes and question-answer sequences. In: Presser S, Rothgeb JM, Couper MP, Lessler JT, Martin E, Martin J, Singer E, editors. Methods for Testing and Evaluating Survey Questionnaires. New York: Wiley; pp. 109–130.

Willis GB (2005). Cognitive Interviewing. Thousand Oaks, CA: Sage.

3 Response 1 to Fowler's Chapter: Coding the Behavior of Interviewers and Respondents to Evaluate Survey Questions

NORA CATE SCHAEFFER and JENNIFER DYKEMA
University of Wisconsin–Madison

3.1 WHY DO WE STUDY INTERACTION IN THE SURVEY INTERVIEW?

When Johannes van der Zouwen (2002) posed this question several years ago, he answered that an analysis of interaction can tell us what problems the interactants have in performing their task, how they try to overcome those problems, and what actions of interviewers affect the data. Here we examine a more restricted version of that question, why do we study interaction in the survey interview if our goal is to evaluate survey questions? Fowler (this volume) reviews the history of and contributions made by studies that pursue that goal. Such studies have offered practical recommendations to question writers (e.g., place definitions before the target question), and have also shown links between the behavior of interviewers and respondents and data quality. In this chapter we complement Fowler's discussion by formulating more formally than has been done previously the links among question characteristics, behavior, and measurement.

One general answer to Van der Zouwen's question is that we think behavior in the interview might indicate something useful about measurement. At the least, observing the interviewer's behavior informs us about how standardized he/she is, and observing the respondent's behavior provides clues about his/her cognitive processing. Because we expect both the level of standardization

Question Evaluation Methods: Contributing to the Science of Data Quality, First Edition.
Edited by Jennifer Madans, Kristen Miller, Aaron Maitland, Gordon Willis.
© 2011 John Wiley & Sons, Inc. Published 2011 by John Wiley & Sons, Inc.

and the respondent's cognitive processing to affect the quality of measurement, the behavior of the participants should be associated with the reliability and validity of the resulting data. The characteristics of survey questions should also affect the interaction, because survey questions affect the interviewer's behavior and the cognitive processing of the respondent, which affect the respondent's behavior.

We first discuss a conceptual model of the relationship among question characteristics, the behavior of interviewers and respondents, and measurement quality. To help us assess how studying behavior might contribute to improving survey measurement, we summarize findings from several studies that examine how the behavior of interviewers and respondents is associated with the reliability and validity of data. Next, we highlight conversation analysis and its potential contributions to identifying behaviors that might be useful in evaluating survey questions. Finally, we comment on the question characteristics that affect survey data.

3.2 QUESTIONS, BEHAVIOR, AND QUALITY OF MEASUREMENT: A CONCEPTUAL MODEL

Although interaction coding has been used to study survey questions (see, e.g., Fowler, this volume; Van der Zouwen and Dijkstra, 2002; Van der Zouwen and Smit, 2004), the conceptual model underlying this strategy has not been made explicit.[1] Figure 3.1 provides a simplified view of one portion of the process of answering a survey question. Because our goal is to evaluate survey questions, we ask how the question characteristics might be linked to the behavior of interviewers or respondents, and then to indicators of data quality.

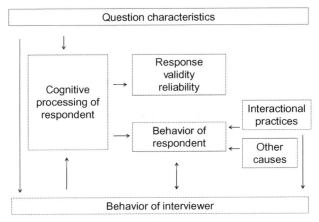

FIGURE 3.1. Conceptual model of the relationship among question characteristics, behavior, and measurement.

One route is that question characteristics affect the interviewer's behavior.[2] The interviewer's behavior (e.g., how she reads the question) affects the respondent's cognitive processing of the question, but it may also affect the respondent's behavior in other ways (e.g., the respondent may mirror the interviewer's pace in ways that reflect interactional rather than cognitive practices). Question characteristics also affect the respondent's cognitive processing directly; for example, the respondent may process a rating scale with numeric category labels differently than a rating scale with verbal category labels. The respondent's cognitive processing has two outcomes of interest here: a response (evaluated by estimating reliability and validity) and other observable behaviors (e.g., disfluencies, repairs, and use of mitigators). Thus, we think of these other behaviors of the respondent as potentially associated with the quality of the answer, in part because both the answer and the respondent's behavior are influenced by the respondent's cognitive processing of the question as presented by the interviewer, and because they influence the subsequent behavior of the interviewer (Dijkstra and Ongena, 2006).

Although the model in Figure 3.1 refers to behavior, much of the model is fundamentally psychological because the key intervening variable is the respondent's cognitive processing of the survey question. This approach ignores the many ways in which conversational practices, rather than psychological processes, shape behavior. In fact, the respondent's behavior is subject to influences other than that of the process of answering the question, including conversational and interactional practices. For example, a respondent may preface an answer with a mitigator such as "about" because it is one of a set of conversational practices that signal that an answer is an approximation. "About" thus may convey information about the cognitive processes used to produce the answer (estimation) or the respondent's evaluation of its accuracy (an approximation), but perhaps not in the same way for all respondents. Other behaviors of the respondent, for example tokens such as "hmm," may convey information about measurement that is less specific than the information mitigators convey because the tokens occur in a wide range of cognitive processing situations, reflect personal habits, or for other reasons.

In addition to "indicating" cognitive processes, the respondent's behavior may also occasion behavior by the interviewer that has consequences for data quality. For example, a mitigator displays the interactional status of the answer as an approximation for the interviewer to pursue or not. Similarly, reports, hesitations, tokens, repairs, and similar behaviors may be treated by interviewers as relevant to the task of producing an answer in ways that may affect data quality, for example, by increasing interviewer variance (by providing occasions for probing) or decreasing accuracy (see, e.g., the discussion of preemptive probes in Schaeffer and Maynard, 2002, and of "choosing" in Dijkstra and Ongena, 2006).

As these comments indicate, the participants influence each other through time: Figure 3.1 recognizes that the interviewer's behavior influences the respondent's and vice versa. This influence could be simple, as when the presence of another person (e.g., the interviewer) evokes the speaking aloud of mitigators, or complex, as when a history of answers that do not meet the requirements of the question and the interviewer's probes lead to a final recorded answer.

The conceptual model in Figure 3.1 implies that the correlation between an indicator of data quality and a given measure of behavior may vary across questions (depending on the characteristics of the question) and behaviors (depending on how the behaviors are related to cognitive processing and interactional practices). For example, interruptions may be relatively frequent and associated with lower reliability for questions in which a definition follows something that can be heard as a question, but relatively infrequent and not associated with lower reliability for items in which the definition is placed before the target question. Although the interviewer's behavior may influence the respondent's answer and its quality, we do not necessarily expect the interviewer to be the most important influence. Similarly, although the respondent's behavior reflects some features of the respondent's cognitive processing, not all cognitive processes, problems in cognitive processing, or errors in cognitive processing will have observable behavioral correlates. Additionally, problems or errors in cognitive processing that do affect behavior will not have perfect behavioral correlates of which they are the sole cause.

To summarize, to use the behavior of interviewers and respondents to evaluate survey questions, we need to be able to measure relevant characteristics of questions and behaviors of interviewers and respondents—where "relevant" means features that affect (in the case of interviewers) or reflect (in the case of respondents) the quality of answers. Thus, we must look for evidence that the behaviors vary with the characteristics of survey questions and are associated with data quality.

3.3 CODING THE BEHAVIOR OF INTERVIEWERS AND RESPONDENTS

3.3.1 The Interviewer's Behavior

Codes for interviewers' behaviors have often been oriented toward standardization, as illustrated by Brenner's (1982) elaborate analysis of the rules underlying standardization, which is designed to reduce interviewer variability (see review in Schaeffer et al., 2010). Although practices of standardization vary, the rule to read the question as worded, except in specified circumstances, appears in all systems of standardization with which we are familiar (Viterna and Maynard 2002).

Few studies have had data that allow an assessment of how the interviewer's behavior is related to data quality, as indicated by interviewer variance, re-interviews, or a record check of the accuracy of the answer, although there have been a few experimental manipulations of probing (e.g., Dijkstra 1987;

TABLE 3.1. Associations between Interviewers' Behaviors and Measurement Quality

Reliability	
No relationship between question reading and clarity and pace of question delivery on interviewer variance in analysis of 25 items	Groves and Magilavy, 1986
Probing associated with increased interviewer effects in analysis of 130 items	Mangione et al., 1992
No effect of exact reading of question on test-retest index of inconsistency in analysis of 34 items	Hess et al., 1999
Validity	
When probing and administration of feedback had significant effects, both were associated with greater inaccuracy	Belli and Lepkowski, 1996[1]
Substantive change in question reading had no effect in 9 of 10 items, increased accuracy in 1	Dykema et al., 1997
Major change in question reading had no effect in 9 of 11 items, decreased accuracy for 1, increased accuracy for 1	Dykema and Schaeffer, 2005[2]
Follow-up by interviewer associated with decreased accuracy (8 of 11 items for adequate follow-up, 10 of 11 items for inadequate follow-up)	Dykema and Schaeffer, 2005

[1]Several studies report results from the Health Field Study (HFS); some focus on the total number of health visits, which was obtained by asking four questions, the sum of which was verified by the interviewer (e.g., Belli and Lepkowski, 1996; Mathiowetz, 1999; see also Belli, Lepkowski, and Kabeto 2001). Results from multiple outcomes in the HFS are reported in Dykema et al. (1997), so we selectively report results from some other HFS analyses in Tables 3.1 and 3.2.

[2]For details about one of the items in Dykema and Schaeffer, 2005, see Schaeffer and Dykema, 2004.

Smit, Dijkstra, and Van der Zouwen 1997). Table 3.1 summarizes some results, focusing on studies that use surveys and assess reliability or validity. Hess et al. (1999) found that the interviewer's reading of the question did not affect reliability. Similarly, misreadings appear to have only a small effect on validity, although behaviors at specific questions have been shown to decrease or increase reporting accuracy (Dykema et al., 1997; Dykema and Schaeffer, 2005). However, the interviewers in these studies were standardized, and most deviations from standardization may have had a low likelihood of affecting the data. It has also proven difficult to measure substantive changes to the reading of a question that can be coded reliably across a wide range of questions; thus, many reading changes captured by codes may not affect the respondent's understanding of key survey concepts. Therefore, we might interpret these results as conditional on a particular regime of interviewing practice (see, e.g., Belli, Lee, Stafford, and Chou, 2004). When the interviewer's reading of the question does affect the accuracy of the answer in either direction, it is worth investigating further whether interviewers are improving faulty questions or encountering situations that the question does not anticipate.

3.3.2   The Respondent's Behavior

Table 3.2 summarizes associations between the respondent's behavior and indicators of data quality. Response latency is associated with accuracy, such that answers produced after longer response latencies are less accurate, for most cases in which there is a relationship.[4] The size of the cognitive task also affects response latency (Draisma and Dijkstra, 2004; Dykema and Schaeffer, 2005). For example, respondents with more challenging digit ordering tasks had longer response latencies than those with simpler tasks (Schaeffer et al., 2008).

Behaviors that might indicate uncertainty (e.g., "doubt" words, qualifiers) are associated with decreased reliability (Hess et al., 1999) and sometimes with less accurate answers (Dykema et al., 1997; Mathiowetz, 1999; Draisma and Dijkstra, 2004; Dykema and Schaeffer, 2005). However, it is not clear which of the various behaviors (e.g., use of qualifiers, mitigators, "don't know," "doubt words," or "uncertainty") indicate distinct varieties of uncertainty or whether contradictory findings are due to variations in the reliability with which the behaviors are coded, to variations in the operationalizations, or to variations in model specification. For example, Dykema et al. (1997) find no relationship between "uncertainty" and accuracy for 10 items, but a relationship between "qualification" and accuracy for two of those items. Mathiowetz (1999), analyzing the same study, finds a relationship between uncertainty and a composite of four items. To the extent that operationalizations of "uncertainty" overlap or are redundant, it is not clear which is the strongest. Additional studies have shown how some of these behaviors might be linked to the respondent's difficulties in mapping experience onto survey concepts, which could be related to measurement error (e.g., Van der Zouwen and Smit 2004; Schober and Bloom, 2004; Holbrook, Cho, and Johnson 2006; Garbarski et al., 2011, forthcoming).

Of the other behaviors that have been examined, there is fairly consistent evidence that answers with more words (Draisma and Dijkstra, 2004), more exchange levels (Dykema and Schaeffer, 2005), or interruptions are associated with reduced accuracy. Filled pauses (which might be associated with uncertainty) are not related to accuracy (Draisma and Dijkstra, 2004), and seeking clarification is associated with increased accuracy for one item (Dykema and Schaeffer, 2005).

3.4   CONVERSATION ANALYTIC STUDIES OF INTERACTION IN THE INTERVIEW

The selection of behaviors in recent studies is inspired by an orientation to sources in addition to standardization, including linguistics (see, e.g., Schober and Bloom, 2004). One alternative way of identifying behaviors of interest uses conversation analysis, which provides both a theoretical orientation and

TABLE 3.2. Associations between Respondents' Behaviors and Measurement Quality[1]

Reliability	
Percent qualified answers associated with decreased index of consistency from test-retest in analysis of 34 items	Hess et al., 1999
Validity—Response latency	
Shortest for correct answer (analysis of 11 items in each of 2 data sets)	Draisma and Dijkstra, 2004
Pause of 2 seconds or more associated with reduced accuracy for 6 of 11 items	Dykema and Schaeffer, 2005
Longer response latency associated with better performance on digit ordering task, poorer performance on word fluency task	Schaeffer et al., 2008
Validity—Constructs related to uncertainty	
Uncertainty not associated with accuracy in analysis of 10 items	Dykema et al., 1997
Qualification associated with reduced accuracy for 2 of 10 items	Dykema et al., 1997
"Don't know" associated with reduced accuracy for 3 of 10 items	Dykema et al., 1997
Uncertainty associated with reduced accuracy for total of responses to 4 individual items	Mathiowetz, 1999
Doubt words more likely with incorrect answers	Draisma and Dijkstra, 2004
Switching answer more likely with incorrect answer	Draisma and Dijkstra, 2004
Qualification associated with reduced accuracy for 8 of 11 items	Dykema and Schaeffer, 2005
Validity—Other behaviors	
Larger number of words associated with incorrect answers	Draisma and Dijkstra, 2004
Filled pauses not associated with accuracy	Draisma and Dijkstra, 2004
Problematic sequences (which often begin with a "mismatch" answer) associated with lower accuracy than non-problematic sequences	Dijkstra and Ongena, 2006
Elaboration associated with reduced accuracy for 1 of 11 items	Dykema and Schaeffer, 2005
Seeking clarification associated with increased accuracy for 1 of 11 items	Dykema and Schaeffer, 2005
More than one exchange level associated with reduced accuracy for 9 of 11 items, with increased accuracy for 1 of 11 items	Dykema and Schaeffer, 2005
Validity—interruption	
Interruption associated with decreased accuracy for 1 of 10 items	Dykema et al., 1997
Interruption associated with increased accuracy for 1 of 11 items	Dykema and Schaeffer, 2005

[1]See footnotes for Table 3.1. Additional results from the environmental studies discussed in Draisma and Dijkstra (2004) are presented in Dijkstra and Ongena (2006).

EXCERPT 3.1. WISCONSIN LONGITUDINAL STUDY, QUESTION 2

FI: During the past four weeks, have you been able to see well enough to read ordinary newsprint without glasses or contact lenses? (1.2)

FR: Ah (0.1), just reading glasses. (0.7)

FI: Okay. I'm just going to reread the question.

FR: Okay.

FI: During the past four weeks, have you been able to see well enough to read ordinary newsprint (0.1) without glasses or contact lenses? (0.3)

FR: Ah (0.3), no.

FI: Okay.

---

a method of analysis to examine in detail what actually happens between the interviewer and respondent (e.g., Schaeffer, 1991). Studies of interviewing that use conversation analysis can identify behaviors with possible implications for measurement quality. In addition, descriptions of questions in action can both identify failures in question design and suggest solutions.

In the "paradigmatic" question-answer sequence presumed by standardization, the interviewer reads the question, the respondent provides an adequate answer, and the interviewer (optionally) provides a receipt of the answer (Schaeffer and Maynard, 1996).[5] One type of deviation from this sequence is a "report," a label borrowed from Drew (1984) to describe an answer that does not use the format projected by the question, as illustrated in Excerpt 3.1 (see Schaeffer and Maynard, 1996, 2002).

Speakers use reports to perform a variety of tasks (Schaeffer and Maynard, 2008). The report in Excerpt 3.1 suggests that the respondent is not sure how to classify her experience into the offered categories, a kind of "mapping" problem. Garbarski et al. (2010) report similar possible mapping problems for the self-reported health item, which requires respondents to combine information about disease and functioning to arrive at an overall judgment about their health (see Excerpt 3.2). In Excerpt 3.3, the respondent may be unsure where to place the threshold that separates "no" from "yes": Does "yes" mean that one can hear all of the time, most of the time, some of the time, or just at all? The generalizing speculation in Excerpt 3.4—"I think everybody gets irritable"—is an indirect answer to a question that asks for a potentially embarrassing admission. Such speculations may help identify threatening questions or respondents who are less able to admit engaging in threatening behaviors. These examples suggest that reports may locate problems and suggest possible solutions.

EXCERPT 3.2. WISCONSIN LONGITUDINAL STUDY, QUESTION 1

I: In general would you say your health is excellent, very good, good, fair, or poor?
      (1.5)
R: I think it's great.
R: Except for arthritis Sir Arthur visits every day it makes it
I: Okay.
R: You gotta move around or it makes it ache.
I: Okay, so would you say excellent or very good or good?
R: I I think it's very good.
I: Okay. (.4)
R: I five years ago I had surgery for colon cancer and seems they got it all. (.4)
I: Mm good.
R: And last January I had a (.4)really neat(.5)deal with uh (1.2)um(.5) what (.7)colosticitis pancreatitis and(.7) and paralytic alias that was really great so they took out my gall bladder. (.5)
I: Uh huh. (.5)
R: Everything's fine.
I: Okay. (1.0)

EXCERPT 3.3. WISCONSIN LONGITUDINAL STUDY, QUESTION 7

FI: Alright, without a hearing aid and while in a group conversation with at least three other people, have you been able to hear what is said?
       (1.8)
MR: Most of the time.
       (1.2)
FI: Is that a yes then or
MR: Yes.
       (0.1)
FI: Okay.

3.5   CHARACTERISTICS OF SURVEY QUESTIONS

To be diagnostic for improving survey questions, behavior must be associated both with measurement error and with question characteristics. A simple example illustrates the relationship among question characteristics,

---

EXCERPT 3.4. WISCONSIN LONGITUDINAL STUDY, QUESTION 35

MI: Okay. And during the past four weeks, did you ever feel fretful, angry, irritable, anxious, or depressed?
(0.7)

MR: Oh, I think everybody gets irritable.
(0.3)

MI: So, I should say yes to that question?

MR: Yes.

MI: Okay.

---

interactional practices, and data quality: If a definition is provided after an answerable question, the respondent is likely to "interrupt" the reading of the question before the definition is read (Oksenberg et al., 1991; Schaeffer, 1991; Van der Zouwen and Smit, 2004). This may increase interviewer variability, because only some interviewers will forge ahead with the definition, and respondents will then answer with varying levels of information (see Collins, 1980; Van der Zouwen and Smit, 2004; Fowler, this volume).

Characteristics of questions affect other behaviors of interviewers. Questions with field-coded answers may require probing (associated with interviewer variability), and longer questions are more likely to be misread (Fowler, this volume). Interviewer effects have been found with sensitive, nonfactual, and open questions (Schnell and Kreuter, 2005), although such effects have not always been found (see Groves and Magilavy, 1986; Mangione et al., 1992; O'Muircheartaigh and Campanelli, 1998), and with questions that require interviewers to judge whether the question applies to the respondent (Collins, 1980).

Some documented links between question characteristics and the respondent's behavior are summarized by Fowler (this volume). Results reported there indicate that questions that do not project the response format, do not project that response alternatives will be offered, or do not define ambiguous terms generate predictable behaviors by respondents (and subsequent followup by interviewers). In addition, longer response latencies are observed when a negative response alternative is presented before a positive response alternative (Holbrook et al., 2000) and when questions are double-barreled or use negative constructions (Bassili and Scott, 1996). (See Uhrig and Sala [2011] for discussion of how dependent interviewing affects behavior.)

One way researchers generate such findings is to locate question characteristics that appear to cause the problem, revise the questions, and demonstrate that the behavior is reduced (e.g., Oksenberg et al., 1991). A challenge in extending such investigations is the lack of a comprehensive analysis of structural or other (e.g., grammatical) characteristics of survey questions. Further, these studies often lack criteria for determining if the revised question improved data quality.

There are several explicit, and even more implicit (and usually partial and highly focused), taxonomies of the features of survey questions. (For a discussion of taxonomies of question characteristics, see Schaeffer and Dykema, 2010; for illustrations, see Forsyth et al., 2004; Saris and Gallhofer, 2007; Fowler and Cosenza, 2008.) However, these taxonomies encompass several levels of analysis. For example, whether the content of the question is sensitive may be relevant in different ways than technical features of questions such as the number of scale points. Moreover, there is variability in whether the presence or the absence of a feature (such as a transitional statement) is noted and in when a characteristic is only conditionally relevant (e.g., a transitional statement is needed only when the topic changes). As illustration, Table 3.3

TABLE 3.3. Illustrative Descriptions of Question Complexity

Graesser et al., 2006, p. 10	Complex syntax: The grammatical composition is embedded, dense, structurally ambiguous, or not well formed syntactically. Syntactic parsers are used in conjunction with Brill's (1995) part-of-speech classifier when syntactic analyses are performed. A sentence is flagged as having complex syntax if one of the following conditions is met: (a) the number of words before the main verb of the main clause is greater than a threshold, (b) the number of modifiers of a noun in a noun phrase is greater than a threshold, or (c) the mean number of higher-level constituents per word is greater than a threshold. A sentence with many higher-level constituents per word is structurally complex, has many levels of structure, or is not syntactically well formed.
Knauper et al., 1997	Question complexity: whether or not Q is (mainly syntactical) complex
Forsyth et al., 2004	Complex, awkward syntax [Question structure]
Willis, 2005	Wording: Question is lengthy, awkward, ungrammatical, or contains complicated syntax. [Clarity]
Saris and Gallhofer, 2007	In introduction: Interrogative sentence Number of subordinate clauses In request: Number of interrogative sentence Number of subordinate clauses Number of syllables per word
Yan and Tourangeau, 2008	Number of clauses (excluding response categories)

Note: Text in square brackets indicates authors' label for the construct.

TABLE 3.4. Illustrative Analysis of Some Dimensions on which Two Types of Survey Questions Vary (Based on Schaeffer and Presser, 2003)

Questions about Events and Behaviors	Rating Scales about Subjective Things
Relevance: Filter for opportunity or relevance	
Name of the event	Name of the object
Definition of the event	Name of the evaluative dimension
Placement of definition	Relevance: Filter for object or dimension
Reference period	Polarity
Absolute or relative	Bipolar
Boundaries	Unfolding (valence, then intensity)
Level of detail	Middle category
Placement of reference period	Labeling of middle category
Response dimension[1]	Unipolar
Occurrence	Presence/absence filter
Clarify threshold	Number of categories
Frequency	Labels of categories
Occurrence filter or not	Numeric only
Absolute frequencies	End point only labeled
Open	Completely labeled
Metric	Knowledge or certainty
Closed	State uncertainty
Number and labeling of categories	Task uncertainty
Relative frequencies	
Number and labeling of categories	
Duration	
Rate	
Timing	

[1]Detail shown for occurrence and frequency only.

compares the way several studies define or operationalize a question characteristic, complexity. It is apparent that a variety of approaches is possible, and without studies that make direct comparisons among them, we cannot tell which operationalization is superior.

An illustrative structural analysis of some dimensions on which survey questions vary is shown in Table 3.4. This example, which is restricted to a few common types of questions, has a decision tree implicit in portions of the taxonomy. For example, in writing questions about events or behaviors, when one includes a reference period, decisions must be made about whether the metric will be absolute (e.g., the last calendar year) or relative (e.g., the previous 12 months), which boundaries will be specified (e.g., beginning or ending or both) and at what level of detail (e.g., what combination of day and date), and where the reference period will be placed in the question. In writing a question about frequency, some of the choices are nested: to use an occurrence filter or not; whether to request absolute (e.g., counts of the number of times) or relative (e.g., adverbial descriptions of "how often") frequencies; and if absolute frequencies are used, the question may be open or closed.

3.6 CONCLUSION

Coding the interaction between interviewers and respondents has contributed to deepening our understanding of the structure of survey questions and the interactional expressions of cognitive processing and measurement error. In many cases, however, we have had to rely on an orientation to the paradigmatic question-answer sequence as a criterion for determining when survey questions are problematic and when they have been improved, because only a few studies provide strong criteria such as test-retest reliabilities, interviewer effects, or record checks. Furthermore, the variety of conceptualizations and operationalizations of both behaviors and question characteristics makes it difficult to distinguish null findings from poor measurement. Interpretation of behaviors is complicated by the lack of explicit and exclusive relationships between specific behaviors and either question characteristics or measurement error.

If interaction coding is going to be used routinely to evaluate and improve survey questions, future research needs to identify the key set of behaviors that can be coded reliably and that lead to improvements in measurement that make the effort cost-effective. This set of core codes would identify behaviors that occur frequently enough to be worth training coders to use them (or that signal serious measurement problems when they do occur), that are specific enough that it is possible to locate and correct the problem causing the behavior, and that improve measurement when the problem is corrected.

ACKNOWLEDGMENTS

Some of the research reported here was supported in part by the National Institute of Child Health and Human Development (Center Grant R24 HD047873), by the National Institute on Aging (Center Grant P30 AG017266) and the Wisconsin Longitudinal Study: Tracking the Life Course P01 AG-021079, by the Wisconsin Center for Demography and Ecology, by the Wisconsin Center for Demography of Health and Aging, by grants from the Graduate School Research Committee to Schaeffer and Maynard, and by the University of Wisconsin Survey Center (UWSC), which is supported by the College of Letters and Science. Nora Cate Schaeffer and Douglas W. Maynard are coinvestigators on the Wisconsin Longitudinal Study Supplement "Cognition and Interaction in Surveys of Older Adults."

NOTES

1 The model of Van der Zouwen and Smit (2004) focuses on the stages of the question-answer process captured by specific codes.

2 The figure omits the interviewer's cognitive processing and the respondent's true value, which can both influence the respondent's cognitive processing and interact with the characteristics of the survey question to affect the respondent's behavior in the interview (e.g., Schaeffer, 1994 and Dykema and Schaeffer, 2000).

3 Tables 3.1 and 3.2 focus on studies that examine interaction in survey interviews using explicit measurement criteria, and most of the studies include multiple survey items. There are also relevant studies that use non-survey experiments or more distant proxies for reliability and validity (e.g., Schober and Bloom, 2004; Ongena and Dijkstra 2009).

4 However, an experimental study with a self-administered instrument suggests that the relationship may be curvilinear (those who answer very quickly may also be less accurate) (Ehlen et al., 2007).

5 Van der Zouwen and Smit (2004, p. 111) discuss labels for "paradigmatic" sequences used in earlier studies.

REFERENCES

Bassili JN, Scott BS (1996). Response latency as a signal to question problems. Survey research. Public Opinion Quarterly; 60:390–399.

Belli, RF, Lepkowski, JM (1996). Behavior of Survey Actors and the Accuracy of Response. In: edited by R. Warneke. Hyattsville, MD, Department of Health and Human Services, Public Health Service, Centers for Disease Control and Prevention, National Center for Health Statistics, pp. 69–74.

Belli RF, Lepkowski JM, Kabeto MU (2001). The respective roles of cognitive processing difficulty and conversational rapport on the accuracy of retrospective reports of doctors' office visits. In: Seventh Health Survey Research Methods Conference Proceedings (University of Illinois–Chicago, Chicago), pp. 197–203.

Belli RF, Lee EH, Stafford FP, & Chou C-H (2004). Calendar and question-list survey methods: Association between interviewer behaviors and data quality. Journal of Official Statistics; 20(2):143–184.

Brenner M (1982). Response effects of "role-restricted" characteristics of the interviewer. In: Dijkstra W, Van der Zouwen J, editors. Response Behavior in the Survey Interview. London: Academic; pp. 131–165.

Brill, E. (1995). Transformation-based error-driven learning and natural language processing: A case study in part-of-speech tagging. Computational Linguistics; 21: 543–566.

Collins M (1980). Interviewer variability: a review of the problem. Journal of the Market Research Society; 22(2):77–95.

Dijkstra W (1987). Interviewing style and respondent behavior: An experimental study of the survey interview. Sociological Methods and Research; 16(2): 309–334.

Dijkstra W, Ongena Y (2006). Question-answer sequences in survey-interviews. Quality & Quantity; 40(6):983–1011.

Draisma S, Dijkstra W (2004). Response latency and (para) linguistic expression as indicators of response error. In: Presser S, Rothgeb JM, Couper MP, Lessler JT, Martin E, Martin J, Singer E, editors. Methods for Testing and Evaluating Survey Questionnaires. New York: Wiley; pp. 131–148.

Drew P (1984). Speakers' reportings in invitation sequences. In: Atkinson JM, Heritage J, editors. Structures of Social Action: Studies in Conversation Analysis. Cambridge: Cambridge University Press; pp. 129–151.

Dykema J, Schaeffer NC (2000). Events, instruments, and reporting errors. American Sociological Review; 65(4):619–629.

Dykema J, Schaeffer NC (2005). An investigation of the impact of departures from standardized interviewing on response errors in self-reports about child support and other family-related variables. Paper presented at the annual meeting of the American Association for Public Opinion Research, May, Miami Beach, FL.

Dykema J, Lepkowski JM, Blixt S (1997). The effect of interviewer and respondent behavior on data quality: analysis of interaction coding in a validation study. In: Lyberg L, Biemer P, Collins M, De Leeuw E, Dippo C, Schwarz N, Trewin D, editors. Survey Measurement and Process Quality. New York: Wiley-Interscience; pp. 287–310.

Ehlen P, Schober MF, Conrad FG (2007). Modeling speech disfluency to predict conceptual misalignment in speech survey interfaces. Discourse Processes; 44(3): 245–265.

Forsyth B, Rothgeb JM, Willis GB (2004). Does pretesting make a difference? An experimental test. In: Presser S, Rothgeb JM, Couper MP, Lessler JT, Martin E, Martin J, Singer E, editors. Methods for Testing and Evaluating Survey Questionnaires. New York: Wiley; pp. 525–546.

Fowler FJ, Cosenza C (2008). Writing effective questions. In: De Leeuw ED, Hox JJ, Dillman DA, editors. International Handbook of Survey Methodology. pp. 136–160.

Garbarski D, Schaeffer NC, Dykema J (2011, forthcoming). Are interactional behaviors exhibited when the self-reported health question is asked associated with health status? Social Science Research.

Graesser, AC, Cai, Louwerse, M, Daniel, F (2006). Question understanding AID (QUAID): A web facility that tests question comprehensibility. Public Opinion Quarterly; 70(1): 3–22.

Groves RM, Magilavy LJ (1986). Measuring and explaining interviewer effects in centralized telephone surveys. Public Opinion Quarterly; 50(2):251–266.

Hess J, Singer E, Bushery JM (1999). Predicting test-retest reliability from behavior coding. International Journal of Public Opinion Research; 11(4):346–360.

Holbrook A, Cho YI, & Johnson T (2006). The impact of question and respondent characteristics on comprehension and mapping difficulties. Public Opinion Quarterly; 70(4):565–595.

Holbrook AL, Krosnick JA, Carson RT, Mitchell RC (2000). Violating conversational conventions disrupts cognitive processing of attitude questions. Journal of Experimental Social Psychology; 36:465–494.

Knauper B, Belli RF, Hill DH, Herzog AR (1997). Question difficulty and respondents' cognitive ability: the effect on data quality. Journal of Official Statistics; 13(2): 181–199.

Mangione TW, Fowler FJ, Louis TA (1992). Question characteristics and interviewer effects. Journal of Official Statistics; 8(3):293–307.

Mathiowetz NA (1999). Expressions of respondent uncertainty as indicators of data quality. International Journal of Public Opinion Research; 11(3):289–296.

Oksenberg L, Cannell CF, Kalton G (1991). New strategies for pretesting survey questions. Journal of Official Statistics; 7(3):349–365.

O'Muircheartaigh C, Campanelli P (1998). The relative impact of interviewer effects and sample design effects on survey precision. Journal of the Royal Statistical Society, Series A; 161:63–77.

Ongena, YP, Dijkstra, W (2009). Preventing mismatch answers in standardized survey interviews. Quality & Quantity; 44(4):641–659.

Saris WE, Gallhofer IN (2007). Design, Evaluation, and Analysis of Questionnaires for Survey Research. New York: Wiley.

Schaeffer NC (1991). Conversation with a purpose—or conversation? Interaction in the standardized interview. In: Biemer PP, Groves RM, Lyberg LE, Mathiowetz NA, Sudman S, editors. Measurement Errors in Surveys. New York: John Wiley & Sons; pp. 367–392.

Schaeffer NC (1994). Errors of experience: response errors in reports about child support and their implications for questionnaire design. In: Schwarz N, Sudman S, editors. Autobiographical Memory and the Validity of Retrospective Reports. New York: Springer Verlag; pp. 142–160.

Schaeffer NC, Dykema J (2004). A multiple-method approach to improving the clarity of closely related concepts: distinguishing legal and physical custody of children. In: Presser S, Rothgeb JM, Couper MP, Lessler JT, Martin E, Martin J, Singer E, editors. Methods for Testing and Evaluating Survey Questionnaires. New York: Wiley; pp. 475–502.

Schaeffer NC, Dykema J (2010). Characteristics of survey questions: a review. American Association for Public Opinion Research, May, Chicago, IL.

Schaeffer NC, Maynard DW (1996). From paradigm to prototype and back again: interactive aspects of cognitive processing in survey interviews. In: Schwarz N, Sudman S, editors. Answering Questions: Methodology for Determining Cognitive and Communicative Processes in Survey Research. San Francisco: Jossey-Bass; pp. 65–88.

Schaeffer NC, Maynard DW (2002). Occasions for intervention: interactional resources for comprehension in standardized survey interviews. In: Maynard DW, Houtkoop-Steenstra H, Schaeffer NC, Van der Zouwen J, editors. Standardization and Tacit Knowledge: Interaction and Practice in the Survey Interview. New York: Wiley; pp. 261–280.

Schaeffer NC, Maynard DW (2008). The contemporary standardized survey interview for social research. In: Conrad FG, Schober MF, editors. Envisioning the Survey Interview of the Future. Hoboken, NJ: Wiley; pp. 31–57.

Schaeffer NC, Presser S (2003). The science of asking questions. Annual Review of Sociology; 29:65–88.

Schaeffer NC, Dykema J, Garbarski D, Maynard DW (2008). Verbal and paralinguistic behaviors in cognitive assessments in a survey interview. In: American Statistical Association 2008 Proceedings of the Section on Survey Research Methods. Washington, DC: American Statistical Association; pp. 4344–4351.

Schaeffer NC, Dykema J, Maynard DW (2010). Interviewers and interviewing. In: Marsden PV, Wright JD, editors. Handbook of Survey Research, 2nd ed. Biggleswade, UK: Emerald Group; pp. 437–470.

Schnell R, Kreuter F (2005). Separating interviewer and sampling-point effects. Journal of Official Statistics; 21(3):389–410.

Schober MF, Bloom JE (2004). Discourse cues that respondents have misunderstood survey questions. Discourse Processes; 38(3):287–308.

Smit JH, Dijkstra W, Van der Zouwen J (1997). Suggestive interviewer behaviour in surveys: An experimental study. Journal of Official Statistics; 13(1):19–28.

Uhrig SN, Sala E (2011, forthcoming). When change matters: An analysis of survey interaction in dependent interviewing on the British Household Panel Study. Sociological Methods & Research.

Van der Zouwen J (2002). Why study interaction in survey interviews? In: Maynard DW, Houtkoop-Steenstra H, Schaeffer NC, Van der Zouwen J, editors. Standardization and Tacit Knowledge: Interaction and Practice in the Survey Interview. New York: Wiley; pp. 47–66.

Van der Zouwen J, Dijkstra W (2002). Testing questionnaires using interaction coding. In: Maynard DW, Houtkoop-Steenstra H, Schaeffer NC, Van der Zouwen J, editors. Standardization and Tacit Knowledge: Interaction and Practice in the Survey Interview. New York: Wiley; pp. 427–448.

Van der Zouwen J, Smit JH (2004). Evaluating survey questions by analyzing patterns of behavior codes and question-answer sequences: a diagnostic approach. In: Presser S, Rothgeb JM, Couper MP, Lessler JT, Martin E, Martin J, Singer E, editors. Methods for Testing and Evaluating Survey Questionnaires. New York: Wiley; pp. 109–130.

Viterna, JS, Maynard, DW (2002). How Uniform Is Standardization? Variation within and across survey research centers regarding protocols for interviewing. In: Maynard DW, Houtkoop-Steenstra H, Schaeffer NC, Van der Zouwen J, editors. Standardization and Tacit Knowledge: Interaction and Practice in the Survey Interview. New York: Wiley; p. 365–401.

Willis GB (2005). Cognitive Interviewing: A Tool for Improving Questionnaire Design. Thousand Oaks, CA: Sage.

Yan T, Tourangeau R (2008). Fast times and easy questions: the effects of age, experience and question complexity on web survey response times. Applied Cognitive Psychology; 22(1):51–68.

4 Response 2 to Fowler's Chapter: Coding the Behavior of Interviewers and Respondents to Evaluate Survey Questions

ALISÚ SCHOUA-GLUSBERG

Research Support Services, Chicago

4.1 INTRODUCTION

This chapter is in response to Jack Fowler's chapter on behavior coding as a method for evaluating survey questions. In his chapter, Fowler describes the method clearly and succinctly, talks about its origins and history, how it has evolved, how it is used, what it tells us about questions, as well as the strengths and possible weaknesses of the method. While this response will discuss a few issues in that chapter, it will primarily focus on one specific use of behavior coding not mentioned in Fowler's chapter, but increasingly discussed in the literature: the use of behavior coding to evaluate questions in multilingual/multicultural surveys.

4.2 IN RESPONSE TO FOWLER'S CHAPTER

If there is one aspect that is missing in discussions about behavior coding, and indeed in Fowler's chapter as well, it is that of the sociocultural context in which the survey interview takes place. Behavior coding primarily examines respondent behavior as a result of question-related issues, such as whether the question includes an explanatory statement or whether the question is worded in a way that does not suggest a list of response options will be offered. The discussions barely touch on issues of interviewer–respondent interaction, perceived social distance and hierarchy, or the interview as a special type of scripted conversation.

Question Evaluation Methods: Contributing to the Science of Data Quality, First Edition.
Edited by Jennifer Madans, Kristen Miller, Aaron Maitland, Gordon Willis.
© 2011 John Wiley & Sons, Inc. Published 2011 by John Wiley & Sons, Inc.

In his chapter, Fowler states that there are three premises upon which behavior coding is based. First, deviations from the ideal question-and-answer process pose a threat to how well answers to questions measure target constructs. Second, the way a question is structured or worded can have a direct effect on how closely the question-and-answer process approximates the ideal. Finally, the presence of these problems can be observed or inferred by systematically reviewing the behavior of interviewers and respondents.

Absent from these three premises is the recognition that the interview process does not occur in a vacuum. In fact, an interview may be affected not only by question wording or characteristics, but by a host of other variables. Respondents' different degrees of familiarity with the survey process and lack of knowledge about the ideal interaction expected from them in the survey interview can detract from the goals of the survey. There are cases in which respondents are asked questions they have never considered before and, for that reason, may ask for clarification or make extraneous comments before answering. There are also cases where respondents may be new to surveys and take a long time to learn that they are expected to select a response choice from among those offered rather than come up with their own.

By the same token, some of the problems behavior coding seeks to identify cannot always be observed—for example, when respondents provide an answer even if they did not interpret the question as intended. Misinterpretation can be attributed to acquiescence or satisficing, but it may also be due to unintentionally interpreting the question differently than intended in the case of ambiguous questions. In his chapter, Fowler recognizes this, and provides a good example with the case of the question, "Did you eat breakfast this morning?", but there needs to be more emphasis on this important shortcoming of behavior coding.

In his chapter, Fowler acknowledges that when behavior coding identifies problems in a question, it does not reveal the causes of the problems it detects. He offers many ways to gain insight into these causes and suggests that if the questions were cognitively tested in advance, that testing may indicate where the problems originate. We would argue, however, that if the cause of the problems is identified at that early stage, the items should be reworded or redesigned before the pretest or pilot interviews and would not become an issue later during behavior coding.

Fowler points out that "an additional attractive feature of behavior coding is that it provides objective, quantitative data" while most other methods for question evaluation yield qualitative results. And indeed, survey researchers as quantitative researchers tend to be more comfortable with data that appear as numeric codes rather than with text or descriptive observations. It is, however, questionable whether the quantitative data resulting from coding observations—that is qualitative data—are more objective than the observational data itself. Ending up with quantifiable codes may make some researchers more comfortable, but these codes should not be viewed as more objective than the observational data from which it is derived.

4.3   THE USE OF BEHAVIOR CODING TO EVALUATE QUESTIONS IN MULTILINGUAL SURVEYS

Those who are involved in multilingual or multicultural surveys are faced with the need to evaluate how questions work for respondents from different cultures. Not only do we ask ourselves whether the questions are performing well, but we also ask whether they are being interpreted the same way by respondents from different languages and cultures, and whether they are performing satisfactorily for all groups in the study. In search of question evaluation methods that will address these concerns, we have begun using question evaluation methods, such as cognitive interviewing and behavior coding, in cross-cultural studies. These methods were originally designed in the United States and Northern Europe. It is only within the last 5 years that studies focusing on cognitive interviewing and attempting to evaluate how well such methods work in different cultures have begun to emerge (Gerber and Pan, 2004; Goerman, 2006; Pan et al., 2010). Behavior coding needs to be examined the same way if we are to continue using it in cross-cultural surveys as it has been used in recent years (e.g., Edwards et al., 2004; Cho et al., 2006; Hunter and Landreth, 2006; Kudela et al., 2006; Childs et al., 2007).

Pan et al. (2010) explores how cognitive interviewing works for several languages/cultures through the lenses of discourse analysis and the ethnography of communication. Their conclusion specifically addresses how their findings about communicative styles and cultural norms can inform behavior coding. They state, " ... in our transcript analysis we looked at the extent to which respondents across different cultures challenge interviewers or their questions. We identified patterns of differences across language groups. This has strong implications for behavior coding because many behavior-coding studies assume that respondents behave in the same way when a question is problematic, regardless of the language of the interview."

In survey research we make assumptions that are challenged repeatedly by qualitative research findings, particularly in cross-cultural/multilingual studies. We believe that by asking identically worded questions, we are presenting the same stimulus to respondents, and therefore their answers can be compared because they are all answers to the same question. As Fowler's chapter states, behavior coding is based on this basic precept of survey research: "When the survey process involves an interviewer and when the process goes in the ideal way, the interviewer asks the respondent a question exactly as written (so that each respondent is answering the same question). The respondent understands the question in the way the researcher intended." However, this is a model, and reality is often more complex.

Cognitive testing offers a glimpse into how different individuals can interpret the same question differently. In some cases this can be attributed to idiosyncratic reasons, but often there are also demographic or sociocultural factors to consider. In a recent test of questions on Reactions to Race, Ridolfo and Schoua-Glusberg (forthcoming) found that low education monolingual

Spanish-speaking Hispanic immigrants interpreted a question about whether others see them as Hispanic or Latino as a question on whether they are discriminated against because of being Hispanic/Latino.

Respondents from other ethnic groups interpreted the item as asking literally if others "see" or "perceive" them as Hispanic/Latino. This difference in interpretation may be due to different life experiences or to the priming effect of the introduction to the question, which reads as follows:

> This next set of questions asks about your health and how other people identify you and treat you. Please remember that your answers to these questions are strictly confidential.

> How do other people usually see you in this country? Would you say people see you as Hispanic or Latino?

In the Ridolfo and Schoua-Glusberg case, the introduction stated that the following set of questions would be about how "other people identify you and treat you." This in turn, could have predisposed the respondents to interpret the "how others see you" in subsequent questions as asking about discrimination.

Behavior coding, on the other hand, observes behavior and infers certain things about the performance of survey questions from those behaviors, specifically from interviewer–respondent communications. One factor to keep in mind is that cultures differ in communication styles, and this has important implications for survey interviews. How do we interpret behavior coding findings in the absence of an understanding of specific cultural communication norms?

Two of the premises upon which behavior coding is based, according to Fowler, relate to the process of the interview and to the interaction between interviewer and respondent. These cannot be taken as givens or even accepted as basic premises in the multicultural, multilingual context. The ideal question-and-answer process may not work equally well cross-culturally. In fact, limiting the interviewer–respondent exchange as we do in surveys may indeed pose threats to eliciting useful answers. Let us consider, for example, interviewer behavior in delivering questions as worded. In the course of translating cognitive testing protocols for a 2005 study evaluating American Community Survey respondent materials in multiple languages, Korean translators raised the issue that, as interviewers, they would need to address a respondent differently depending on the respondent's age and other perceived characteristics. This would not only affect the introductory statements or transition phrases across sections, but the wording of the questions themselves. After hours of discussion on how to standardize the translation, the translators concluded—and we agreed—that the interviewers would need to adapt the text depending on who they were interviewing. Edwards et al. (2004) report a similar finding for Koreans and point out that diverging from the script might be the culturally appropriate thing to do. In behavior coding, if we expect interviewers to read as worded, a Korean interviewer adjusting to appropriate cultural norms of communication would rate very low in "reading as worded." Those who rate highly for "reading as worded" may be perceived as culturally inappropriate by respondents, and this may affect the respondents' answers in unknown ways.

What about cases in which respondents give an answer that, while meeting the objectives, is qualified by wording that suggests "less than complete certainty about whether the answer is correct" (Fowler, this volume)? Pan et al. (2010) found that Korean respondents prefaced their answers often with "it seems" or "I guess" in cognitive interviews, either possibly as a reflection of culturally appropriate modesty in answering or potentially as an indication of satisficing behavior. In this case as well, it might be that the culturally appropriate behavior gets coded incorrectly as a problem with the question.

And what about respondents asking the interviewer to repeat all or part of the question or asking for clarification? Cultures with experience in surveys and interviews may exhibit one type of interaction or interviewer–respondent behavior in this regard, while cultures without interview experience may act differently. For example, Edwards et al. (2004) point out that the Spanish-language interviews had "the lowest rate of respondents interrupting with answers and providing qualified answers, and the highest rate of providing adequate answers despite having relatively high rates of inadequate answers and don't know's." These differences, they hypothesize, perhaps point to a more passive role by Spanish-language respondents.

Do particular ethnic/cultural groups engage in satisficing more than others and how is that reflected in the interaction with the interviewer? Smith (2003) notes, for example, respondents from some countries prefer to guess rather than give don't know answers. It is not a stretch to think that such respondents are more likely to choose an answer among those offered than ask for clarification. Their problems in interpreting the question will go unnoticed in behavior coding.

4.4 WHAT TO DO?

Reports on behavior coding in bilingual or multilingual studies normally do not make reference to how different communication styles may affect the use of this method. Some of these reports treat all language versions as the same, in that they compare how each question performs for different cultural or language groups as though the question were really identical, and not a translated version. Others focus on the usefulness of behavior coding to help determine equivalence in questions across different language versions, that is, in a way as a translation assessment tool. In either approach, they do not address the validity of the method in cross-cultural studies, or how cultural communication styles may affect the observed interviewer–respondent interaction.

In their findings, Edwards et al. (2004) present several such issues as possible explanations for findings that differ by language group in a multilingual behavior coding effort. Others make only passing references to these matters. For instance, Cho et al. (2006) mention in passing: "It is also unclear whether the inherent assumption that respondents from varying cultural backgrounds overtly express the behaviors being coded in a similar manner and to a similar degree is appropriate." Childs et al. (2007), in searching to explain a higher rate of extraneous comments from Spanish-speaking respondents, propose that cultural conversational norms may be responsible for making Spanish-speaking respondents feel that a discussion is warranted instead of giving the brief answer that is expected of them.

In the absence of basic research that examines how different cultural communication styles affect the building of appropriate codes to detect question problems in different cultures/languages, all these authors are left with the task of explaining differences across groups in their findings by hypothesizing about the possible causes. I propose that the findings in these studies ought to be about question performance and about how each question performs culturally and linguistically. They should not just be an opportunity to learn about how other cultures communicate, mainly because some differences may not be shown by this approach.

To employ behavior coding effectively in cross-cultural and multilingual studies, we need to calibrate the methodology by using it within each language/culture first, before we can compare the findings across cultures or languages. How do interviewers and respondents behave in a language and culture when a survey question is working exactly as intended? How do they behave when it is not? What ought to be the telltale signs in that culture that a question is not working as intended? Can every type of respondent be addressed the same way, or do other considerations apply that make reading the question exactly as worded inappropriate? Is not reading as worded more appropriate culturally than sticking to the script? These and other such questions may inform basic research on how behavior coding can be most useful in different cultural contexts. Only once we learn more about these factors within cultures can we make useful comparisons across cultures.

In the meantime, in the absence of that basic research about how best to use behavior coding across cultures, researchers should be wary of relying just on the judgment and insight of bilingual interviewers and coders, an approach often followed in studies where researchers are not familiar with the language/culture of some or all respondents. Fowler mentions this approach as one that may provide insight into behavior coding findings. Indeed, several of the studies reviewed debriefed interviewers and coders as to the possible causes of the problems (e.g., Edwards et al., 2004; Kudela et al., 2006). While this is common practice, it should be employed with caution, as interviewers and coders may often be poor reporters of what is happening in the interview interactions. We should remember that bilingual interviewers are different from the target population in terms of acculturation, if nothing else, and that they vary in their degree of comfort and familiarity with the target language and culture. In addition, they are, at best, members of the target culture, not experts, and they may often make generalizations about communication in the target culture without basis.

4.5 CONCLUSION

To conclude, this chapter reflects the perspective of an anthropologist trained in ethnoscience ethnography, which defines culture as what members of a group see themselves doing. Simply observing from the perspective of our own culture may not provide us with the full picture. We also need to understand how the members of a specific cultural group understand the interview situation and explain why they react in a certain way. Behavior coding in multicultural contexts predetermines, from the point of view of the Anglo-Saxon culture, what are appropriate ways in which interviewers and respondents must interact if a question is performing well. It leaves no room for other cultures to explain to us why or how they are interpreting the question, or even how it is appropriate to communicate in that situation.

REFERENCES

Childs J, Landreth A, Goerman P, Norris D, Dajani A (2007). Evaluating the English and the Spanish Version of the Non-Response Follow-Up (NRFU). Behavior Coding Analysis Report. Statistical Research Division Research Report Series, Survey Methodology #2007-16. U.S. Census Bureau. Available online at http://www.census.gov/srd/papers/pdf/ssm2007-16.pdf.

Cho Y, Fuller A, File T, Holbrook A, Johnson TP (2006). Culture and survey question answering: a behavior coding approach. American Statistical Association 2006 Proceedings of the Section on Survey Research Methods. Washington, DC: American Statistical Association; pp. 4082–4089.

Edwards WS, Fry S, Zahnd E, Lordi N, Willis G, Grant D (2004). Behavior coding across multiple languages: the 2003 California health interview survey as a case study. American Statistical Association 2004 Proceedings of the Section on Survey Research Methods. Washington, DC: American Statistical Association; pp. 4766–4772.

Gerber E, Pan Y (2004). Developing cognitive pretesting for survey translations. Paper presented at the Second International Workshop on Comparative Survey Design and Implementation (CSDI), Paris, France, April 1–3, 2004.

Goerman P (2006). Adapting Cognitive Interview Techniques for Use in Pretesting Spanish Language Survey Instruments. U.S. Census Bureau Research Report Series, Survey Methodology #2006-3. U.S. Census Bureau. Available online at http://www.census.gov/srd/papers/pdf/rsm2006-03.pdf.

Hunter J, Landreth A (2006). Evaluating Bilingual Versions of the NRFU for the 2004 Census Test. Behavior Coding Analysis Report. Statistical Research Division

Research Report Series, Survey Methodology #2006-7. U.S. Census Bureau. Available online at http://www.census.gov/srd/papers/pdf/ssm2006-07.pdf.

Kudela M, Forsyth B, Levin K, Lawrence D, Willis G (2006). Cognitive interviewing versus behavior coding. Paper presented to AAPOR, Montreal, Canada, May 2006.

Pan Y, Landreth A, Park H, Hinsdale-Shouse M, Schoua-Glusberg A (2010). Cognitive interviewing in non-English languages: a cross-cultural perspective. In: Harkness JA, Braun M, Edwards B, Johnson TP, Lyberg L, Mohler PP, Pennell B-E, Smith TW, editors. Survey Methods in Multinational, Multiregional, and Multicultural Contexts. Hoboken, NJ: John Wiley & Sons, Inc.

Ridolfo H, Schoua-Glusberg A (forthcoming). Analyzing cognitive interview data using the constant comparative method of analysis to understand cross-cultural patterns in survey data. Field Methods.

Smith TW (2003). Developing comparable questions in cross-national surveys. In: Harkness JA, Van de Vijver Fons JR, Mohler P, editors. Cross-Cultural Survey Methods. New York: John Wiley & Sons; pp. 69–91.

PART II
Cognitive Interviewing

5 Cognitive Interviewing

KRISTEN MILLER

National Center for Health Statistics

5.1 INTRODUCTION

Virtually every article within the last decade on cognitive interviewing methodology begins with a statement about its popularity and widespread use among survey research organizations. Arguably, it is the primary method among the U.S. federal statistical agencies to evaluate survey questions. Invariably, however, authors soon follow up with the assertion that there is no common or set standard for conducting such studies. Although the method is a central piece in the assurance of data quality for the federal government, there are no recognized best practice guidelines as there are for other facets of survey methodology such as data editing and response rate calculations (Office of Management and Budget Standards and Guidelines for Statistical Surveys, 2006). Office of Management and Budget (OMB) guidelines stipulate only that new survey questions should be tested "using methods such as cognitive testing, focus groups, and usability testing" and that the results should be "incorporated" into the final design. It does not stipulate how to conduct such tests, what the test results should look like, or how they should be incorporated.

Indeed, as previously recognized by practitioners of the method, how it is actually implemented can vary from organization to organization, and even among researchers within one organization. Although there has been debate over interviewing practices throughout the method's 25-year history (Beatty, 2004; Conrad and Blair, 2004, 2009; Willis, 2005), this has made little impact on the uniformity or standards of the actual practice. Moreover, it is not a common practice for federal agencies to make cognitive test studies accessible—which

Question Evaluation Methods: *Contributing to the Science of Data Quality,* First Edition.
Edited by Jennifer Madans, Kristen Miller, Aaron Maitland, Gordon Willis.
© 2011 John Wiley & Sons, Inc. Published 2011 by John Wiley & Sons, Inc.

would make both the findings and the specific methods public. This lack of transparency makes general knowledge of the actual methodological practice impractical. The Q-Bank effort is a significant step in changing this general practice and, with hope, will provide a better understanding of how different agencies go about conducting and implementing cognitive interviewing studies.

Regardless of these larger issues, cognitive interviewing methodology has significant and unique advantages to benefit survey research. Like no other question evaluation method, cognitive interviewing offers a detailed depiction of meanings and processes used by respondents to answer questions—processes that ultimately make up the survey data. As such, the method offers an insight that can transform understanding of question validity and response error, which are typically conceptualized homogeneously, as simply "error" or "no error," and operationalized as relational variables between other existing measures. Just as important, in comparison to the amount of dollars invested in survey research, the method is inexpensive and feasible, and is perhaps the reason why it is one of the most used methods for pretesting questions.

It is also important to recognize that there is considerable worthy criticism, both within the field of cognitive testing as well as from the broader field of survey methodology. Critics from within the method tend to focus on interviewing practices and the fact that certain techniques can impact the type of information produced in the interview (Beatty, 2004; Conrad and Blair, 2004). Those from the broader area of survey methodology more typically critique it for its qualitative characteristics and the apparent lack of scientific merit. For example, the method has been called anecdotal (Madans, 2004) and impressionistic (Conrad and Blair, 1996; Tourangeau et al., 2000), and has been critiqued for its inability to produce falsifiable findings (Tucker, 1997). Because of the significant potential (not to mention its widespread use), it is important to address these issues. This chapter will, first, describe the method and its critique. It will then describe a perspective practiced within the question evaluation program at the National Center for Health Statistics (NCHS) that addresses these concerns. Finally, it will describe specific areas that require particular attention in order to improve the overall credibility of cognitive interviewing methodology.

5.2 THEORETICAL UNDERPINNINGS, OBJECTIVES, AND PRACTICE

As it is true for all scientific pursuits, methodological practices are inextricably linked to theoretical traditions and preexisting sets of assumptions (Kuhn, 1962). This is also true for the method of cognitive testing. This first section will describe the method as it is rooted in the theoretical tradition of cognitive psychology and the movement known as CASM (Cognitive Aspects of Survey Methodology). It will then describe an approach that integrates a sociological perspective into the general understanding of the question response process.

In so doing, the chapter will contrast the two perspectives to illustrate how they differently inform characteristics of the method and its practice. Those characteristics include the overall objective, the purpose, and protocol for the interview, sample criteria and respondent characteristics, analysis and documentation of findings, as well as evaluation of the method itself.

5.3   THE CASM MOVEMENT AND THE PRACTICE OF COGNITIVE TESTING

The origins of cognitive interviewing are rooted within the intersection of survey methodology and cognitive psychology, whereby respondents' cognitive processes for answering survey questions are made explicit. Cognitive processes, it is believed, drive the survey interview and are represented in psychological models that depict the question response process. A commonly cited model contains four stages: comprehension, retrieval, judgment, and response (Tourangeau et al., 2000; also, see Willis, 2005 for a thorough discussion). For every survey question that is asked, each respondent, regardless of their demographic or personal background, must proceed through the cognitive steps to answer the question. In this regard, the cognitive process for question response can be understood as a universal process that occurs in the individual mind of each and every respondent. Furthermore, a primary contention is that knowledge of cognition is central to understanding sources of response error and how to design questions to reduce that error. The psychological model serves as a basis for understanding how and specifically where in the process error can be introduced (Tourangeau et al., 2000; Willis, 2004, 2005). Understanding cognitive process, then, provides the basis for developing a better theory for survey question design as well as offering a practical basis for reducing response error in survey data. Additionally, it was (and still is) believed that by studying the cognitive processes of question response, much can be learned about cognitive processes in general. The CASM movement is seen as mutually beneficial to both survey methods as well as to the field of cognitive psychology.

It is through this perspective that the practice of cognitive interviewing evolved as a pretest method for survey research organizations needing to "test" their questions prior to fielding. The method is characterized by conducting in-depth interviews with a small sample of respondents to reveal the cognitive processes—for making those hidden processes explicit—with the sole purpose of identifying sources of error and "flaws" in survey questions. The method, then, emerged as a practical application that has veered from the purer psychological pursuit of examining cognitive processes. As Willis (2004) argues, "this activity is not carried out primarily for the purpose of developing general principles of questionnaire design, but rather, to evaluate targeted survey questions, with the goal of modifying these questions when indicated" (p. 23). The interview structure, then, consists of respondents first answering a draft survey question and then providing textual information to reveal the cognitive processes involved in answering the test question. Specifically, cognitive interview respondents are asked to describe how and why they answered the question as they did. Through the interviewing process, various types of question-response problems that would not normally be identified in a traditional survey interview, such as interpretive errors and recall accuracy, are uncovered. DeMaio and Rothgeb (1996) have referred to these types of less evident problems as "silent misunderstandings." In theory, identified response errors can be linked to a flaw in a question that may ultimately warrant modification.

In practice, cognitive interviews themselves are not standardized, although there continues to be debate over whether or not at least some standardization would improve information about question flaws (Conrad and Blair, 2009, 1996; Willis, 2005). While it is difficult to say exactly how interviews may vary throughout the federal statistical enterprise, it is fair to say that interviews can vary depending on the way interviewers actually go about eliciting the textual follow-up from respondents, whether it is through "think aloud" or probing techniques, either prescripted or emergent probing, the particular types of probes used, and whether that probing is concurrent or retrospective. Variation across interviews is also dependent on the respondents' life experiences, particularly as it relates to the subject matter, and also their abilities to verbalize their thoughts. (Again, for more discussion, see Willis, 2005.) Indeed, a principal critique of the method centers on whether respondents are even capable of reporting their cognitive processes and whether their attempts might artificially alter subsequent thoughts (Willis, 2004; Conrad and Blair, 2009).

The number of respondents interviewed in typical cognitive testing studies can also vary substantially within and across survey organizations. Again, with no established best practice guidelines, there is no identified requisite number to perform a quality test. Willis (2005) advises that, in an ideal situation, interviews are conducted until all the major problems have been detected and satisfactorily addressed. And while he also suggests that, in reality, the sample size is often dictated by the survey schedule, he offers a sample size of 5 to 15 as a typical range. It is also well known in the field that studies are frequently conducted consisting of nine interviews, or at least several rounds of nine interviews, not because of any theoretical or methodological rationale, but because federal studies of nine and under do not require OMB authorization. Because cognitive processes are assumed to be universal processes that do not vary substantially across respondents, a large sample is simply not necessary. Furthermore, it is not warranted because the ultimate goal is not to describe statistical trends or to make estimations, but rather to detect flaws in questions. Additionally, it is argued that, by forgoing large numbers, findings are more detailed which is particularly useful for determining the exact nature of an identified problem. Nevertheless, as many as 150 interviews has been reported for any one particular testing project (Harris-Kojetin et al., 1999), and while it is impossible to know for sure, anecdotally, it does appear that there is a growing trend to increase interview numbers.

For traditional cognitive testing studies, respondents are not selected through a random process, but rather are selected for specific characteristics such as race or health status or some other attribute that is relevant to the type of questions being tested. There is, then, no claim that all social and demographic groups are represented, and practitioners readily acknowledge that cognitive interviewing does not produce statistically generalizable findings. However, because the response process is seen as a core process of human cognition, it is unnecessary to assemble a sample with a full spectrum of demographic characteristics. And, again, because the goal is to identify the presence of problems, as opposed to making estimations or causal statements, a randomly drawn sample is not demanded. In fact, as Willis (2005) notes, it is actually beneficial to purposefully select respondents because this can ensure proper coverage of the questionnaire with respondents who are more likely to expose question problems.

Unlike sampling and interviewing procedures, specific explanations of how cognitive interview data are analyzed are often minimal. While there is much discussion about probing techniques, specifically, what are the best types of probes to illicit the most problems (ones that are "true" problems and not "false alarms"), there is little discussion regarding how the textual data are actually analyzed to derive at question problems. There is an implicit assumption that question flaws simply emerge from the interview text, revealing themselves as well as the particular character of the problem. Furthermore, there is an implicit assumption that the flaws appear in an objective form, devoid of need for study or explanation. The fact that there would be an analytic step of interpreting the text is consistently glossed over. This assumption is carried over into the presentation of cognitive test findings. In some cases, reports are not even written, but rather meetings are held whereby interviewers verbalize the problems that they identified when conducting the interviews. When reports are written, often they are thin with little documentation as to what was articulated in the interviews and the processes by which conclusions were made. Reports more typically assert the problems that were found within the confines of a few sentences or paragraph and, if no problems are found, discussion is often skipped for that question.

The lack of a well-articulated analytic method, indeed, has left the methodology vulnerable to criticism for being impressionistic, lacking objectivity and, ultimately, for being unscientific. As a way to compensate for the deficit, some researchers have developed coding schemes by which the interview texts are coded and then analyzed using quantitative methods (Conrad and Blair, 1996; Tourangeau et al., 2000). While at first glance this might appear to be an objective and systematic way to make conclusions, the text must still be interpreted in order to assign a code, and this coding technique still assumes that the flaws (i.e., the codes) will simply emerge from the interview text. With this assumption overlooked, the method is no more objective or systematic than in previous versions of the methodology.

Rather than establishing a rigorous method of analysis, more attention has been given to the interview itself and the credibility and objectivity of respondents' textual reports. If a cognitive testing study produces problems that are not "real flaws" or if real flaws are missed, then the source of the problem is traced back to the interview and the specific type of probing techniques used. In Conrad and Blair's (2009) piece, for example, an experiment was designed to evaluate cognitive interviewing techniques in which interviewers performed two types of interviews: a more standardized version in which specific follow-up probes were asked only when respondents indicated having a problem, and the other in which interviewers were unrestricted in the timing and types of probes they could ask. With the resulting interview text, independent judges were then asked to assign their own codes. Finally, inter-coder reliability measures between the interviewer and the judges were used to evaluate the two interviewing techniques, presuming that the technique with the higher convergence rate is the superior practice. The experiment design alone assumes no kind of analysis in cognitive interview studies, and the authors move directly from an examination of the inter-coder reliability score to conclusions about interview format. The paper presents no discussion of analysis for either the interviewers or the judges; specifically, it does not describe the processes used to sift through the text, how the text was actually examined or pieced together to form conclusions, as well as the types of decisions used by the interviewers and judges to make sense of contradictory information, attend to gaps in respondents' reports, or whether to disregard information that they might have deemed irrelevant or potentially dubious. Interestingly, while they offer the credentials and experience of their cognitive interviewers, this is absent for the coding judges. While it was assumed to be a noteworthy piece of information for the interviewers who had a role in creating the text, it was not deemed relevant for those interpreting the text.

Given the objective to discover question problems, like the study above, cognitive interviewing methodology is routinely evaluated by the numbers as well as the consistency of problems it can identify. For example, in their attempt to identify efficient respondent selection strategies for cognitive testing studies, Ackermann and Blair (2006) argued that the best strategy would be the one that selected respondents who were able to produce the most problems in the least amount of interviews. Rothgeb et al. (2001) compared the number and types of problems identified in one questionnaire through cognitive interviewing techniques with those evaluations identified by expert review and forms appraisal. And, DeMaio and Landreth (2004) compared the number and types of problems identified by different interview teams using not only different interviewing techniques, but also such factors as recruiting method, remuneration amount, and summarization method. It is an underlying assumption that the method, in its ideal form, would produce results that are consistent across researchers and teams of researchers, and the fact that this does not occur is deemed problematic and an indication of a flawed method. After a reading of various cognitive interviewing studies,

Presser et al. (2004) make this assumption explicit: "In sum, research on cognitive interviews has begun to reveal how the methods used to conduct the interviews shape the data produced. Yet much more work is needed to provide a foundation for optimal cognitive interviewing."

It is important to recognize that underneath this critique of the method is a tacit understanding: Because of the implicit belief of universality in the question response process across respondents, there exists an essential truth regarding the quality of a survey question and that such a truth can and should be discovered. To be sure, the underlying objective of identifying problems is such an integral piece of the method, that there is little recognized value of cognitive testing when no problems are found. Significantly, this objective has broad implication for the method and how it is practiced. More succinctly, the theoretical position asserting a universal response process produces the ultimate goal of cognitive interviewing (i.e., to identify question flaws) and then defines method characteristics such as interview structure, the probing techniques used, respondent characteristics, the required number of interviews, and even the standards for evaluating the method.

5.4   SHIFTING PARADIGMS: TOWARD A MORE INTEGRATIVE APPROACH TO QUESTION RESPONSE

An alternative viewpoint redefines the theoretical underpinnings of cognitive test methodology to incorporate a sociological perspective. While the CASM perspective argues that it is important to understand cognition, this approach maintains that interpretation and the construction of meaning are critical for understanding the question response process. And, in effect, the previously articulated models of question response process are significant, not because of the cognitive processes per se, but because they represent the types of interpretive processes used by respondents to answer survey questions. Furthermore, as it is a firmly established sociological principle, this perspective asserts that interpretations do not merely and spontaneously occur within an asocial vacuum, but rather are inextricably linked to institutional and interactional constructs within a social milieu (Berger and Luckmann, 1966). Therefore, rather than conceptualizing the question response process within the confines of an individual's mind and their cognitive process, the shift recognizes that the question response process itself is set within a sociocultural context—not simply within the context of the survey interview—but (and perhaps even more significantly to survey data quality) within the context of that respondent's life circumstance and the perspective that he or she brings to the survey interview. That is, how respondents go about the four stages—of comprehending, recalling, judging, and responding—is informed by the sociocultural context in which they live. Significantly, respondents' social location, including such significant factors as their socioeconomic status, education, and age, can impact how respondents go about interpreting and processing survey questions (Miller, 2003).

While the CASM perspective lends itself to focus on aspects of memory and the retrieval of information, this alternative approach emphasizes the interpretive aspects of the question response process. This holds true for not just the initial stage of comprehension, but also acknowledges an interpretive element in each of the other stages. Obviously, understanding what a question is asking requires interpretive processing, but what the respondent goes about remembering is also tied to interpretive processes. It would not matter if, for example, a respondent can accurately remember the number of alcoholic beverages he drank in the last week, if he defines his nightly glass of wine with dinner as "part of the meal" and not as "drinking an alcoholic beverage." In the same way, Willson (2006), in her project testing an alternative health module, found that respondents' personal understandings of themselves as either "users" or "nonusers" of alternative health medicine informed what they defined as "taking an herbal supplement." Those nonusers, as opposed to users, were not inclined to think of such practices as drinking a cup of chamomile tea to relax as "taking an herbal supplement," even though the question clearly directed respondents to do so.

How respondents go about processing the retrieved information as well as to map that assessment onto the response categories are also rooted within interpretation. For example, for the question, Overall, during the past week, how much physical pain or discomfort did you have?, Miller (2008) found variability in the ways in which respondents explained their answer. Respondents spoke about multiple dimensions of their pain that they factored into their answer, including episodic frequency, intensity, length of time for each episode, impact on their life, and seriousness of condition. Formulating an answer, then, required respondents to first identify and label those dimensions that they deemed relevant, and then to develop a method by which to formulate or calculate an answer that would accurately portray (at least to them) their experience of pain in the past week—all of which are interpretive tasks. Some respondents, for example, only chose to include the most severe episodes such that it impacted their daily routine. On the other hand, some respondents counted each time that they could remember feeling any type of pain regardless of intensity, and still others averaged across the week. As for providing an actual answer, matching their own assessment with the provided response categories (which, in this case, were "none at all," "a little," "moderate," "a lot," or "extreme") is a fully interpretive process. As is often the case for survey question response options, the categories consisted of vague quantifiers that, in and of themselves, are essentially meaningless. The categories only take on the intended meaning if the respondent intuits the degree of severity through the order that the categories are presented.

This shift in thinking not only impacts the basic assumptions regarding the question response process, but it also has implication for question design and the overall objective of cognitive interviewing methodology. Table 5.1

TABLE 5.1. Contrasting Paradigms for Question Design and Cognitive Testing Methodology

	CASM Paradigm	Integrative Paradigm
Underlying assumptions	Cognitive processes are the focal point of question response	Interpretive processes are the focal point of question response
	Response processes are universal and do not vary across respondents	Because response processes are tied to social context, they can vary tremendously across respondents
	Essential truth exists about the quality of a question	No essential truth exists. Meaning of all questions has potential to vary depending on respondents' historical and life circumstances
Principles for question design	Rules for question design are factual and universal	Dubious of the universality of accepted rules of question design
Unit of analysis	The question and its flaws	The interaction between the respondent and the survey question
Primary purpose of cognitive interviewing studies	Fixing question flaws	Documenting theoretical framework regarding the question's performance and the phenomena captured by the question

summarizes the theoretical differences between the two paradigms. While the previous perspective assumes a (more or less) universal response process for respondents and, therefore, a universal truth about a question and its flaws, this perspective assumes that the essential meaning of all questions has potential to vary depending on respondents' historical and life circumstances. A question may perform one way among some groups of respondents in particular contexts and another way among other groups in different contexts. Taken to its logical conclusion, the paradigm casts doubt on traditional or elemental principles of question design. Instead, it contends that question design must be based on an empirical investigation of the interpretive processes used by respondents as they relate the survey question to their own life experience. Rather than to identify and correct specific problems, the objective of this approach is to identify and explain the patterns of interpretation throughout the entire question response process. More specifically, the purpose is to generate a theoretical framework for understanding a question's performance. This theoretical framework, then, delineates the phenomena captured by the question. Obviously, some of these patterns may be unintended and, if left without modification, would lead to error in the survey data. In this regard, such studies are important for pretesting. But to the extent that this type of study captures the interpretive processes by which respondents construct their answers, findings are also relevant to the interpretation of survey data (even if the question is not ultimately modified) and can be useful to data users in making sense of relationships found in the data (Ridolfo and Schoua-Glusberg, forthcoming). After all, if data users do not know what a question is actually capturing, it is not possible for them to purposefully use the data in a valid manner.

Another significant point of departure is in the way that the interview itself is conceptualized as well as the criteria for what constitutes a good interview. Interviews within the CASM paradigm, in their ideal form, consist of respondents' verbal reports that reveal their cognitive processes. Given this supposition, it is important that the interviewer not contaminate respondents' memory or ask probes that would cause reactivity. As Conrad and Blair (2009) argue, this type of interviewing can introduce "error" into the cognitive interviews.

The integrative approach, however, understands the interview as a narrative endeavor, specifically, of respondents relaying how and why they answered the question as they did. Most succinctly, in the interview's ideal form, the resulting text contains respondents' explanation of their interpretive process. As it is the aim of a narrative investigation to examine the interpretation of experience (Josselson and Lieblich, 1995), in the context of the cognitive interview it is to examine how individual respondents relate survey questions to their own life experience and circumstance. Again, as it has been long recognized, those in different sociocultural locations can interpret and process questions differently than those in entirely different locations. But, it is only through narrative investigation that it is possible to determine the ways in which (and the extent to which) life circumstances mediate the interpretive aspects of answering survey questions. Consequently, the primary goal of the interviewer is to capture as much of the narrative as possible and to ask whatever questions are necessary to fill in gaps or to address apparent contradictions. To be sure, as required by the CASM approach, the interview text is not the essence of respondents' cognitive processes; at the very most, it is respondents' interpretations of their cognitive processes. The integrative approach acknowledges this and, in fact, argues that this is, in actuality, what is required for a truer understanding of the question response process: respondents' interpretive processes. To the extent that the interview captures respondents' narrativity, when analyzed it provides a detailed portrayal of the question response process and offers unique insight into the performance and validity of survey questions. It is important, however, that these data are understood exactly as they are—respondents' interpretations of their interpretive processes. And, it is critical that these interpretive texts are analyzed using appropriate qualitative methodologies. Table 5.2 summarizes these epistemological and methodological differences.

While criticism of the CASM response model as overly individualistic is not new (Forgas, 1981; Schwarz, 1994; Sudman et al., 1996), it was not until the last decade that the theoretical basis for the critiques began finding its way into

TABLE 5.2. Contrasting Paradigms: Epistemological and Methodological Differences

	CASM Paradigm	Integrative Paradigm
Essence of cognitive interview data	A representation of respondents' cognitive processes	Respondents' interpretations of the question response process
Role of interview	Interviewing reveals hidden problems in questions	Interviewing generates narrative text that when analyzed illustrates interpretive processes as well as how respondents' sociocultural contexts inform those processes
Methodological focus	Focus on probing techniques, specifically, the best probes to reveal true problems	Focus on how to conduct analysis and how to interpret cognitive interview data
Sample size	Interviewing complete when all problems are identified. Implicit desire to maximize the discovery of most problems from one single interview	Interviewing complete when all interpretive patterns are identified. Implicit desire to increase sample size to understand multiple perspectives
Focus analysis of	Identifying problems within individual interviews; little attention to interpretation or differences across interviews	Assessment of data quality and identifying interpretive patterns, particularly, variations or contradictions within or across interviews

the discourse on cognitive interviewing methodology. Eleanor Gerber (1999), an anthropologist by training, was perhaps the first practitioner of cognitive interviewing methodology to advocate for a more integrative approach. And, at the 2002 meetings for the American Association for Public Opinion Research, she and colleagues held an entire cognitive interviewing session devoted to the integration of sociocultural phenomena into the question response model. Around the same time (and perhaps in response to Gerber), Willis (2004) began advocating for the application of theory from other disciplines, including anthropology, sociology, and social psychological theory as a way of compensating for the theoretical limitations of cognitive psychology. At that time, it should also be noted that Willis began calling for better documentation of the methodologies and techniques used in cognitive testing studies.

Additionally, with the growing use of cross-national surveys and the need to produce harmonized questionnaires for populations across various language and cultural backgrounds, focus on sociocultural context has increased dramatically (Harkness et al., 2003). This trend has also placed increasing attention on the actual practice of cognitive interviewing methodology and its ability to attend to these emerging needs. This is certainly true for the question evaluation program at the NCHS. The increased amount of international testing work, particularly in the area of disability measurement (an especially complex concept steeped in social connotation), has forced the development of a testing practice that is more compatible with an integrative approach. Many of the methodologies developed for those international projects were ultimately adopted for the practice of cognitive test studies in general. To be sure, the theoretical approach embraced by the NCHS evaluation program is an integrated one, and this is reflected in its methodological practice and the reports it produces. The remainder of this chapter will focus on the methodology as it is practiced at the NCHS, focusing on methodological advantages and strategies for improving credibility.

5.5 ONTOLOGICAL AND EPISTEMOLOGICAL ADVANTAGES TO SURVEY RESEARCH

As described earlier, there is considerable criticism of cognitive interviewing methodology, particularly as it is practiced in the production of federal statistics. There are two reasons that support the merit of this criticism. First, the method as it is practiced does not hold up to the widely accepted criteria for trustworthiness of a quantitative investigation (Cook and Campbell, 1979). The method, however, cannot (and really should not) conform to these standards because the empirical investigation is in its essence an interpretive investigation and, therefore, necessarily a qualitative investigation. Second, and perhaps more importantly, the method, as it is currently practiced, does not entirely hold up to the criteria for trustworthiness of a qualitative investigation. Those criteria include credibility, transferability, dependability, and confirmability, and are the parallel axioms of internal and external validity, reliability, and objectivity (Lincoln and Guba, 1985). It is a mistake to sidestep methodological rigor simply because data are qualitative. When applied, cognitive interviewing methodology can offer unique and significant advantages to survey research. That is, cognitive interviewing, if it is practiced as a qualitative methodology, is valuable to survey research specifically because of the ontological and epistemological characteristics inherent to qualitative methodology.

Over the past decade, there have been many different arguments and examples shown to illustrate the advantages of a mixed-method approach in social science research (Green and Caracelli, 1997; Tashakkori and Teddlie, 1998, 2002). In this same way, there are numerous ways in which cognitive interviewing as a qualitative method can advantage survey research. While many of the benefits are yet to materialize, this chapter will begin by laying out three strengths that, in and of themselves, are of benefit to survey research methodology. They include (1) its interpretive quality, (2) its ability to capture and portray complex phenomena, and (3) the fact that its findings are grounded within empirical inquiry.

5.6   INTERPRETIVE QUALITY

As previously argued, to the extent that interviews capture respondent narrative, cognitive interviewing methodology is necessarily an interpretive endeavor that relays how and why respondents answered questions the way they did. It is only through this type of investigation that it is possible to determine the ways in which (and the extent to which) life circumstances mediate the interpretive aspects of answering survey questions. Unlike any other question evaluation method, this method can portray the interpretive processes that ultimately make up survey data. Because the information that survey respondents provide is quantified, it is easy to lose sight of the fact that survey data itself is a culmination of interpretive processes. Cognitive interviewing methodology, as it is practiced as a qualitative methodology, reveals these processes as well as the type of information that is transported through statistics.

To illustrate this point, the excerpt below is taken from a cognitive testing report which summarizes the findings of a project that examined the performance of a questionnaire on people's experiences with terrorism. A primary finding was that respondents' particular interpretations of "terrorism" were significant to determining how they would answer the subsequent questions. However, it was also found that respondents' interpretations varied a great deal—some respondents even included such things as gang violence and neighborhood crime. In order to write a successful survey question (i.e., one that would elicit responses framed within the intended definition), it was important to make sense of these seemingly "off-the-wall" interpretations and to fully understand the various dimensions in which terrorism was being defined. The inclusion of the below excerpt is intended to illustrate, first, those interpretive processes and, second, the level of insight that cognitive interviewing methodology can reveal about them. This passage is in reference to a set of questions that begin with the following introduction: "The next questions are about your direct experiences with terrorism. When I use the word "terrorism," I mean an attack against the American people, such as the September 11 attacks in 2001 and the anthrax letters in 2002. With that definition in mind ... were you ever within 5 miles of a terrorist attack at the time it occurred?" (For the entire set of questions and the full discussion, see Miller and Willson, 2004)

> Several core dimensions that constituted a general definition of terrorism were identified. Respondents' definitions of terrorism were based, not so much on the

act itself, but on the characteristics of the perpetrator (e.g., Are they foreigners? What is their motivation?) and on the characteristics of the victim (e.g., Is more than one person killed? Is the attack random?). In conceptualizing a definition of terrorism, respondents considered the characterization of victims in terms of their numbers (an individual vs. a group) and their arbitrariness (random vs. specific). Perpetrators were considered in terms of being insiders (i.e., Americans) or outsiders (i.e., foreigners), and also have two dimensions, affiliation (organized group vs. loner) and motivation (political/ideological vs. "temporary insanity"). Tables 5.3–5.5 graphically depict the many dimensions that respondents considered when framing a definition of terrorism.

TABLE 5.3. Dimensions of Terrorism—Victim

	Victims	
	Individual	Group
Random		
Specific		

TABLE 5.4. Dimensions of Terrorism—Perpetrators (Outsiders)

	Perpetrator (Outsider)	
	Organized	Loner
Political		
Insane		

TABLE 5.5. Dimensions of Terrorism—Perpetrators (Insiders)

	Perpetrator (Insider)	
	Organized	Loner
Political		
Insane		

Without a definitional statement outlining these dimensions, respondents needed to generate their own personal definition of terrorism so that they could then formulate a response to the question. This created a great deal of burden for respondents (who sometimes asked the interviewer to provide "more information" or answered with a "that depends"). Perhaps more serious, these makeshift interpretations of terrorist attack varied immensely across respondents. For example, one respondent saw police brutality as a form of terrorism, another included burglary in her definition, and still another decided that one person could potentially "terrorize" another.

Although examples of terrorism were included in the question as a way to help respondents understand what types of events to focus on, alone they did not go far enough in clarifying the term terrorism. For instance, despite the fact that the Anthrax letters of 2002 were included in the introductory statement, at least three respondents did not include anthrax in their answers, even though one received his mail from, and lived within walking distance of the Brentwood Post Office. This is because they did not define those events as a terrorist act ("we never found out who it was"). Lacking knowledge of the perpetrator made it difficult for some respondents to judge whether or not those letters constituted terrorism.

It is important to note that, in this capacity, the ultimate goal of such a project is to identify all of the types of interpretive patterns occurring throughout the question response process. In this case, the identified patterns of interpretation pertained to specific characteristics of perpetrators and victims. A credible statement could not be made, however, about the actual prevalence of those patterns. For example, there were some respondents whose interpretive pattern was inconsistent with the intent of the question (e.g., the respondent who counted burglary). With this data alone, it is not possible to determine the prevalence of this pattern as well as to make a statement regarding the amount of response error (in statistical terms) that would appear should this question be fielded without modification. This type of inquiry would require a mixed-method design.

5.7 CAPABILITY OF CAPTURING COMPLEXITY

The above excerpt, additionally, begins to convey the fact that cognitive interviewing methods can offer a complex picture of the interpretive processes. The level of complexity grows when it is understood that social context informs those processes, and that those from different social circumstances might construct different interpretations of the same concept. As a qualitative methodology, cognitive interviewing is able to capture this level of complexity because it can reveal how and why context informs interpretation. While quantitative studies show relationships through tests of statistical significance, qualitative studies show relationships through the explanation of how and why, and the trustworthiness of such a study is relayed through the level of detail and the believability of that description. The below excerpt illustrates how the method is able to reveal this connection to social context—in this case social context represents respondents' access to and the quality of dental health care (Miller, 2006). The question that this passage refers to is: "Another common problem with the mouth is gum disease. By gum disease we mean any kind of problem with the gums around your teeth that lasts for at least two weeks—except for problems caused by injury or problems caused by partials or dentures. Do you think you have gum disease? Yes or No."

The differences in the way respondents came up with an answer depended primarily on whether or not they had seen a dentist in the recent past. Those respondents who had not been to the dentist in many years experienced the most burden responding. To answer, those respondents were required to use the provided description of gum disease while considering their own symptoms and then speculate on whether or not they actually had the condition. A few of these respondents provided particularly tentative answers (saying "possibly" or answering don't know) because, despite the introductory clause, they were not entirely sure what symptoms should count as gum disease. One man who had difficulty answering, for example, had not seen a dentist in at least five years and because he was "not a professional," said he needed a dentist's opinion to answer with confidence. In the end, he answered explaining that, because the question specifically asked do you think you have gum disease? as opposed to do you have gum disease?, he was able "to make a guess;" he answered yes based on the fact that his gums bleed when he brushes his teeth. Another respondent stated that she was not sure what to answer because she recently had experienced some swelling around a particular tooth, but that the condition seemed to have improved. It was not clear to her whether or not she should count this flare-up as gum disease.

On the other hand, those respondents who were easily able to answer the question were those respondents who had been to the dentist and who had been told directly by a dental professional that they had no problems. For example, as one woman who had seen a dentist within the past month explained: "I really don't know what that [gum disease] is. All I know is I don't have it because my dentist would have told me if I did." For these respondents, it did not matter if they understood the introductory definition because a dental professional had indirectly weighed in on this assessment.

Those respondents who had been diagnosed with some kind of condition did not necessarily have an easy time answering the question. Some respondents had been treated, for example with scaling or by having teeth pulled, but were unsure if their condition was cured with the treatment. Additionally, unless a respondent was told specifically by their dentist that they had gum disease (as opposed to having another condition such as gingivitis or in need of a deep cleaning), they needed to decide what should or should not count as gum disease. For example, one respondent who had 16 teeth pulled due to decay—as recently as two months ago, stated that her dentist never told her that she had gum disease (only that her teeth needed to be pulled) and she ultimately answered no to the question.

When the interpretive processes of question response are connected in this way to context and social location, it is then possible to see how some respondents may introduce one type of error while other respondents might introduce another type of error. In so far as the method demonstrates this link, it reveals a complex depiction of response error that is not necessarily linear. It also suggests that, when modeling response error, it would be prudent to compensate for this level of complexity.

In this same way, the method can reveal multidimensionality of measurement in general, not simply the error component of a measure. To be sure, even valid questions can capture subtly different phenomena across respondents. To illustrate this point, an excerpt is taken from a project to test an NCHS follow-up question to the racial identity question (Miller and Willson, 2002). If respondents provided more than one racial category, they were to be asked: "Which one of these groups, that is [READ GROUPS], would you say BEST represents {your/name's} race?" For this project, multiracial respondents (i.e., those whose parents were of different races) were purposefully selected in order to identify the processes by which respondents constructed an answer. Indeed, a few (i.e., 3 out of 24) respondents were simply unable to do so and could not answer the question. They explained that they felt as though the question was asking them to choose one parent over the other and would not do so. However, the vast majority of respondents were able to construct a viable interpretation, though interpretations across these respondents consisted of different dimensions:

Those 21 participants, who responded to the question, based their answer on one of four different dimensions of racial identity:

1. socially, by the way in which they believed others most often perceived them (This was based on their perception of their physical characteristics or what others have said to them.)
   "If I were light-skinned and had long straight hair, I might tell you white best represents [me]. . . . When I look in the mirror, I see a black man."
2. culturally, by the particular community in which they felt a stronger sense of belonging,
   "I was raised that way [as an African American]. I was taught the things that you teach your son and daughter to look out being not white in this world. Growing up in the South, you know, you had to understand what you do and what you don't do to keep the family surviving, from your house being burnt down."
3. administratively, by the way in which they (or their parents) most often reported their race in administrative or official capacities, such as with birth certificates, driver's licenses, and employment or school applications,
   "In reality, it's who do you say you are, like [what] you say on your forms."
4. and ancestrally, by the group which composed the largest percentage of their genealogy.
   "If you look at the [family] tree, I'm like over 85 percent white."

Participants who had no trouble responding immediately interpreted the question as inquiring into one of these dimensions. Those who had difficulty but who ultimately provided an answer also based their answer along one of these dimensions. These participants, however, required additional time and effort (in some cases a substantial amount) to make sense of the question within a social, cultural, administrative, or ancestral framework. Those who refused to provide an answer did not conceptualize the question as inquiring into one of these dimensions, but rather stated that they were unsure, if not skeptical, of the question's intent because they saw it as dismissive of their multiple-race backgrounds. These three participants viewed the question as an unreasonable attempt to classify respondents within standardized race categories.

The above excerpt illustrates how cognitive interviewing methodology, as it generates a theoretical framework (i.e., the four dimensions of racial identity), can illustrate how it is possible to have differing interpretations of the same question that are also all valid. The advantage of such a framework is that question designers and subject matter experts can make better decisions regarding the adequacy of a question as it pertains to their research agenda. Additionally, interpretive trends of long-standing questions can be tracked across the years, such that those questions can be modified if historical and contextual changes alter the interpretive processes. Furthermore, the documentation can be utilized for survey harmonization or as an aid to translators needing to know the full interpretive scope of a question to create a comparable translation.

5.8   FINDINGS ARE GROUNDED

The last advantage, as is also illustrated in these previous excerpts, is that findings from these types of qualitative studies are inductively constructed, that is, conclusions derive directly from the observed narrativity. Glaser and Strauss' (1967) classic work, *The Discovery of Grounded Theory: Strategies for Qualitative Research*, forms the basis for this line of inquiry. Grounded theory is a systematic qualitative research methodology in the social sciences that emphasizes the generation of theory from data. It is a research method that operates in a reverse fashion from traditional, quantitative research. Rather than beginning with the development of a hypothesis, the first step is data collection. From the data collected, the key points are marked with a series of codes and then grouped into similar concepts. From these concepts, categories are formed which form the basis for theory generation. (For detailed description of the method, see Strauss and Corbin, 1990; Charmaz, 2006.) To the extent that theory is induced from observation, it is said to be grounded in the empirical data. This is the opposite of a deductive, quantitative research model in which a theoretical framework is first developed and then applied. These opposing epistemological positions form the basis for a beneficial mixed-method framework for cognitive interviewing methodology and survey research. Because interpretation of statistical data must be inferred (through assumptions built into the study design as well as into the actual meaning of identified relationships), cognitive interviewing methodology provides a grounded theoretical framework for making such inferences.

The previous few examples have begun to illustrate how such an inductive process can benefit survey research. For example, as in the terrorism project, patterns of interpretation found in the cognitive interviews generated a theoretical framework for understanding respondent-applied definitions of terrorism. This understanding was then used to inform a modified, and hopefully improved, version of the question. Additionally, the mixed-method approach is beneficial when the method is brought together with field testing. This strategy was used in a Washington Group evaluation project to evaluate six core disability questions developed for use on censuses around the world (Miller et al., 2010). Cognitive testing was conducted first at the NCHS, and then the interpretive patterns identified were used to develop a field test questionnaire that was then administered to a total of 1290 respondents in 15 countries in Central and South America, Asia, and Africa. The protocol was specifically designed to capture: (1) whether core questions were administered with relative ease; (2) how core questions were interpreted by respondents; (3) the factors considered by respondents when forming answers to core questions; and (4) the degree of consistency between responses to core questions and a set of more detailed functioning questions. Demographic and general health questions allowed for an examination of comparability, specifically, whether test questions performed consistently across all respondents, or if nationality, education, gender or socioeconomic status impacted the ways in which respondents interpreted or considered each core question. From this design, it was possible to identify patterns of response error in the survey data as well as to determine which respondents had fallen into those error patterns. The findings revealed, first, how the newly developed questions worked as a measure of disability, and then, second, whether that measure was comparable across socio-demographic groups.

5.9   ELEMENTS OF A CREDIBLE COGNITIVE INTERVIEWING STUDY

Lincoln and Guba (1985) argue that axioms governing quantitative methods should not be applied to qualitative methods and, instead, offer an alternative set of axioms. Those criteria include credibility, transferability, dependability, and confirmability. They also lay out methodological techniques for ensuring these principles. Indeed, there is a massive qualitative literature that articulates methodological techniques for assuring the credibility of qualitative investigation. While it would be fruitful to fully discuss each component as they relate to cognitive interviewing (as well as to include other critical components, such as subjectivity and reflexivity), there is not enough space in the confines of this chapter. However, three particular areas that are especially deficit in the current practice of cognitive interviewing have been identified and require at least mention. Those components include standards for evidence, a deliberate method of analysis, and transparency. The NCHS question evaluation staff are specifically thinking about and developing strategies to make improvements in these areas. In the end, it is believed that advancements in these particular areas strengthen the overall credibility of cognitive interviewing methodology and, in turn, its usefulness to survey research.

5.9.1   Standards for Evidence

As previously articulated, the narrative interview is the ideal format for identifying the interpretive processes of question response. However, it is critical
to recognize that, in the course of most cognitive interviews, respondents do
not provide narrative throughout the entire interview—regardless of the
probing techniques used (Willson, 2009). Other than the narrative, in the
course of conducting NCHS testing projects, at least three other types of
information that can be embedded in the interview text have been identified:
out-of-context interpretations, out-of-context speculations, and respondents'
question design opinions and preferences. Significantly, these types of information are not equivalent and do not convey the same kind of meaning. Most
importantly, they are not necessarily comparable, and when taken together
they can portray opposing realities about the performance of a survey question. For example, when answering a question about number of lifetime sexual
partners, a respondent can count their partners based on one definition of sex,
describe an entirely different definition of sex when asked, "what does the
word 'sex' mean to you?", and then speculate that "most people" consider sex
to be an entirely different concept. If different researchers base their findings
on different standards for evidence, they will most likely reach different conclusions about the performance of a question.

It is essential to recognize and evaluate these different (and sometimes
opposing) types of interview data, giving highest credibility to narrative information. In the interview, contradictions should be followed up and, to the best
of the interviewer's ability, resolved. Additionally, discrepant information must
be attended to when performing analysis. Because different types of information can lead to different conclusions, it is important to acknowledge these
discrepancies as well as to consider and weigh out what is actually known from
the evidence provided. As a best practice, these decisions should be made
transparent and documented in the final report.

5.10   DELIBERATE METHOD OF ANALYSIS

As previously articulated, there is minimal discussion regarding the method
of analysis for cognitive interviews. And, as argued above, analysis serves to
identify and reconcile dubious or inconsistent information as well as to identify
patterns and relationships and form theories about the interpretive processes.
It is this step in the cognitive interviewing project that makes cognitive interviewing relevant to survey research. The fact that analysis is not discussed or
debated openly in the literature, nevertheless, does not mean that analyses are
not occurring. However, because there is little in terms of documentation,
methods of analyses are rarely apparent and, therefore, may or may not be
thoughtful or deliberate.

As with quantitative methods, there are numerous ways in which an analyst
can go about examining cognitive interview data. The method of analysis must
be considered in the context and purpose of the cognitive interviewing project.

Miller (2007) and Miller et al. (forthcoming), for example, argue that when conducting a cognitive interviewing study specifically to examine cross-cultural or multinational comparability, it is necessary to perform three levels of analysis, moving from intra-interview comparisons to inter-interview comparisons, and then to country or cultural group comparisons. Willson (2006), on the other hand, to understand the false reports of alternative medicine usage, incorporates a highly detailed analytic scheme involving respondents' awareness of an alternative medicine culture and preconceptions of themselves in relation to this culture.

Most importantly, and just as in quantitative methodology, how an analyst actually goes about conducting analysis of cognitive interviews informs the findings of that study. Researchers using different methods of analysis can, therefore, reach different conclusions. And, incomplete analyses can result in incomplete or erroneous conclusions. For example, in her study of questions designed to collect health behaviors of new mothers before, during, and after pregnancy, Willson (2007) found that, with incomplete analysis, the response errors provided by some new mothers could easily be interpreted as a "telescoping problem." Specifically, while answering questions about smoking and healthy eating *prior* to their pregnancy, some women reported on their behaviors well *after* they were pregnant. However, with an additional analytic step, Willson was able to explain how the problem was specific only to those women who did not plan their pregnancies. Rather than being a general cognitive problem of telescoping, the real problem was related to a built-in assumption that women planned their pregnancies and that they were conscious of and were able to discretely distinguish between the time periods of before and during pregnancy. Critically, the question's "fix" is different depending on how the question design problem is understood. A telescoping problem is solved with emphasis on the time period or drawing on memory cues. However, regardless of the amount of provided cues, the underlying assumption problem would remain and systematically produce erroneous information about the specific group of women who did not plan their pregnancies—most likely those who are teens, have lower socioeconomic status, and older women. Instead, the more appropriate solution is to restructure the questionnaire such that the assumption is eliminated.

Because findings are intrinsically related to the analyses performed, it is important that users of cognitive interviewing studies can assess the completeness and credibility of those findings. Additionally, should the same question be evaluated at another time, it is important to be able to make sense of any differences found in the conclusions. As a best practice, the method of analysis must be explicit in a final report.

5.11 TRANSPARENT PROCESS

Finally, as it has been a running theme throughout this chapter, the method of cognitive testing has lacked transparency. As already noted, historically cognitive testing reports themselves have not been publicly accessible. Without knowing how a study was carried out, how data quality was ensured, as well as how those data were analyzed, it is impossible to defend the credibility of such a study. There can be no credibility to cognitive testing studies that are not accessible. Additionally, survey organizations have little ground to claim that their questions have been tested when there is no documentation to back up that claim. To be sure, a critical value of Q-Bank, and a primary reason that the NCHS has supported the endeavor, is that making reports accessible to the public brings credibility to the testing projects, as well as to the credibility of the method, and ultimately to the credibility of the survey data.

Providing public access to cognitive interviewing reports, however, is only the first (albeit, significant) step. The credibility of a particular study is tied to the level of detail and evidence provided in the testing reports. Thin reports with little description of what was found have questionable credibility. A best practice for reports is to document the method of data collection and the method of analysis that have led to the conclusions. Outside of the report, it is important that an audit trail be established tracing each finding to the original source so that an outside researcher can verify the conclusions. To this end, the NCHS has developed Q-Notes, a cognitive interview analysis program that generates such an audit trail. Additionally, the video of the actual interview is embedded within the application so that the findings can truly be traced to the original source.

5.12   CONCLUSION

In the method's 25-year history, cognitive interviewing methodology has developed from a novel innovation founded on the principles of cognitive psychology to a standard tool used in the production of survey data. In relationship to the cost of fielding a survey, the method is inexpensive, feasible, and offers unique advantages to survey research. Like no other question evaluation method, cognitive interviewing offers a detailed depiction of the meanings and processes that ultimately make up survey data. It offers an immediate understanding of the construct captured by a particular question and provides insight into question validity and response error.

Particularly because of its popularity and widespread use among U.S. federal statistical agencies, it is important to document and make accessible the actual practices of the method and the processes by which findings are drawn. Furthermore, it is necessary that "best practices" be developed through empirical study and open debate. Guidelines should address what constitutes a finding and the criteria for assessing the validity of the cognitive interviewing study itself. It is not possible for survey research to reap the full benefit of the method if criteria for assessing the validity of such evaluation studies are not available. Survey managers cannot be expected to incorporate question evaluation findings if there is no way to account for the veracity of those findings. Because of the reliance on cognitive interviewing methodology in the production of official statistics, it is even more imperative that the next generation take on these epistemological and methodological challenges. Such advancements should produce a truer, more accurate portrayal of the population which is represented through official statistics.

REFERENCES

Ackermann A, Blair J (2006). Efficient respondent selection for cognitive interviewing. Paper presented at the American Association for Public Opinion Research Annual Meeting: Montreal, Canada.

Beatty P (2004). The dynamics of cognitive interviewing. In: Presser S, Rothgeb JM, Couper MP, Lessler JT, Martin E, Martin J, Singer E, editors. Methods for Testing and Evaluating Survey Questionnaires. New York: Wiley; pp. 45–67.

Berger PL, Luckmann T (1966). The Social Construction of Reality: A Treatise on the Sociology of Knowledge. Garden City, NY: Anchor Books.

Charmaz K (2006). Constructing Grounded Theory: A Practical Guide Through Qualitative Analysis. Thousand Oaks, CA: Sage.

Conrad F, Blair J (1996). From impressions to data: increasing the objectivity of cognitive interviews. In: American Statistical Association 1996 Proceedings of the Section on Survey Research Methods. Washington, DC: American Statistical Association; pp. 1–9.

Conrad F, Blair J (2004). Data quality in cognitive interviews: the case of verbal reports. In: Presser S, Rothgeb JM, Couper MP, Lessler JT, Martin E, Martin J, Singer E, editors. Methods for Testing and Evaluating Survey Questionnaires. New York: Wiley; pp. 67–89.

Conrad F, Blair J (2009). Sources of error in cognitive interviews. Public Opinion Quarterly; 73(1):32–55.

Cook T, Campbell D (1979). Quasi Experimentation: Design and Analytical Issues for Field Settings. Chicago, IL: Rand McNally.

DeMaio T, Landreth A (2004). Do different cognitive interview techniques produce different results? In: Presser S, Rothgeb JM, Couper MP, Lessler JT, Martin E, Martin J, Singer E, editors. Methods for Testing and Evaluating Survey Questionnaires. New York: Wiley; pp. 89–108.

DeMaio T, Rothgeb J (1996). Cognitive interviewing techniques: in the lab and in the field. In: Schwarz N, Sudman S, editors. Answering Questions: Methodology for Determining Cognitive and Communicative Processes in Survey Research. San Francisco, CA: Jossey-Bass; pp. 177–195.

Forgas J (1981). Social Cognition: Perspectives on Everyday Understanding. London and New York: Academic Press.

Gerber E (1999). The view from anthropology: ethnography and the cognitive interview. In: Sirken M, Herrmann D, Schechter S, Schwarz N, Tanur J, Tourangeau R, editors. Cognition and Survey Research, Hoboken. New Jersey: John Wiley and Sons; pp. 217–234.

Glaser B, Strauss A (1967). The Discovery of Grounded Theory: Strategies for Qualitative Research. Chicago, IL: Aldine.

Greene JC, Caracelli VJ, editors (1997). Advances in Mixed-Method Evaluation: The Challenges and Benefits of Integrating Diverse Paradigms. New Directions for Program Evaluation, No. 74, San Francisco, CA: Jossey-Bass.

Harkness J, Van de Vijver F, Mohler P, editors (2003). Cross-Cultural Survey Methods. Hoboken, NJ: Wiley.

Harris-Kojetin L, Fowler F, Brown J, Schnaier J, Sweeny S (1999). The use of cognitive testing to develop and evaluate CAHPS(TM) 1.0 core survey items. Medical Care; 37(3):MS10–MS21.

Josselson R, Lieblich A, editors (1995). Interpreting Experience: The Narrative Study of Lives. Beverly Hills, CA: Sage.

Kuhn TS (1962). The Structure of Scientific Revolutions. Chicago, IL: University of Chicago Press.

Lincoln YS, Guba EG (1985). Naturalistic Inquiry. Beverly Hills, CA: Sage.

Madans J (2004). Discussant commentary: tools, policy and procedures for survey improvement. 2004 Federal Committee on Statistical Methodology Statistical Policy Seminar. December 15–16, 2004.

Miller K (2003). Conducting cognitive interviews to understand question-response limitations among poorer and less educated respondents. American Journal of Health Behavior; 27(S3):264–272.

Miller K (2006). Results of Cognitive Testing for Oral Health Questions. Available online at http://www.cdc.gov/qbank.

Miller K (2007). Design and analysis of cognitive interviews for cross-national testing. 2007 European Survey Research Association Annual Meeting. Prague, Czechoslovakia.

Miller K (2008). Results of the Comparative Cognitive Test Workgroup Budapest Initiative Module. Available online at http://www.cdc.gov/qbank.

Miller K, Willson S (2002). Results of Cognitive Testing for the NCHS Best Race Question. Available online at http://www.cdc.gov/qbank.

Miller K, Willson S (2004). Results of Cognitive Testing for Terrorism Experience Questions. Available online at http://www.cdc.gov/qbank.

Miller K, Mont D, Maitland A, Altman B, Madans J (2010). Results of a cross-national structured cognitive interviewing protocol to test measures of disability. Quality and Quantity; 45(4):801–815.

Miller K, Fitzgerald R, Padilla JL, Willson S, Widdop S, Caspar R, Dimov M, Gray M, Nunes C, Pruefer P, Schoebi N, Schoua-Glusberg A (forthcoming). Design and analysis of cognitive interviews for comparative multi-national testing. Field Methods.

Office of Management and Budget Standards and Guidelines for Statistical Surveys (2006). Available online at http://www.whitehouse.gov/omb/inforeg/statpolicy/standards_stat_surveys.pdf.

Presser S, Couper MP, Lessler JT, Martin E, Martin J, Rothgeb JM, Singer E (2004). Methods for testing and evaluating survey questions. Public Opinion Quarterly; 68(1):109–130.

Ridolfo H, Schoua-Glusberg A (forthcoming). Analysis of cognitive interview data using the constant comparative method of analysis to understand cross-cultural patterns in survey data. Field Methods.

Rothgeb J, Willis G, Forsyth B (2001). Questionnaire pretesting methods: do different techniques and different organizations produce similar results? In: American Statistical Association 2001 Proceedings of the Section on Survey Research Methods. Washington, DC: American Statistical Association.

Schwarz N (1994). Judgment in a social context: biases, shortcomings, and the logic of conversation. In: Zanna M, editor. Advances in Experimental Social Psychology, Vol. 26. San Diego, CA: Academic Press; pp. 123–162.

Strauss A, Corbin J (1990). Basics of Qualitative Research: Grounded Theory Procedures and Techniques. Newbury Park, CA: Sage.

Sudman S, Bradburn N, Schwarz N (1996). Thinking about Answers: The Application of Cognitive Processes to Survey Methodology. San Francisco, CA: Jossey-Bass.

Tashakkori A, Teddlie C (1998). Mixed Methodology: Combining Qualitative and Quantitative Approaches. Thousand Oaks, CA: Sage.

Tashakkori A, Teddlie C (2002). Handbook of Mixed Methods in Social and Behavioral Research. Thousand Oaks, CA: Sage.

Tourangeau R, Rips L, Rasinki K (2000). The Psychology of Survey Response. Cambridge: Cambridge University Press.

Tucker C (1997). Methodological issues surrounding the application of cognitive psychology in survey research. Bulletin of Sociological Methodology/Bulletin de Methodologie Sociologique June 1997. 55(1):67–92.

Willis G (2004). Cognitive interviewing revisited: a useful technique, in theory? Presser S, Rothgeb JM, Couper MP, Lessler JT, Martin E, Martin J, Singer E, editors. Methods for Testing and Evaluating Survey Questionnaires. New York: Wiley; pp. 23–44.

Willis G (2005). Cognitive Interviewing: A Tool for Improving Questionnaire Design. Thousand Oaks, CA: Sage.

Willson S (2006). Cognitive Interviewing Evaluation of the 2007 Complementary and Alternative Medicine Module for the National Health Interview Survey. Hyattsville, MD: National Center for Health Statistics. Available online at http://www.cdc.gov/qbank.

Willson S (2007). Cognitive Interviewing Evaluation PRAMS. Hyattsville, MD: National Center for Health Statistics. Available online at http://www.cdc.gov/qbank.

Willson S (2009). Data quality in cognitive interviewing: a new perspective on standardizing probes in multi-language and multi-cultural projects. Paper presented at QUEST meetings, Bergen, Norway.

6 Response 1 to Miller's Chapter: Cognitive Interviewing

GORDON WILLIS

National Cancer Institute

6.1 INTRODUCTION

Miller's chapter (this volume) advocates a new paradigm for the practice of cognitive interviewing. In her view, the "Cognitive Aspects of Survey Methodology (CASM) Paradigm" has enough serious limitations that it should be eclipsed by her proposed "Integrative paradigm." Overall, I find many of the arguments supporting the development of the Integrative paradigm to be useful, and in fact these ideas appear to reflect the culmination of an evolutionary development in the author's thinking. These have, in fact, influenced my own views, especially in the area of analysis of cognitive interviews. Nevertheless, a close review of the proposed model begs several questions:

(a) Does the "CASM paradigm" as described adequately reflect current conceptualizations of cognitive interviewing?

(b) Of the listed features of the Integrative paradigm, which of these truly represent a novel conceptualization, as opposed to the re-branding or extension of existing ideas?

(c) Similarly, of the listed features of the Integrative paradigm, which of these (whether novel or not) are either empirically or logically supported?

(d) To what extent does the inclusion of "new thinking" demand the *supplanting* of an existing paradigm, as opposed to *augmentation* of that paradigm?

Question Evaluation Methods: *Contributing to the Science of Data Quality*, First Edition.
Edited by Jennifer Madans, Kristen Miller, Aaron Maitland, Gordon Willis.
© 2011 John Wiley & Sons, Inc. Published 2011 by John Wiley & Sons, Inc.

I will address these questions for each of a number of ideas expressed in the chapter, focusing mainly on ideas summarized in two key tables (see Chapter 5, Tables 5.1 and 5.2) that strive to compare and contrast the CASM and Integrative paradigms. Concerning the characterization of the "CASM paradigm," I will focus on the field as it exists today. It is important to note that the CASM field has itself evolved since the time of the CASM I conference held almost 30 years ago—and that it does not necessarily encompass the same views as those it began with, in actual practice (and as well, may vary in significant ways between practitioners).

6.2   ARE WE LIMITED TO "COGNITIVE PROCESSES"?

First, concerning the underlying assumption Miller makes that "Cognitive processes are the focal point of question response," I have taken pains elsewhere (Willis, 2005) to point out that the cognitive model, as initially conceptualized by Tourangeau (1984) and colleagues, was simply the starting point from which cognitive interviewing has developed. This developmental course, which has taken us from a purely "cognitive" orientation, has a long and somewhat torturous history. At this point, it seems somewhat pedestrian, and uncontroversial, to state that cognitive interviews produce numerous results that are not well conceptualized as especially "cognitive" in nature. Questions that make false assumptions from their inception, those that have been mistranslated into target languages, and failures of items to satisfy measurement objectives, are not especially cognitive in nature, according to most conceptions of the term "cognition."

There was some initial resistance to this stance: When Willis et al. (1991) claimed that identified problems could be divided, roughly, into "cognitive" and "noncognitive" in nature, this notion was not well received by influential parties who had helped to midwife the CASM movement. There has, however, been a steady shift toward the accounting for perspectives other than the purely "cognitive." As such, the "CASM" field has become increasingly multidisciplinary, and has incorporated concepts from linguistics, sociology, anthropology, linguistics, and survey methods generally. As part of the conduct of a recent DC-AAPOR short course (Willis, 2009), I asked the audience (consisting of seasoned practitioners of cognitive interviewing, both within and outside of the federal government) the following: (1) How many of them had an academic background in cognitive psychology? and (2) How many made explicit use of the four-stage CASM-based cognitive model of the survey response process when engaged in their everyday work? Few (approximately 5 of around 30) featured a classically cognitive academic background; and significantly, although the group felt that the cognitive model has utility as an organizing framework, it simply does not characterize their view toward the actual conduct of cognitive interviews.

Although the cognitive orientation remains important, it simply does not dominate the practice of cognitive interviewing. In fact, returning again to the formative era of cognitive interviewing, it was likely never the case that those who were "down in the trenches" were constrained by the cognitive model. Trish Royston (1989), who was instrumental in the development of cognitive probing practices, opted to delete the cognitive label in her writings and to instead christen the practice as "intensive interviewing"—reasoning that relatively little of what went on in the cognitive lab was "cognitive" in the academic sense. It may be that if Royston's preferred term had become the preferred one, the field would at this time be freer of the nomenclature straightjacket implied by the "cognitive" label. Although proponents have found it useful to retain this term, I would reiterate that the field encompasses a conceptualization that is broader and more integrative, at least in terms of the affiliated interdisciplines.

6.3   UNIVERSALITY OF RESPONSE PROCESSES

Concerning the second listed underlying assumption of the CASM model, that "Response processes are universal and do not vary across respondents"—I will argue that this has never been a key viewpoint of the CASM approach, and certainly is not represented in the field as it currently exists. In fact, the explicit consideration of individual differences as a vital factor in the survey response process actually formed the basis for the approach. The seminal CASM-oriented paper by Loftus (1984), within the CASM I proceedings, investigated the degree to which individuals had a tendency to retrieve information on doctor visits in a forward chronological order, a backward order, or neither of these. She found support for the operation of disparate cognitive strategies, supporting the notion that response processes are clearly *not* universal. A later (but still relatively "historic") volume that represented the CASM field, by Herrmann et al. (1996), also explicitly concerned individual differences with respect to a range of processes relevant to the survey response process (see in particular the chapter by Willis on respondent selection of response strategies). Perhaps the most compelling example of CASM-oriented authors' regard for variation of cognitive processes across respondents is reflected in Schechter (1994), which was devoted to the in-depth investigation of the range of interpretations taken by the single survey question: *Would you say that your health in general is excellent, very good, good, fair, or poor?*. Similarly, current recommendations concerning the analysis of cognitive interviewing, as what could be called the "successive reduction" procedure by Willis (2005), again take pains to examine individual differences.

Miller's claim concerning invariance of processing appears to rely on the fact that CASM researchers often study small sets of subjects, under the assumption that once a finding is revealed, it will extend to the larger group—and that question functioning is therefore uniform in that sense. The actual reasoning encompassed by this approach is somewhat more subtle, however. What practitioners normally assert is that (1) multiple issues worthy of detection do exist, across different individuals, (2) these exist at varying frequencies, and that (3) the most common issues will likely emerge with smaller samples. There are of course additional types of issues that emerge when additional individuals (and categories of individuals) are interviewed, so that in an absolute sense, more testing is better. However, given usual constraints on sample sizes for cognitive tests, we make the assumption that a small number of interviews can detect the most common, and potentially serious, issues (i.e., potential sources of response error or source of variance) relatively quickly. This assumption may be incorrect. It is not, however, predicated on the belief that cognitive processes are invariant or uniform—simply that there are at least some aspects of question functioning that are shared among multiple respondents.

To summarize, the recognition that response processes vary across respondents has been a core feature of both the initial CASM conceptualization, and of successive variations of the cognitive interviewing field. It seems that what Miller means to say, given her emphasis on social context, is that there should be more understanding of the *correlates and determinants* of individual differences. This is a point well taken, as we sometimes have been content to chronicle the range of interpretations represented by a survey question, but without attempting to tie these disparate interpretations further to other constructs (whether viewed as demographic, situational, or cultural). As a result, cognitive interviewers have been somewhat limited in understanding sociocultural contributions to individual differences. Her suggestion to expand our viewpoint to encompass these factors is therefore a useful contribution.

6.4   PURPOSE OF THE COGNITIVE INTERVIEW

A third feature of cognitive interviewing described in Miller's paradigm table—and one that is perennially of interest—concerns the key issue of the "Primary purpose of cognitive interviewing studies." Miller describes this, under the existing (flawed) CASM paradigm, as "Fixing question flaws," whereas the novel (improved) paradigm stresses "the question's performance and the phenomena captured by the question." My first reaction is that this statement is ironic, given that the explicit initial focus of CASM was specifically on the issue of "understanding survey response processes" without much regard for "problems" in survey questions. There was much initial focus on understanding human cognition itself, with survey questions as the vehicle, but without an overwhelming emphasis on question defects. Miller has altered the perspective in a subtle way—that is, more toward the *survey item* as the unit of analysis than on the *individual who is processing the item*, as reflected in her phrasing concerning "question performance." However, the thrust of her argument is really to "return to our roots" in this regard, and to suggest that cognitive interviewers have been unwise in departing from the initial objective regarding scientific understanding, in favor of a view of the cognitive interview as a type of "fix-it tool" to directly solve our questionnaire design problems.

I consider this view—that we have abandoned the goal of understanding question function—as somewhat too strident. Admittedly, there has been a strong and continual applied focus on finding problems and then fixing them (Willis, 2005). A review of existing cognitive testing reports, however (as part of the process of writing that volume), revealed the inclusion of comments that simply strive to chronicle question function, without necessarily regarding these findings as "problems." As an example, virtually any report that includes the general health item ("Would you say your health in general is excellent . . . poor?") has commented on the range of interpretations of both the item itself, and its response categories—but without necessarily concluding that these represent a problem, or that the item should be altered (in particular, see Schechter, 1993).

Upon closer inspection, the issue of whether a survey question presents a "problem" is itself somewhat more murky, and nuanced, than is commonly appreciated. Often, the statement that a question is problematic is tempered by the inclusion of a conditional argument: Depending on the base measurement objective of the item, the item *may* be problematic—or it may not be. That is, the characteristic of "problem" is not immutable, but only exists in the context of the investigator's intent (see Beatty and Willis, 2007, for an extensive discussion of this point). As an example from Willis (2005), the tested item: "Have you had your blood tested for the AIDS virus?" was found to have (at the least) two key interpretations: (1) The respondent took active steps to have this done, versus (2) This was simply done (whether actively or passively—such as part of a military induction examination). Beyond documenting this variation, the investigators were interested in whether this in fact was a problem. If the client cares about the identified ambiguity, and prefers one or the other interpretation, then the item should be seen as exhibiting a problem, with respect to its objectives. In this case, it was determined that the client did in fact prefer one interpretation ("To your knowledge, has your blood been tested for the AIDS virus?"). The point is that we often first strive to understand question functioning, and to then determine whether the manner in which the question functions presents a "problem," in the context of the investigation. As such, Miller is certainly right that the classical approach is to search for problems—but this is often done in the context of understanding question functioning, as she advocates.

Further, I argue that the issue of whether our paradigm leads us to look for problems, as opposed to more generally understanding question function, may be misguided—as it is normally not the underlying *paradigm* that guides this, but the practical dictates of the situation. In the great majority of cases, clients (whom cognitive interviewers serve, after all) request that we identify potential sources of response error—in effect, a discrepancy between what is desired, and what is obtained. We therefore investigate question function (as Miller advocates), but focus on the product desired by the ultimate consumer of the results (the client). Changing paradigm in this case is not conceptually difficult—as we are clearly enabled in this regard. The challenge, rather, is that if we only report on question function—without relating this to the issue of "Is this a problem for the survey?"—we may no longer satisfy the needs of those parties who enable the practice of cognitive interviewing in the first place (by paying for it). To some degree, this tension may simply reflect the dissimilar interests of those such as Miller who prefer a purer, scientifically defensible approach, and those who are constrained by the practical factors that impinge upon the science. I believe that it is possible to satisfy both requirements—but only through compromise.

Looking to the next table of Chapter 5 (Table 5.2), a further presumed assumption of the CASM paradigm, under *Essence of cognitive interview data* is "A representation of respondents' cognitive processes," whereas the Integrative paradigm concerns "Respondents' interpretations of the question response process." Concerning the former, I have already made the case that the CASM orientation, at least in current form, concerns a much wider array of issues than simply cognitive processes. Whenever we find that a question sounds strange when read aloud by the interviewer, we are engaged in a "finding" that is not meaningfully "cognitive" in the manner defined here. Or, noting that we have asked about weight, but neglected height, and that we therefore will not be able to compute a Body Mass Index, again only involves the *cognitive interviewer's* cognitive processes (which may be alternately be labeled as "expert review," whether that practice is encouraged or not). To those who may object that these examples do not constitute products of "cognitive interviewing," my response is that, from the point of view of those who have actually conducted cognitive interviews, they most certainly do, as these are the types of problems that repeatedly emerge, despite our best efforts to avoid them in the initial design phase—whether our not our theory or paradigm demands that we seek them.

Turning to the proposed refinement reflected in the Integrative paradigm— that the essence of the cognitive interview is "Respondents' interpretations of the question response process"—I admit some bafflement concerning what is meant. If it means that we desire that respondents reflect upon their own response process, then this seems far too much to ask—as I do not believe that we should ask them to account for their own response processes (and ironically, this demand has sometimes been criticized as an unrealistic feature of the classic cognitive interview). More simply, respondents should provide information relevant to these processes. If Miller's intent is to say instead that we should study "Respondent interpretations" of the items—as she discusses extensively within the text—then that seems much more sensible.

A CASM proponent might argue that such a strong focus on one process— question comprehension/understanding/interpretation (however termed)—is somewhat limited, in that this only concerns one cognitive stage (and ignores retrieval, decision, and response processes, as discussed by Tourangeau, 1984).

Admittedly, however, a focus on interpretation processes does have considerable support in the empirical literature on cognitive interviewing. Analyses that have characterized question function (and in particular, question flaws or "problems") according to categorized type typically reveal that a great deal of the "action" concerns this category (Presser and Blair, 1994; Willis, 2005). Gerber and Wellens (1997) concluded that cognitive interviewing is, in essence, "the study of meaning," and of the manner in which meaning varies across individuals and social contexts. In fact, given the existing strong emphasis on interpretive processes, I take issue with Miller's statement that " . . . the CASM perspective lends itself to focus on aspects of memory and the retrieval of information." The CASM approach does encompass retrieval processes, but it is certainly not dominated by them, and as empirical studies have demonstrated, if it has a major focus, it clearly is the manner in which survey questions are *interpreted*.

Interestingly, as cognitive interviewing has moved to become more broad in scope (to encompass a range of issues that are not necessarily "cognitive"), it has simultaneously moved more narrowly toward an emphasis on one slice of this total pie—concerning interpretive processes—which does appear to constitute an enormously sized slice. It is this piece, I believe, that has become more interesting as we have increasingly included the viewpoints of sociologists, anthropologists, and linguists—and widened our conceptualization of item interpretation beyond its initial CASM-based definition. Hence, I agree with Miller that that there is considerable evidence that "interpretive processes" should be emphasized. This view does not represent a paradigm shift, but rather the advocacy of a shift within the field, relative to its roots.

6.5 SAMPLE SIZE

Further down the list, within Table 5.2, *Sample size* is explicitly addressed. The table indicates that, according to the CASM approach, interviewing is complete when all problems are identified, and for the Integrative paradigm, interviewing is complete when all interpretive patterns are identified. However, I am not sure that the CASM movement has ever come to grips with the issue of sample size, on theoretical grounds. Miller is certainly correct in one sense: For example, work by Conrad and Blair (2009) on the effect of sample size on cognitive interviewing does rely on the assumption that practitioners concur that our objective is to find as many (true) problems as we can.

The difficulty here is that, in applied practice, theoretical issues involving the relationship between sample size and problem detection almost never arise, as we tend to be limited either by administrative factors (e.g., OMB Paperwork Reduction Act), or by constraints of time, funding, and other resources. Hence, as I have stated repeatedly to audiences in cognitive interviewing short courses, I have *never* been in the situation in which we tested a series of questions until all the problems were found (or, using Miller's terms, until all interpretive processes had been identified)—we simply do not have this luxury in the world of real-life, "assembly-line" cognitive testing.

As such, I argue that sample sizes usually chosen for cognitive interviewing projects are not determined by paradigmatic outlook as much as they are by constraints imposed by existing circumstances. The relevant issue, therefore, is not whether the CASM paradigm and the Integrative paradigm differ with respect to sample size requirements (as both seem to invoke a "more is better" approach). Rather, we should ask whether the very limited sample sizes we often used are in fact sufficient to attain our testing objectives. To the extent that our objective is to identify problems in survey questions, the dominant view of cognitive interviewers can be summarized as "more testing is better, but some is better than none" (see Willis, 2005, or Beatty and Willis, 2007). Miller does seem to have raised the stakes in this regard, by suggesting instead that *we need to conduct more interviews, to be effective.*

Again, this is not necessarily a conclusion that derives from her paradigm— as Fred Conrad and Johnny Blair have made the same argument for some time. But whereas Conrad and Blair's argument is basically that, for a uniform set of individuals, more testing reveals more problems, Miller's view has a different basis, that I believe brings out a more serious issue that we increasingly face: To the extent that there is significant variability across subpopulation groups of interest to survey data collectors in interpretive practices, we naturally need to increase the scope of our testing in order to capture these differences. Whereas the Conrad–Blair argument is quantitative in nature (the number of interviews correlates with the number of problems identified), Miller is instead invoking a qualitative argument: Put simply, if our surveys endeavor to increasingly cover cross-cultural, multilingual, and even cross-national respondent groups, we need to include members of those groups in our testing. Overall, sample sizes naturally need to be increased to reflect the increased range of interpretive processes represented.

Miller may object that her approach is not limited to cross-cultural investigations. That is, the need to increase sample size to fully explore variation in interpretive processes becomes explicit and obvious within such investigations, but remains true for any study, whether or not it is explicitly cross-cultural. I am inclined to agree with this assessment. Where Miller especially hits home is with respect to the argument that we have insufficiently investigated whether cognitive testing, as normally practiced, reaches *saturation* (as the term is used by qualitative researchers), in which *all* dominant interpretations have been identified. To the degree that a single round of nine cognitive interviews is likely insufficient for reaching saturation, this is a difficult argument to dispel. So, I agree that cognitive interviewers should expand what they are already doing (i.e., studying variation in interpretive processes, and especially when engaged in explicitly comparative research).

However, a caveat I would place on the recommendation to greatly increase sample sizes for cognitive interviews is that, as many good recommendations, it may serve us better in theory (or as a feature of "best practice") than it does in "real-life" practice. Given the constraints (1) "We want you to test this survey" and (2) "It needs to go into the field in a month," it may be somewhat unrealistic to demand testing that achieves saturation with respect to interpretive practices. Of course, proponents of the Integrative paradigm are free to argue that cognitive testing is only beneficial if one conducts it according to ideal practices—and that otherwise, testing may even be counterproductive, and should not be carried out (i.e., either "Do it right" or "Don't do it at all"). This is a vital discussion to have but cannot be resolved based on abstract discussions of theoretical paradigms. Rather, critical conclusions in this regard must be driven by an empirical examination of data quality (or some measure of effectiveness of cognitive testing) under varying conditions of resource expenditure. In any event, the fundamental question of "How much testing is enough?" has been asked for some time. It *is* of considerable importance, but it is not uniquely spun from the Integrative (or any other specific) paradigmatic model.

6.6   METHODOLOGICAL FOCUS OF COGNITIVE INTERVIEWS

I next return to a particularly significant element of Table 5.2—concerning *Methodological focus*. Here the CASM model and the Integrative model purportedly depart strongly—as Miller states that the former focuses on verbal probing techniques, and the latter on analysis of conducted interviews. This distinction has considerable potential merit—but is somewhat exaggerated. First, an extreme claim that cognitive interviewing as currently conducted ignores analysis can be quickly refuted, as the one major book on (presumably current) cognitive interviewing techniques (Willis, 2005) contains a chapter devoted to analysis, and delves into issues such as how to compare and contrast results from different interviews, and how to combine these into a report that illustrates a coherent picture. In fact, although the terms are not used in that book, the notions of "grounded theory" and "constant comparison," from the qualitative research tradition, are at least somewhat represented by the recommended approach.

Similarly, the statement that according to the CASM tradition, "question flaws simply emerge from the interview text," independent of further interpretation, is not one that I have ever heard a proponent of cognitive interviewing make. Miller is certainly correct that cognitive testing reports are often very thin, in describing the path between data and conclusions made—and this is clearly a problem to be rectified. However, this does not in itself indicate that there was *no* such path or set of interpretive processes—only that these are not sufficiently transparent.

With respect to the opposite end of the continuum, if Miller's view is that "probing does not matter" and that analysis itself can carry the weight of data quality—that would constitute a somewhat radical statement, which I doubt she truly believes. In fact, Miller has elsewhere (Willis and Miller, 2008) contributed some excellent examples where defective probing has produced uninterpretable or misguided results: For example, where interviewers failed to probe adequately, they were unable to establish whether a subject understood what 100 yards is (i.e., simply accepting "Yes, I know what 100 yards is" is too thin a description). In contrast, probing to obtain "thick description" (of marching back and forth on a 100-yard football field as part of band practice) resulted in more convincing information.

Miller states that what is ultimately important is to obtain a full "narrative" from the subject to obtain a rich and useful description. This view is certainly laudable—and I view it as completely consistent (if not essentially identical to) the notion of "elaborative probing" (Willis, 2005). Elaborative probing consists largely of inducing subjects (in whatever manner) to provide a full explanation of the phenomenon under study, akin to the narrative. The consumer of this information then relates this rich set of information back to the response given to the survey question, to determine whether that answer is supported by the narrative, and to obtain a clear understanding of the subject's understanding (i.e., interpretive processes). Interestingly, although this approach is not well articulated under the original CASM framework, Willis (2005) found it to be well represented in the cognitive interviewing field, as testing reports are replete with such examples. This is perhaps a telling example of how the actual conduct of cognitive interviewing (what we do) departs from academic descriptions of the processes (what we say we do). Although this practice appears to have been in existence, if not dominant, through the past 20 years, it was only recently that it has been described as such, based on a review of actual cognitive testing results.

Miller stresses that such narrative is important—and in fact appears to be suggesting that such elaborative types of probing (in order to produce the necessary narrative) be our sole source of information. Although it may be too extreme to state that this is the *only* type of probing that we should do, or the only paradigm that should apply, I have some sympathy for this overall viewpoint, given that request for elaboration—getting the story in the subject's own words—is relatively unbiasing in comparison to alternative forms of probing. Further, as pointed out by Jaki Stanley (personal communication, September 18, 2009), this approach closely matches the philosophy underlying think-aloud, as it emanated originally from the CASM movement.

However, Miller does not indicate *how* one probes in order to obtain the narrative—and perhaps just as importantly, how one should *not* probe. Just as she claims that analysis does not spring automatically from the data, I will claim, similarly, that the data do not spring automatically from the conduct of a cognitive interview. That is, the interviewer engages in some type of activity, in order to produce those data—and *how* this is accomplished is not only relevant, but vital. Especially for those who have trained new cognitive interviewers, or who have been involved in the cross-cultural research field, it is apparent that the ability to probe effectively enough to obtain a rich narrative is a skill that takes considerable practice and judgment. Further, examples of poor probing techniques, and resultant probing pitfalls, are legion (see Willis, 2005, chapter 8), and can lead to seriously erroneous results. This remains true no matter how careful or sophisticated an analysis is conducted (as garbage in cannot produce anything other than garbage out). *Hence, I caution practitioners not to underestimate the importance of good probing practice—as this is a large component of the fundamental practice of cognitive interviewing.*

This is not to say that probing is more important than analysis. The objection I have to Miller's treatment of this subject is that it views probing-versus-analysis as something of a zero-sum game, in which a practitioner is forced to choose sides and decide that one side of the equation is more important than the other. Instead—if only to avoid the inevitable to-and-fro swinging of the conceptual pendulum—I suggest that she develops her very useful dictate to focus more intensely, and intensively, on analysis—without denigrating the value of careful probing. Her arguments in favor of more systematic analysis are in fact compelling. In particular, the recommendation to be more transparent with respect to cognitive interviewing results—by including these within the Q-Bank database of testing results (Miller, 2005)—is laudable. Further, the recommendation to incorporate a levels-of-analysis approach within comparative cognitive interviewing investigations (so that we can assess trends across groups, as well as within them) serves to inform the manner in which we should conduct analysis.

6.7 RESPONSES TO SEVERAL OTHER POINTS IN THE CHAPTER

Although Tables 5.1 and 5.2 constitute the major organizing framework for Miller's Integrative paradigm, she both precedes and follows with several other ideas that are worthy of comment.

6.7.1 The Need to "Know What a Question Is Actually Capturing"

This theme is articulated throughout the chapter, and is presented as a strength of the Integrative paradigm, and as a weakness of the CASM view. It is difficult to know exactly what the phrase is meant to encompass, due to intrinsic vagueness of the description, but given the associated emphasis on interpretive processes, I take it that this refers to the capacity of testing to *identify the full range of interpretations that are assigned to the item by survey respondents who are eligible to be administered the question.* Extending from my discussion above, traditional (or current) CASM-oriented practitioners do at least attempt to focus on "what a question is actually capturing," as that is normally viewed as part of our job. What we have not done, however, and which seems to be the "value added" by Miller's argument, is to chronicle or meaningfully articulate a *full* rendering of what a question is capturing.

Again, this argument recapitulates, perhaps in an ironic way, some elements of the distant past history of the cognitive approach. Perhaps the most extreme proponent of attempts to determine *everything* an item captures was Belson (1981), who favored an aggressive, exhaustive form of intensive probing in which the individual could be grilled for virtually an hour on the same item (to paraphrase his approach: "Ok, now I have all your main ideas about TV watching—Now I want the rest of your ideas . . ."). Later developments in the field have backed off from this extreme (as it would now be referred to as "over-probing"). Miller seems to be suggesting a return to this philosophical objective (to get all the potential information about how an item could be interpreted), but in a very different manner, as she focuses on *more* subjects, as opposed to *more* interrogation of the same subject.

Hence, there is an important distinction to be made, according to within-subject versus between-subject variation in "what the question captures." Miller's contribution is to recapitulate the intent embodied by Belson's approach (i.e., fully understanding what the item measures), but in a way that focuses on the manner in which the item's "sphere of capture" exists *between* individuals who represent varying social contexts. I find this to be eminently logical—but is achieved at a cost (increased sample size) that is only some-times achievable, at least given the parameters that we currently accept as minimal requirements for the conduct of cognitive interviews. An open ques-tion, of course, is whether we should reconceptualize those parameters, by raising our minimal standards.

6.7.2   Use of Case Studies

Near the end of the chapter, Miller has chosen to present a few case studies of questions that were evaluated, supposedly according to the Integrative para-digm (regarding, e.g., dimensions of terrorism and gum disease). Paradoxically, these examples seem somewhat weak exemplars of her best points. Despite the explicit emphasis on sample size, neither example reports how many sub-jects were tested (as she points out, a key defect of cognitive testing reports generally). Further, the prose-based summary of the results of testing that are provided strike me as similar to the types of findings that are represented in the dozens of "CASM-oriented" cognitive testing reports that I reviewed in order to summarize then-current analysis practice, when writing Willis (2005).

Interestingly, although Miller has earlier disavowed an emphasis on prob-lems with items, she includes this type of description here, with such statements as "Those respondents who had been diagnosed with some kind of condition did not necessarily have an easy time answering the question." Hence, the description is not as clearly of "interpretive processes," as she claims, as it is of difficulties with items (according to the classical paradigm, and which seems sensible). On the other hand, the description of who had problems, and under what circumstances, may be somewhat of an extension of our usual efforts to determine *why, and for whom,* questions present problems. However, this does not strike me as evidence of a fundamental shift that constitutes a new paradigm.

I find it disappointing that she did not instead choose to include in this chapter one of the best ideas I have seen developed recently through the elaboration of analysis procedures: The recommendation to include a "charting" approach to cognitive interview results, in which the findings from each interview, with respect to a series of explicit cognitive testing objectives, are carefully depicted on a matrix (the chart). Such a depiction makes clear what data were collected in each interview, and what was actually gleaned from each. Such an organizing framework helps to "keep us honest" in exhibiting what was sought, and what was found, within each interview, in a way that reduces reliance on extraneous factors (such as our opinions on what "should" or "should not" work). This organizing framework also tailors the interpretation in a way that facilitates a grounded theory approach, and enables a constant comparative procedure for assessment. The charting approach seems to well represent her emphasis on the need to evolve our analysis, and represents a novel contribution.

6.8 CONCLUSION

My conclusion, overall, is that Miller's ideas—especially concerning the need to modify our approach to analysis—have considerable merit. The overall critique I have is that the Integrative paradigm, in a more general sense, is either not novel, or makes several assertions that are not supported by the existing literature. It seems questionable to assert that the field is in need of a new paradigm that should supplant an older one. I advocate a less radical assertion: The field would be enhanced by an approach that recommends *augmentation* rather than *substitution*. I resist a view that presupposes that the status quo is clearly inferior (e.g., bloodletting as our major medical intervention) and needs to be supplanted by a superior paradigm (e.g., modern medical care involving hygiene and the administration of antibiotics).

Instead, I think we should conceptualize cognitive interviewing techniques according to a "toolbox" model, containing a number of tools that can be used as circumstances dictate. As such, the question of "Which paradigm is better?" is misguided, and we should therefore redefine our objective, and ask: "Under what circumstances and conditions is one approach likely to be more effective than another?" Hence, there are likely to be times when a fully Integrative paradigm, relying on large sample sizes, and that exhaustively investigates the full pattern of interpretive processes associated with a set of survey items, is the optimal approach. However, this does not mean that the "CASM" perspective—in the various guises in which it now exists—does not have an equally valid application, under different circumstances.

More specifically, it may be that a charting approach is necessary for cross-cultural investigations where there is considerable value in organizing the analysis. On the other hand, where a set of seasoned cognitive interviewers are given a week to complete a project, they might still be effective when applying more limited "tried and true," extant techniques. It is the job of methodologists to determine the conditions under which each of these systems best applies. Just as it makes little sense to conclude that Newtonian mechanics is "wrong" because quantum theory or general relativity theory apply best under particular circumstances (respectively, at the very microscopic vs. the very macroscopic), I also believe that there is room for a variety of perspectives that should be adopted, where they are found to best apply.

REFERENCES

Beatty P, Willis G (2007). Research synthesis: the practice of cognitive interviewing. Public Opinion Quarterly; 71:287–311.

Belson WA (1981). The design and understanding of survey questions. Aldershot, UK: Gower.

Conrad F, Blair J (2009). Sources of error in cognitive interviews. Public Opinion Quarterly; 73(1):32–55.

Gerber ER, Wellens TR (1997). Perspectives on pretesting: cognition in the cognitive interview? Bulletin de Methodologie Sociologique; 55:18–39.

Herrmann D, McEvoy C, Hertzog C, Hertel P, Johnson M (1996). Basic and Applied Memory Research: Practical Applications, Vol. 2. Mahwah, NJ: Erlbaum.

Loftus E (1984). Protocol analysis of responses to survey recall questions. In: Jabine TB, Straf ML, Tanur JM, Tourangeau R, editors. Cognitive Aspects of Survey Methodology: Building a Bridge between Disciplines. Washington, DC: National Academy Press; p. 61.

Miller K (2005). Q-bank: development of a tested-question database. In: American Statistical Association 2005 Proceedings of the Section on Survey Research Methods. Washington, DC: American Statistical Association.

Presser S, Blair J (1994). Survey pretesting: do different methods produce different results? In: Marsden PV, editor. Sociological Methodology, Vol. 24. Washington, DC: American Sociological Association; pp. 73–104.

Royston PN (1989). Using intensive interviews to evaluate questions. In: Fowler FJ, editor. Health survey research methods (DHHS Publication No. PHS 89–3447. Washington, DC: U.S. Government Printing Office; pp. 3–7.

Schechter S (1993). Investigation into the Cognitive Processes of Answering Self-Assessed Health Status Questions (Cognitive Methods Staff Working Paper No. 2). Hyattsville, MD: Centers for Disease Control and Prevention/National Center for Health Statistics.

Schechter S, editor (1994). Proceedings of the 1993 NCHS Conference on the Cognitive Aspects of Self-Reported Health Status (Cognitive Methods Staff Working Paper No. 10). Hyattsville, MD: Centers for Disease Control and Prevention/National Center for Health Statistics.

Tourangeau R (1984). Cognitive science and survey methods: a cognitive perspective. In: Jabine T, Straf M, Tanur J, Tourangeau R, editors. Cognitive Aspects of Survey Design: Building a Bridge between Disciplines. Washington, DC: National Academy Press; pp. 73–100.

Willis GB (2005). Cognitive Interviewing: A Tool for Improving Questionnaire Design. Thousand Oaks, CA: Sage.

Willis GB (2009). Current issues in cognitive interviewing. Short Course presented at the meeting of DC-AAPOR, Bureau of Labor Statistics, Washington, DC.

Willis GB, Miller KM (2008). Analyzing cognitive interviews. Short Course presented at the meeting of the Southern Association for Public Opinion Research, Durham, NC.

Willis GB, Royston P, Bercini D (1991). The use of verbal report methods in the development and testing of survey questionnaires. Applied Cognitive Psychology; 5:251–267.

7 Response 2 to Miller's Chapter: Cognitive Interviewing

FREDERICK G. CONRAD

University of Michigan

7.1 I AM LARGELY SYMPATHETIC TO THE VIEW PROPOSED BY MILLER

The crux of Kristen Miller's proposal about cognitive interviewing is what she calls the "Integrative Paradigm." The key idea is that respondents actively interpret survey questions because the meaning intended by the author is not directly transferred into respondents' heads upon hearing the questions. Instead respondents make sense of the question by bringing to bear their life circumstances and social situation, all of which are experienced outside the pretesting laboratory. This is important because it reminds us that word meanings are not as stable and as universal as we typically assume and that there may be systematic differences in the experiences and, thus, interpretations of different groups of respondents. Miller illustrates the idea nicely in Chapter 5. Answering a question about whether they have suffered from gum disease is easy for respondents who had recently seen the dentist: Either the dentist told them they had gum disease or did not but would have if they had had the disease so no mention was taken as evidence they were disease-free. For those who had not recently been to the dentist, the task was harder. The chapter argues convincingly through many clear examples that the information brought to bear in interpreting questions is individualized, or at least differs by groups of individuals, and is far from universal.

I am sympathetic to this view having shown with Michael Schober (e.g., Schober and Conrad, 1997, 2002; Conrad and Schober, 2000) that accurate

Question Evaluation Methods: Contributing to the Science of Data Quality, First Edition.
Edited by Jennifer Madans, Kristen Miller, Aaron Maitland, Gordon Willis.
© 2011 John Wiley & Sons, Inc. Published 2011 by John Wiley & Sons, Inc.

interpretation of survey questions may require not only active interpretation by the respondent but also collaboration between respondent and interviewer. Interviewers may need to help respondents classify their circumstances. For example, when asked "Do you have more than one job?" a respondent who babysits for three families might not know whether to consider this one job (babysitting) or three (one job with each family). An interviewer who is able to define "more than one job" can resolve the ambiguity in how the concept corresponds to the respondent's situation. So this involves more than just active interpretation by the respondent; it involves collaboration between the respondent and interviewer.

Miller gives almost as much emphasis to the interactive character of cognitive interviews as to the role of respondents' individual circumstances. The idea, I believe, is that the narrative created through the interaction between interviewer and respondent contains rich evidence of the respondent's interpretive processes. And much of this material would not surface under less interactive conditions. I am sympathetic to this as well. I believe there is a story in the details of interaction that is almost always lost when one looks at just the outcome of an exchange, for example, answers to survey questions (see, e.g., the analysis of interaction between respondent and interviewer in cognitive interviews in Conrad and Blair, 2009, pp. 44–47). In fact, I subscribe to the view that there is important information about respondents' interpretive processes not only in their words but also in paralinguistic aspects of their speech, for example, disfluencies. We (Schober et al., under review) have found that respondents produce more *um*s and *uh*s while answering the question when they ultimately changed their answer (an unreliable answer) than when they did not change it. And the details of interpretation are not restricted to spoken information: In certain face-to-face interviews, respondents looked away from the interviewer while answering more often and for longer durations when they ultimately changed their answer than when they did not. So I applaud Miller's commitment to mining the interaction for insight into respondent interpretation processes and encourage her to go further, looking beyond respondents' words to the verbal and visual paradata they produce while answering questions.

Miller mentions several times that an analysis of respondents' interpretive processes can not only inform the (re)design of survey questions but can serve as auxiliary information for users of survey data about the origins of the published estimates. In effect, by considering cognitive interview analyses, data users might gain insight into how respondents probably interpreted the questions when providing the answers from which a particular statistic is derived. I think this is a promising idea and can imagine integrating cognitive interview results with published survey results. For example, one can imagine data users being able to roll their mouse over an entry to obtain this kind of description, much like making question wording and other metadata available to data users.

Appealing as this idea is, it makes it hard to defer the representativeness issue, namely how representative are the cognitive interview findings of the population whose attributes are estimated by the quantitative survey data. If particular respondent subgroups, for example, those who have not been to the dentist recently, answer questions about gum disease by different means than another group, for example, those who have been to the dentist recently, the data use requires knowledge of the subgroups' proportions in the population if the goal is truly to better understand the quantitative results. Yet, I suspect Miller would be the first to point out that cognitive interviewing is not conducted to, nor is it able to, support population estimates.

Miller is rightly critical of the largely black box approach to analysis of respondents' narratives (or "verbal reports," depending on your perspective) in current cognitive interview practice. The interviewers or analysts are often assumed to be experts, and their analytic activities are essentially trade secrets that are not to be shared. But given the centrality of the analyses to converting respondents' narratives into findings, they really should be made available and explicit. Miller's criticism implies that current practice is not only private but also procedurally unspecified. She argues that practitioners should be explicit about "the processes used to sift through the text, how the text was actually examined or pieced together to form conclusions, as well as the types of decisions used by the interviewers and judges to make sense of contradictory information, attend to gaps in respondents' reports, or whether to disregard information that they might have deemed irrelevant or potentially dubious." This degree of specificity and prescription is one approach. It is also possible that analysts need more discretion in how best to interpret what respondents have said, given that they are well-versed in what is known about survey response processes.[1] In any event, greater clarity and precision in reporting analytic processes would be a positive step. I assume Miller has produced this kind of detailed specification for the narrative analyses whose results she reports. Unfortunately, those procedural details are not presented in her chapter. One gets the sense that she advocates study-specific analytic approaches because her examples from different studies seem to result from different types of analyses. Perhaps the analyses are developed from the ground up as are the theories by her view. This is important material that could serve as a model for how the field should design and document its analytic practices so its absence is unfortunate.

Finally, I am sympathetic to Miller's view that larger numbers of interviews than are now typical will help ensure that different perspectives are sampled and considered in the analysis (see Chapter 5, Table 5.2). Johnny Blair and I (Blair et al., 2006; Blair and Conrad, in press) have shown that larger samples of cognitive interviews turn up increasingly larger numbers of problems. And the problems that were turned up for the first time in larger samples (our study goes up to samples of size 90) include almost as many serious or "high-impact" problems as are observed in small samples. This suggests that small numbers of cognitive interviews miss many problems, including serious problems, that is, those judged most likely to distort the answers and inferences based on them. So, I strongly endorse larger samples. Note, though, that larger convenience samples do not address the representativeness issue; they just increase the chances of observing some problems.

7.2   NOT SUCH A PARADIGM SHIFT

Although much of Miller's proposal resonates for me, I do not think it represents quite the paradigm shift that she suggests it does. More specifically, current practice has room for different life experiences and for the interactive origins of the data.

If a respondent in a typical cognitive interview were to refer to her personal history as a source of difficulty understanding or answering a question, there is nothing to prevent this from being included in an analysis of how the question functions for people like her. And there is nothing to prevent such an analysis from concluding that some respondents, with some personal histories, interpret the question in a way that creates ambiguity or otherwise causes trouble, while other respondents with other histories interpret it differently, possibly without trouble. I cannot imagine a practitioner of cognitive interviewing would give less weight to interpretation difficulties due to personal history than, for example, to those due to semantic or grammatical ambiguity, which presumably are experienced by respondents irrespective of their particular histories.

With respect to the interactive character of cognitive interviews, most versions of the technique involve conversation between an interviewer and respondent over multiple turns. Generally, the interviewer triggers additional respondent turns by probing one way or another. While there is a literature, referred to by Miller, that is concerned with the way different probing techniques produce different types of responses, the general point is that the raw data produced by cognitive interviews and which is typically analyzed is *the sequence of turns* comprising the exchange between interviewer and respondent. So while the analyses typically do not trace the impact of one turn upon the next—which would make for an even more interactive analysis—all of the information arising from the interaction is potential data about how respondents interpret and answer (or do not answer) survey questions. Thus, I believe that typical cognitive interviews (in the right hands) already mine much of the information in the interaction between respondent and interviewer.

7.3   SOME MISUNDERSTANDINGS, I THINK

As I have indicated, I think Miller's view and the view she critiques are closer together than she might suggest. I believe the apparent distance between the approaches is exaggerated by some misunderstandings.

First, Miller suggests that the theoretical perspective from which typical cognitive interview research is conducted holds that all respondents go through the same set of response processes answering a question. The implication she draws from this is that problems with a question are assumed to be relatively universal which in turn suggests that a survey question operates the same way for all people under all circumstances. I do not think this an accurate characterization of the information processing approach. No matter what one's intellectual orientation is, it is hard to quarrel with the idea that respondents need to understand what they are being asked and, given their understanding, bring to mind relevant information to produce a response. Yet this view does not require all respondents to experience the questions in the same way. This view is about *processes* not *content*; the information or content which gives rise to comprehension (generally assumed to be the first of these processes) can, on the one hand, come from extra-laboratory, personal circumstances (like the dental example above) or, on the other hand, from knowledge of the world shared by most adults (e.g., that a particular word has more than one conventional meaning, rendering the question ambiguous). In the first case, respondents should have very different experiences; in the latter more similar. The point is that the same set of processes can produce very different results (e.g., problems with a question) depending on the information upon which the processes act. Thus, the information processing and integrative approaches are actually quite compatible: The former is concerned with how the information is processed and the latter with the information itself.

Second, Miller suggests that the primary goals of the Cognitive Aspects of Survey Methodology (CASM) (information processing) and integrative perspectives are different, but I think there is more overlap than she suggests. She characterizes the primary goal of the CASM approach as "fixing question flaws" (e.g., Chapter 5, Table 5.1). I think this is right in principle, although the cognitive interview process can really only identify flaws or problems with questions, and it is up to clever designers to overcome the problems by revising the questions. In contrast, she describes the goal of the integrative approach as "Documenting [the] theoretical framework regarding the question's performance and the phenomena captured by the question" (Chapter 5, Table 5.1). I believe that in this statement she is advocating the documentation of how the question works. But, a major goal of the integrative approach is surely to detect problems with questions so that they can be improved. Miller seems to believe that authors of questions intend respondents to interpret the questions in particular ways. Thus, when respondents do not glean the intended meaning, there is a problem. Consider the following excerpt:

> Rather than to identify and correct specific problems, the objective of this approach is to identify and explain the patterns of interpretation throughout the entire question response process. More specifically, the purpose is to generate a theoretical framework for understanding a question's performance. This theoretical framework, then, delineates the phenomena captured by the question. Obviously, some of these patterns may be unintended and, if left without

modification, would lead to "error" in the survey data. In this regard, such studies are important for pre-testing.

Similarly, Miller seems to treat the goal of cognitive interviewing under the integrative approach as fixing questions so they produce the intended interpretation for most respondents:

> In order to write a successful survey question (i.e., one that would elicit responses framed within the intended definition), it was important to make sense of these seemingly off-the-wall interpretations and to fully understand the various dimensions in which terrorism was being defined.

So while I understand that the approach elucidates how different respondents react to a question and think about the concepts in the question, the reasons why one would use this approach seems to be in the spirit of bringing respondents' interpretation into alignment with what was intended by the question authors, and this is done by rewording (i.e., fixing) the question.

It may be helpful to think about the role played by cognitive interviewing at different points in the questionnaire development process. In the earliest stages there often are not formal questions and when there are, the intentions are not always well-defined so it does not make sense to think about "fixing" questions at this point. Instead, cognitive interview data are most valuable in helping questionnaire designers become immersed in the thinking of respondents, including different groups of respondents. Later in the question development process, when questions have been honed and their intended meanings are clear to question authors, cognitive interviewing that detects problems communicating those intentions is particularly helpful. So understanding how respondents conceptualize the issues in a domain, independent of how well they grasp the intended meaning, may be most helpful in the early stages of questionnaire development and fixing problems with questions may be more helpful when the questionnaire is more mature.

Third, Miller suggests that most of the action in answering survey questions lies in interpretation processes, that is, recall, estimation, and response mapping are largely shaped by interpretation: "How respondents go about processing the retrieved information as well as to map that assessment onto the response categories are also rooted within interpretation." She reports that when respondents answered a question about how often they had experienced pain in the past week, their decisions about what kinds of episodes to include (most intense, any pain, average pain) influenced their answers to a great extent. Surely, retrieval processes are constrained by what people decide they are being asked to search for, but people certainly carry out retrieval processes that are separable from interpretation, and they experience problems and make errors when engaged in these processes (e.g., forgetting and telescoping). So interpretation is far from the whole story. I believe Miller's emphasis on interpretation reflects the strength of cognitive interviewing in tapping into interpretive processes and its weakness in tapping into other processes, for example, retrieval, more than which processes actually contribute to a survey response.

Johnny Blair and I (Conrad and Blair, 2004) investigated which types of problems occurred most often (i.e., most problems per question) in a set of cognitive interviews. The problems that were most prevalent were related to comprehension: lexical (0.13) and logical (0.11) versus temporal (0.02) and computational (0.04) problems. The first two categories apply to problems that would seem to occur in what Miller calls "interpretive processes"; the second two categories involve problems that are less about interpretation and more about formulating an answer. This suggests that cognitive interviewing is simply more sensitive to comprehension-related problems than to problems occurring at other points in the response process. Yet we know from 25 years of experimental CASM research that measurement error occurs in retrieval and estimation and judgment processes as well as comprehension. Error occurs in many response processes, but cognitive interviewing is strongest in making explicit the processes that concern understanding.

Fourth, I believe Miller may be confusing certain practices used to *evaluate* cognitive interviewing with practices used to *conduct* cognitive interviewing. More specifically, she argues that some practitioners of CASM-style cognitive interviewing have developed coding systems to make their analyses more systematic but, in fact, they still leave the analytic process, that is, how coders assign codes, unspecified. In fact, we (Conrad and Blair, 2004, 2009) coded the verbal reports in order to make them comparable across two types of cognitive interviews because the study concerned the performance of two different approaches to cognitive interviewing on measures well served by codes, for example, agreement between coders or coders and cognitive interviewers. This is not something that would necessarily be done in practice, that is, in order to revise a questionnaire, and to my knowledge, it rarely is. Moreover, there is no reason one could not specify exactly the criteria required for assigning a verbal report to a problem category, demystifying the analytic approach. We did not do this because one of the interviewing techniques in our comparison was intended to approximate typical practice, and in typical practice interviewers (who are also the analysts) rarely specify their analytic technique. The point is that evaluation studies may involve methods that the investigators would not advocate for production use, and I suspect this distinction may have been lost in Miller's account, creating illusory distance between the approaches she contrasts.

7.4 POINTS OF DISAGREEMENT: REACTIVITY AND RELIABILITY

Miller addresses the issue of data quality in cognitive interviews on several occasions. She argues for the use of evaluation criteria that are specific to qualitative data: credibility, transferability, dependability, and confirmability.

The meanings behind these labels are not entirely self-evident (and unfortunately not defined by Miller), but they do not seem to include a test of agreement between different analysts. I think this is a significant limitation. It seems reasonable that two or more analysts evaluating the same cognitive interviews should reach similar conclusions if the interpretation of any one analyst is going to be given weight. This is not to advocate for routinely computing agreement in the pretesting laboratory, but occasionally this should be done so that one can be confident the method leads to relatively objective and consistent conclusions. The fact that the analyses may be inherently qualitative does not preclude direct comparison. For example, one might solicit expert judgments about the degree to which two analyses overlap or code the two analyses so that they can be compared with measures like kappa. The point is that without confidence in the reproducibility of a set of results either across replications of the study or across analysts, it seems to be a gamble to revise a questionnaire on the basis of those results. Without some sense of agreement, one has to ask if a set of findings is idiosyncratic, that is, a minority view and if it is wrong.

I am also quite concerned about reactivity, that is, the tendency for what a respondent says while providing a verbal report to change the processes about which he or she is reporting. These effects are documented in the psychological literature on verbal reports (see, e.g., Russo et al., 1989 and Schooler et al., 1993). Johnny Blair and I have extended this idea to include the impact of what interviewers say—probes—in response to respondents' verbal reports. We have observed in our 2004 and 2009 publications that probes that are hypothesis driven—interviewers believe *a priori* that respondents may experience a problem and so inquire about (probe for evidence of) that problem—lead to more ambiguous data (lower agreement about the presence of a problem) than probes that explore explicit evidence the respondent is experiencing a problem. I am especially concerned about reactivity in the approach to cognitive interviews advocated by Miller because of its emphasis on interaction. If interviewers are required to actively engage with respondents to generate a rich narrative, there is serious potential for this engagement to affect what respondents say.

Miller has a very different view:

> [T]he primary goal of the interviewer is to capture as much of the narrative as possible and to ask whatever questions are necessary to fill in gaps or to address apparent contradictions. To be sure, as required by the CASM approach, the interview text is not the essence of respondents' cognitive processes; at the very most, it is respondents' interpretations of their cognitive processes. The integrative approach acknowledges this and, in fact, argues that this is, in actuality, what is required for a truer understanding of the question response process: respondents' interpretive processes.

I am not persuaded that this is risk-free. Interviewers can easily focus respondents on certain topics and as a result distract them from others. It is to avoid this kind of influence that researchers (e.g., Ericsson and Simon, 1993) advocate the collection of *concurrent* verbal reports. Reporting while performing the task affords participants (respondents) little opportunity to interpret or rationalize or tidy up their thinking. Their verbal reports provide a relatively direct window onto their thinking. But I also recognize that limiting what interviewers can and cannot say may well reduce the depth of investigation (e.g., we observed fewer problems in our 2004/2009 study when interviewers' probes were restricted and reactivity was presumably reduced). This is a thorny problem, probably inherent in cognitive interviewing, and one that really needs to be examined and discussed further.

7.5   FINAL COMMENTS

I am confident that everyone involved with cognitive interviews wants to improve survey measurement by refining questionnaires. We are on the same team. I have devoted more words to my agreement with Kristen Miller's proposals than to any differences. And, as I have tried to argue, I believe her proposals share much more with what she calls the CASM approach than her chapter would suggest. Where I have expressed clear differences with her are, I believe, some of the areas where we should focus the next generation of methodological research on cognitive interviews.

NOTE

1   Miller is critical of Conrad and Blair (2009) for lack of clarity about our analytic approach and details about the analysts (judges): "The paper presents no discussion of analysis for either the interviewers or the judges." We actually provide this information, albeit in abbreviated form, on p. 40.

REFERENCES

Blair J, Conrad F (in press). Sample size for cognitive interview pretesting. Public Opinion Quarterly.

Blair J, Conrad F, Ackerman AC, Claxton G (2006). The effect of sample size on cognitive interview findings. In: American Statistical Association 2006 Proceedings of the Section on Survey Research Methods. Washington, DC: American Statistical Association.

Conrad F, Blair J (2004). Data quality in cognitive interviews: the case of verbal reports. In: Presser S, Rothgeb JM, Couper MP, Lessler JT, Martin E, Martin J, Singer E, editors. Methods for Testing and Evaluating Survey Questionnaires. New York: Wiley; pp. 67–89.

Conrad FG, Blair J (2009). Sources of error in cognitive interviews. Public Opinion Quarterly; 73:32–55.

Conrad FG, Schober MF (2000). Clarifying question meaning in a household telephone survey. Public Opinion Quarterly; 64:1–28.

Ericsson A, Simon H (1993). Protocol Analysis: Verbal Reports as Data, 2nd ed. Cambridge, MA: MIT Press.

Russo J, Johnson E, Stephens D (1989). The validity of verbal protocols. Memory and Cognition; 17:759–769.

Schober MF, Conrad FG (1997). Does conversational interviewing reduce survey measurement error? Public Opinion Quarterly; 61:576–602.

Schober MF, Conrad FG (2002). A collaborative view of standardized survey interviews. In: Maynard DW, Houtkoop-Steenstra H, Schaeffer NC, van der Zouwen J, editors. Standardization and Tacit Knowledge: Interaction and Practice in the Survey Interview. New York: Wiley; pp. 67–94.

Schober MF, Conrad FG, Dijkstra W, Ongena Y (under review). Ums, uhs, and gaze aversion as evidence of unreliable survey responses.

Schooler JW, Ohlsson S, Brooks K (1993). Thoughts beyond words: when language overshadows insight. Journal of Experimental Psychology: General; 122:166–183.

PART III
Item Response Theory

8 Applying Item Response Theory for Questionnaire Evaluation

BRYCE B. REEVE

University of North Carolina at Chapel Hill

8.1 INTRODUCTION

A well-designed questionnaire requires a multi-method approach in the development, evaluation, and refinement of the instrument. Recent development of questionnaires has emphasized the integration of methods from the fields of the cognitive aspects of survey methodology, qualitative research, survey design and methods, information technology research, and psychometrics. Within the field of psychometrics, item response theory (IRT) has been recognized as a valuable tool to provide a comprehensive examination of item and scale properties based on collected response data.

IRT models describe, in probabilistic terms, the relationship between an individual's response to a survey question and his or her standing on the construct being measured by the questionnaire. These measured constructs include any latent (i.e., unobservable) variable, such as math ability (in educational testing), depressive symptomatology (in psychological assessment), or fatigue level (in health outcomes research), which require multiple survey questions to estimate a person's level on the construct. It is a person's level on the latent construct that affects their responses to each question in the survey. IRT models provide detailed information about both the characteristics of the items and the respondents to allow instrument developers to understand how the set of items capture the construct of interest in their study population.

Question Evaluation Methods: _Contributing to the Science of Data Quality_, First Edition.
Edited by Jennifer Madans, Kristen Miller, Aaron Maitland, Gordon Willis.
© 2011 John Wiley & Sons, Inc. Published 2011 by John Wiley & Sons, Inc.

In this article, we introduce the basics of IRT, which includes an overview of the IRT models, assumptions underlying the models, and a comparison of approaches to survey construction between classical test theory (CTT) and IRT. Furthermore, we focus on the application of IRT models for the evaluation and refinement of questionnaires, examination of subgroup differences in responding to survey items, and the development of item banks that serve as the foundation for computerized-adaptive testing (CAT).

8.2   THE IRT MODEL

Each item (question) in a multi-item scale is modeled with a set of parameters that describe how the item performs for measuring different levels of the measured construct. For example, in the field of psychological assessment, the question "I don't seem to care what happens to me" would have IRT model properties reflecting it is informative for measuring people with high levels of depressive symptoms, and a question such as "I am happy most of the time" would have IRT model properties reflecting it is informative for measuring people with low levels of depressive symptoms. Each item in a scale is fit with its own IRT model.

The probabilistic relationship between an individual's response to an item and their level on the latent construct is expressed by the IRT model item response curves (also referred to as item characteristic curves, category response curves, or item trace lines). For example, Figure 8.1 presents the IRT

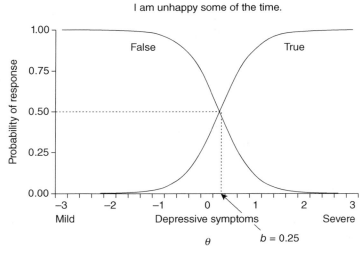

FIGURE 8.1. IRT item response curves representing the probability of a "false" or "true" response to the item "I am unhappy some of the time" conditional on a person's depression level. The threshold ($b=0.25$) indicates the level of depression (θ) needed for a person to have a 50% probability for responding "false" or "true."

response curves for the item "I am unhappy some of the time" which has two responses, "false" and "true." This item is an indicator of a person's level of depressive symptoms that is represented along the horizontal axis in Figure 8.1 and denoted by the Greek symbol theta, θ.

People vary in their level of depressive symptoms from individuals with mild depressive symptoms located on the left side of the θ continuum in Figure 8.1, to people with severe depressive symptoms located on the right side of the axis. Numbers on the θ-axis are expressed in standardized units and, for the illustrations in this discussion, the mean depression level of the study population is set at zero, and the standard deviation is set to one. Thus, a depression score of $\theta = 2.0$ indicates that a person is two standard deviations above the population mean and reports severe depressive symptoms. The vertical axis in Figure 8.1 indicates the probability, bound between zero and one, that a person will select one of the item's response categories. Thus, the two response curves in Figure 8.1 indicate that the probability of responding "false" or "true" to the item "I am unhappy some of the time" is conditional on the respondent's level of depressive symptomatology. Depressed individuals on the right side of the θ continuum will have a high probability for selecting "true," whereas those with lower depressive symptomatology are more likely to select "false" for this item.

The response curves in Figure 8.1 are represented by logistic curves that model the probability P that a person will respond "true" ($X_i = 1$; for the monotonically increasing curve)

$$P(X_i = 1|\theta, a_i, b_i) = \frac{1}{1 + e^{-a_i(\theta - b_i)}}$$

to this item (i) is a function of a respondent's level of depressive symptomatology (θ), the relationship (a) of the item to the measured construct, and the severity or threshold (b) of the item on the θ scale. This model is the IRT two-parameter logistic (2PL) model because the magnitude of the two parameters (a and b) vary for each item in the scale. In IRT, a is referred to as the item discrimination or slope parameter and b is referred to as the item difficulty, severity, or threshold parameter. When an item has just two response categories, the equation for the monotonically decreasing curve (i.e., the "false" response in Fig. 8.1) is a linear transformation of one minus the expression on the right side of the equation above.

The item threshold (b) parameter is the point on the latent scale θ at which a person has a 50% probability of responding "true" to the item. In Figure 8.1, the item's threshold value is $b = 0.25$, which indicates that people with depressive symptom levels a quarter standard deviation above the population mean have a 50% probability of indicating "true" to the question. Note in the equation that the threshold parameter varies for each item (i) in a scale, and it is possible to compare threshold parameters across items to determine items that will more likely be endorsed by people with lower or higher levels of depressive symptoms. For example, the item "I don't seem to care what happens to me" has a threshold parameter of $b = 1.33$, indicating it takes higher depression levels for a person to have a 50% probability for indicating "true" to the item. In general, individuals are less likely to endorse "true" to this severe question.

The discrimination (a) parameter indicates the magnitude of an item's ability to differentiate among people at different levels along the trait continuum. An item optimally discriminates among respondents who are near the item's threshold b. In Figure 8.1, the slope at the inflection point (i.e., the point at which the slope of the curve changes from continuously increasing to continuously decreasing) is $a = 2.83$. The larger the a parameter, the steeper the curve is at the inflection point. In turn, steeper slopes indicate that the trace curve increases relatively rapidly, such that small changes on the latent variable (e.g., small changes in depression level) lead to large changes in item-endorsement probabilities. The a parameter also may be interpreted as describing the relationship between the item and the trait being measured by the scale. Items with larger slope parameters indicate stronger relationships with the latent construct and contribute more to determining a person's score (θ). For example, the item "I cry easily" has a discrimination parameter estimate of $a = 1.11$ and is less related to the construct of depressive symptomatology than the item "I am unhappy some of the time" $(a = 2.83)$.

The equation for the IRT model below includes a third parameter, c, which is referred to as the guessing parameter. This model, the IRT three-parameter logistic (3PL) model, was developed in educational testing to extend the application of IRT to multiple-choice items that may elicit guessing. The guessing parameter is the probability of a positive response to item i if the person does not know the answer. For example, if there are four multiple-choice options, c will often be estimated to be around 0.25 but can vary depending on the attractiveness of alternative choices. When $c = 0$, the 3PL model is equivalent to the IRT 2PL model described above. Including the guessing parameter changes the interpretation of other parameters in the model. The threshold parameter b is the value of theta at which a person has a $(0.5 + 0.5c) \times (100)\%$ probability of responding correctly to the item:

$$P(X_i = 1 | \theta, a_i, b_i, c_i) = c_i + \frac{1 - c_i}{1 + e^{-a_i(\theta - b_i)}}$$

8.3   IRT MODEL INFORMATION FUNCTION

Another important feature of IRT models is the information function (or curve), an index that indicates the range over θ (the construct being measured) for which an item is most useful for discriminating among individuals. In other words, the information function characterizes the precision for measuring persons at different levels of the underlying latent construct, with higher information denoting more precision. Graphs of the information function place

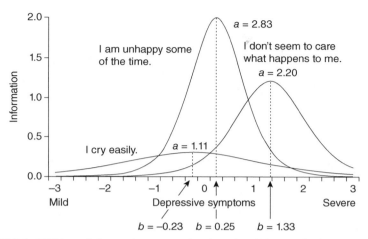

FIGURE 8.2. IRT item information curves for three items in a scale that assesses level of depressive symptoms. A respondent answers "true" or "false to each statement ("I cry easily," "I am unhappy some of the time," and "I don't seem to care what happens to me."). Each information curve is described by the item's discrimination parameter (*a*) and threshold parameter (*b*).

individuals' trait level (θ) on the horizontal axis and information magnitude on the vertical axis. Figure 8.2 presents the item information functions that are associated with three depressive symptom items.

The shape of the item information function is determined by the item parameters (i.e., *a* and *b* parameters). The higher the item's discrimination (*a* parameter), the more peaked the information function will be. Thus, higher discrimination parameters provide more information about individuals whose trait levels (θ) lie near the item's threshold value. The item's threshold parameter(s) (*b* parameter) determines where the item information function is located. In Figure 8.2, the item "I don't seem to care what happens to me" is informative for measuring high levels of depressive symptomatology, while the item "I am unhappy some of the time" is informative for measuring moderate depressive symptomatology, and the item "I cry easily" is not informative for measuring any θ level relative to the other items.

The item information functions are a powerful tool for questionnaire developers. Imagine reducing a 20-item scale into a short form to reduce respondent burden. One can create shorter surveys by selecting the most informative set of items (questions) that are relevant for the population under study. For example, if a researcher is working with a clinically depressed population, one would select items whose information curves are peaked in the high levels of depressive symptoms. On the other hand, if a researcher is working with a population with mild to moderate depressive symptoms, then a different set of items may be selected. Items with low information (low discrimination) may indicate that this particular item has a problem. Among several possibilities,

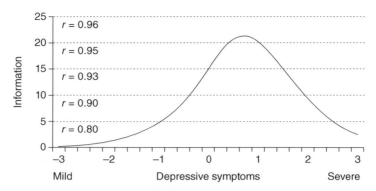

FIGURE 8.3. IRT scale information function for a 57-item questionnaire that assesses level of depressive symptoms. The horizontal dashed lines indicate the level of reliability (*r*) associated with different levels of information.

it may be that: (1) the content of the item does not match the construct measured by the other items in the scale, (2) the item is poorly worded and needs to be rewritten, (3) the item is too complex for the educational level of the population, or (4) the placement of the item in the survey is out of context.

Because of the assumption of local independence (described below), the individual item information functions can be summed across all of the items in a scale to produce the test (or scale) information function. The scale information function for a 57-item depression measure is illustrated in Figure 8.3. Along with the information magnitude indicated along the vertical axis in the graph, the associated reliability (*r*) is provided. Overall, the depression scale is highly reliable for measuring moderate to severe levels of depression (i.e., when the curve is above reliability *r* = 0.90). However, the information function shows that scale precision decreases for measuring persons with low levels of depressive symptomatology.

The IRT scale information curve illustrates one of the unique features of IRT in relation to traditional (or CTT) approaches to evaluating scale reliability. The CTT approach assesses reliability (internal consistency) with coefficient alpha. For the above 57-item questionnaire, the estimated coefficient alpha was 0.92, very high reliability. Such a high coefficient alpha gives questionnaire developers the false impression that their scale is reliable for measuring depressive symptoms no matter who is taking the depression measure. However, the IRT scale information curve in Figure 8.3 shows that the depression scale is highly reliable for assessing moderate to severe levels of depressive symptomatology, but has poor precision for assessing mild levels of depressive symptoms. Thus, under the IRT framework, the precision of a scale varies depending on what levels of the construct one is measuring. For example, a difficult math test may be very precise for differentiating among excellent math students, but the test may have low precision (high measurement error) to differentiate among people with poor to moderate math abilities if they are getting every problem wrong. Thus, the IRT measure of precision (i.e., the information function) better reflects the reliability of a test than the typical single indicator of internal consistency provided under CTT. Further, one can use information functions to look at how much reliability each item in the scale provides for assessing different severity levels along the construct continuum.

8.4   IRT MODEL STANDARD ERROR OF MEASUREMENT (SEM) FUNCTION

At each level of the underlying trait θ, the information function is approximately equal to the expected value of the inverse of the squared standard errors of the θ-estimates (Lord, 1980). The smaller the SEM, the more information or precision the scale provides about θ. For example, if a scale has a test information value of 16 at $\theta = 2.0$, then individuals' scores at this trait level have an SEM of $\left(1/\sqrt{16}\right) = 0.25$, indicating good precision (reliability approximately 0.94) at the level of theta. Figure 8.4 presents the SEM function for the 57-item depression scale, which is simply the inverse of the square root of the scale information function across all levels of the θ continuum. Both the scale information and SEM functions provide the same information about the performance of the scale for assessing different levels of theta, but in different terms—information (precision) and SEM, respectively.

8.5   FAMILY OF IRT MODELS

There are well over 100 varieties of IRT models to handle various data characteristics such as dichotomous and polytomous response data, ordinal and nominal data, and unidimensional and multidimensional data (van der Linden and Hambleton, 1997). There are also nonparametric and parametric IRT

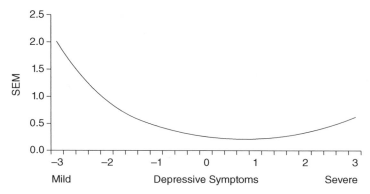

FIGURE 8.4. IRT standard error of measurement (SEM) curve for a 57-item questionnaire that assesses level of depressive symptoms.

models. However, only a few have been used extensively across educational, behavioral, and social science research fields. The common parametric unidimensional models include the 1-, 2-, and 3PL models for dichotomous response data (e.g., questions with responses such as true/false, right/wrong, and yes/no) and the graded response model, partial credit model, rating scale model, nominal model, and generalized-partial credit model for polytomous response data (questions with two or more response categories such as a Likert-type scale).

Beyond the number of response categories the model is designed for, IRT models differ in the type of parameters that are allowed to vary across the items in a scale. All models allow the difficulty (or threshold) parameter to vary from item to item; however, some models constrain every item within a scale to have the same discrimination parameter. For the latter model constraining each item to have the same discrimination, each item is assumed to have an equivalent relationship with the underlying trait being measured by the scale. These models are often referred to belonging to the Rasch family of models named after the Danish mathematician, Georg Rasch, and include the one-parameter logistic model, the rating scale model, and the partial credit model. Other models allow the discrimination parameter to vary from item to item to account for items having a differential relationship with the measured trait. These models include the 2PL model, the graded response model, the nominal model, and the generalized-partial credit model. As noted earlier, there are models that include a third parameter to account for guessing that is often used in educational testing. There are many other models that include additional parameters for different data characteristics such as the testlet response theory model that includes a parameter to account for the clustering effect due to locally dependent items. Several good resources to learn more about the various IRT models include Hambleton et al. (1991), van der Linden and Hambleton (1997), Embretson and Reise (2000), Thissen and Wainer (2001), and Bond and Fox (2007).

IRT models can also be applied to questionnaires that have polytomous response options (i.e., more than two response categories). Figure 8.5 presents IRT category response curves for a question, "In the past 7 days, I felt that nothing could cheer me up" which appears in a depression questionnaire. In the figure, there is a curve associated with each of the five possible responses which models the probability of endorsing the response conditional on a person's level of depressive symptomatology. The IRT Graded Response Model was used to estimate the item properties which includes a single discrimination parameter and four threshold parameters (the number of threshold parameters is equal to the number of response options minus one) which determine the location of the response curves.

The item information function for the polytomous response question, "In the past 7 days, I felt that nothing could cheer me up" is provided in Figure 8.6. The curve indicates this item performs well for discriminating among individuals with high levels of depressive symptoms and performs poorly for

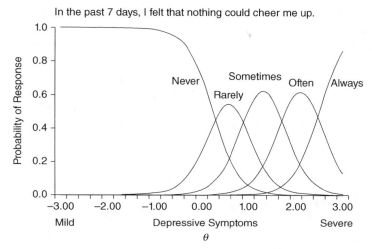

FIGURE 8.5. IRT category response curves representing the probability for selecting one of the five response options ("never," "rarely," "sometimes," "often," and "always") to the statement, "In the past 7 days, I felt that nothing could cheer me up." Each curve is associated with one of the response options. This item comes from a scale that assesses level of depressive symptoms.

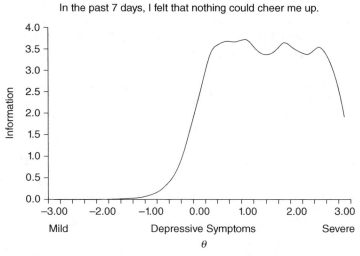

FIGURE 8.6. IRT item information curve associated with the item, "In the past 7 days, I felt that nothing could cheer me up" from a measure that assesses level of depressive symptoms.

discriminating among individuals with low levels of depressive symptoms. Contrast this item information curve with the item information curves shown in Figure 8.2. What do you notice? The information curve in Figure 8.6 is much broader (i.e., it covers much more of the range of θ) than the curves shown in Figure 8.2. This is because the three depression questions in Figure 8.2 only have two possible response categories (true or false) while the single depression question in Figure 8.6 has five response options (never, rarely, sometimes, often, or always). More response categories provide more information about the likelihood of where a person may be located on the underlying trait continuum θ. Two response categories only provide information over a narrow range whether an individual may be above or below a single threshold. However, more response categories are not necessarily better in every case. Each response category has to have meaning to the respondent. A visual analog scale of 0 to 10 may be challenging for some to differentiate a response of 7 versus an 8. For children, we often use less response categories to avoid confusion.

8.6 IRT MODEL ASSUMPTIONS

The parametric, unidimensional IRT models described above make three key assumptions about the data: (1) unidimensionality, (2) local independence, and (3) monotonicity. It is important that these assumptions be evaluated before any IRT model results are interpreted. It should be noted, however, that IRT models are robust (i.e., resistant) to minor deviations from the assumptions and that no real data ever completely meet the assumptions.

The unidimensionality assumption posits that the set of items measures a single continuous latent construct (θ). In other words, a person's level on this single construct gives rise to a person's response to the items in a scale. This assumption does not preclude that the set of items may have a number of minor dimensions (subscales), but it does assume that one dominant dimension explains the underlying structure. The scale unidimensionality can be evaluated by performing an item-level factor analysis that is designed to evaluate the factor structure that underlies the observed covariation among item responses. If multidimensionality exists, the investigator may want to consider dividing the scale into subscales based on both theory and the factor structure provided by the factor analysis, or consider using multidimensional or hierarchical IRT models.

The assumption of local independence means that the only systematic relationship among the items is explained by the conditional relationship with the latent construct (θ). In other words, local independence means that if the trait level is held constant, there should be no association among the item responses. Therefore, differences in item responses are assumed to reflect differences in the underlying trait. Violation of this assumption may result in parameter estimates that differ from what they would be if the data were locally independent; thus, selecting items for scale construction based on these estimates may lead to erroneous decisions (Thissen and Wainer, 2001). For example, items in reading comprehension passages of educational tests are known to be locally dependent because the set of questions are all related to the topic of the passage. Items within a passage will be related to each other more than to items within other passages.

Locally dependent items have high inter-item correlations, and residual correlations in a factor model, that measure constructs tangential to the domain of interest. The impact of local dependence can be measured by observing how the item parameters and person scores change when one or more of the locally dependent items are dropped. One option to control for local dependence is to combine the responses to a set of locally dependent items into a single score, thus creating a testlet. For example, two true/false questions can be combined into one by scoring the items as: 1 = false false; 2 = false true; 3 = true false; 4 = true true. The testlet can then be analyzed with IRT models alongside the other items or testlets in the scale. Testlets are more challenging to create when working with polytomous response items. For locally dependent polytomous response items, it may be better to remove one of the items than create a testlet.

The assumption of monotonicity means that the probability of endorsing or selecting an item response indicative of higher levels of theta should increase as the underlying level of theta increases. Approaches for studying monotonicity include examining graphs of item mean scores conditional on "rest-scores" (i.e., total raw scale score minus the item score) or fitting a nonparametric IRT model to the data that yield initial IRT probability curve estimates.

As noted earlier, the better the data meet these assumptions, the better the IRT model will fit the data and reflect the response behavior. Generally speaking, parametric models attempt to characterize data in a parsimonious fashion, and some extent of misfit is inherent in every unsaturated model. However, considerable misfit in an IRT model implies model misspecification, and inferences based on misspecified models are dubious. As part of the process of model fitting in IRT, it is therefore desirable to employ some diagnostic tool to evaluate the degree of model-data misfit. The fit of the model is often examined through the comparison of model predictions and the observed data (Edelen and Reeve, 2007). A number of fit indices are available; however, there lacks a universally accepted fit statistic because of the many data characteristics (e.g., data distribution, sample size, number of response categories) that can affect the performance of a fit statistic. These fit indices capture the residuals between observed and expected response frequencies. The ultimate issue is to what extent misfit affects model performance in terms of the valid scaling of individual differences (Hambleton and Han, 2005).

8.7   APPLICATIONS OF IRT MODELING FOR QUESTIONNAIRE EVALUATION AND ASSESSMENT

Much of the development and applications of IRT modeling has occurred in educational testing, where IRT is used to help administer and score educational tests like the Scholastic Aptitude Test® (SAT®) and the Graduate Record Examination® (GRE®). Other disciplines have realized the value of these applications and are learning how to adapt these methods for (1) evaluating the properties of existing scales and guiding questionnaire refinement, (2) determining measurement equivalence across research populations, (3) linking two or more questionnaires on a common metric, and (4) developing item banks for CAT applications.

8.7.1 Evaluating Existing Scales and Guiding Questionnaire Refinement

IRT modeling makes an excellent addition to the psychometrician's toolbox for evaluating and revising tests (Edelen and Reeve, 2007). The IRT category response curves help questionnaire developers evaluate how well each of the response categories for each item function for different levels of the measured construct, as well as determine whether more or fewer response categories are needed. For example, in Figure 8.5, each of the five response options for the item "In the past 7 days, I felt that nothing could cheer me up" mapped to a different part of the continuum of depressive symptomatology. Contrast this figure with Figure 8.7. In Figure 8.7, the scale is measuring how the diagnosis and treatment of breast cancer has changed women's psychosocial functioning. The item "Appreciating each day" has six response categories; however, breast cancer survivors are likely to use just four of the response options. Women with $\theta < -0.75$ are more likely to respond "no change" while women between $-0.75 < \theta < 0.25$ are likely to endorse "moderate change" for appreciating each

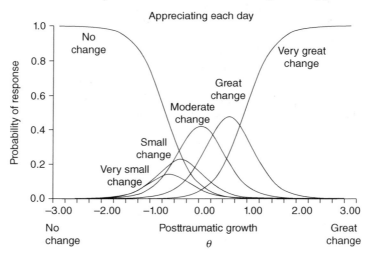

FIGURE 8.7. IRT category response curves representing the probability for selecting one of the six response options ("no change," "very small change," "small change," "moderate change," "great change," and "very great change") to the statement, "appreciating each day." This item comes from a scale that assesses level of posttraumatic growth for women who survived breast cancer.

day after they were diagnosed with breast cancer. The response categories "very small change" and "small change" do not differentiate respondents well over θ and are overshadowed by its neighboring categories. If this behavior was observed for the other items in the scale, then a questionnaire developer may consider dropping one or both response options or rewording them to be more meaningful and better differentiated from the other response categories. Thus, the category response curves provide detail on the optimal number of response categories for each item in a scale.

The IRT information curves serve as a useful tool for instrument developers to evaluate how well an item or scale functions for measuring different levels of the underlying construct. Developers can use the information curves to weed out uninformative questions or to eliminate redundant items that provide duplicate levels of information across the construct continuum. Effects on precision for removing items from the scale can easily be evaluated with the scale information function. Also, information curves allow developers to tailor their instrument to provide high information (i.e., precision) for measuring their study population. For example, if a developer wants high precision to measure a person at any level of depressive symptomatology, then the information function in Figure 8.3 suggests adding more items to the scale (or more response options to existing items) that differentiate among people with mild depressive symptoms. Adding appropriate items or response options will both broaden the range and increase the height of the curve across the underlying construct, and reflect better measurement precision.

8.7.2   Assessing Measurement Equivalence

Questions in a survey are carefully written to ensure that they are tapping into the same construct no matter which population is responding to the questions. For example, considerable care is taken when an instrument is translated from one language to another. Developers use a number of forward-and-backward translations to make sure the items carry the same meaning across translated forms. Despite this careful translation process, it may turn out that although the words are the same (i.e., linguistic equivalence), the two populations may hold culturally different views of the same question. For example, a common finding in depression questionnaires is that Hispanics are more likely to respond positively to a question such as, "I feel like crying" than are non-Hispanics despite controlling for differences between the two populations' depression levels. The difference in response rates between the two groups may result from the Hispanic culture's belief that crying is an acceptable social behavior. Such questions can lead to one group (the Hispanics) having higher (depression) scores than another group (the non-Hispanics) not because they have higher levels of the trait (depression) but because of their cultural beliefs. This is known as differential item functioning (DIF) (Teresi and Fleishman, 2007).

DIF is a condition in which an item functions differently for respondents from one group than for another. In other words, respondents, with similar levels on a latent trait θ but who belong to different populations, may have a different probability of responding to an item. Instruments containing such items may have reduced validity for between-group comparisons because their scores may indicate a variety of attributes other than those the scale is intended to measure (Thissen et al., 1993).

IRT provides an attractive framework for identifying DIF items. IRT item response curves provide a means for comparing the responses of two different groups to the same item. In IRT modeling, item parameters are assumed to be invariant to group membership. Therefore, differences between the curves, estimated separately for each group, indicate that respondents at the same level of the underlying trait, but from different groups, have different probabilities of endorsing the item. More precisely, DIF is said to occur whenever the conditional probability, $P(X)$, of a correct response or endorsement of the item for the same level on the latent variable differs for two groups.

DIF analysis has been used in multiple studies to detect measurement equivalence in item content across cultural groups, males and females, age groups, between two administration modes such as paper-and-pencil versus computer-based questionnaires, and from one language translation of a questionnaire to another. Also, DIF testing can be used for evaluating question ordering effects or question wording effects.

8.7.3   Linking Two or More Scales

It is common in many research settings for several existing quality instruments to measure the same construct. With no clear gold-standard questionnaire to be used across all research applications, researchers will select the questionnaire with which they are most familiar or choose one with the least cost to administer. Combining or comparing results across studies that use different questionnaires in a meta-analytic study is very difficult because the questionnaires may have different lengths, different number of response options, and different types of questions with different psychometric properties. Some standardization of the concepts and metrics is needed to allow investigators to compare instruments and respondents. IRT modeling provides a solution through its ability to link the item properties from different scales on to a common metric (Dorans, 2007).

The goal of instrument linking is not to change the content, but to adjust for the statistical differences in difficulties of the items (McHorney and Cohen, 2000). That is, the scales may be measuring the same construct, but tap into different levels of the underlying construct. Several methodologies exist for linking two (or more) instruments. Ideally, one would administer both instruments to a representative sample and then IRT-calibrate (obtain the properties of) the items simultaneously. However, this method can be a burden on respondents, especially for long scales. As an alternative, a set of items that are common to both instruments can be selected as anchors. The anchor items are used to set the metrics to which items not common to both instruments are scaled. Therefore, instruments with a different number or difficulty of items can be linked by responses to a common set of anchor items.

These applications take advantage of a key feature of IRT models, which is the property of "invariance" (Meredith and Teresi, 2006). If IRT model assumptions are met, item parameters are invariant with respect to the sample of respondents, and respondent scores are invariant with respect to the set of items used in the scale. After the IRT item parameters are estimated (i.e., calibrated), researchers can choose the most salient items to target a person's level of function with the smallest number of items. This method results in different groups receiving different sets of items; however, any given set of items calibrated by the best-fitting IRT model should yield scores that are on a similar metric.

8.7.4   Building Item Banks and CAT

The IRT principle of invariance is the foundation that researchers use to develop CATs (Bjorner et al., 2007; Cella et al., 2007a; Reeve et al., 2007; Thissen et al., 2007). CATs combine the powerful features of IRT with improvements in computer technology to yield tailored instruments that estimate a person's level on a construct (e.g., depression symptomatology) with the fewest number of items. To accomplish this, a CAT has access in its data bank to a large pool of items called an *item bank*. The questions in an item bank have been carefully selected from existing instruments or written by experts to make sure they are valid, reliable, and can be well understood by people with low education levels. Each of the questions in the item bank has been calibrated by IRT models, and this information is used for selecting items in the CAT.

Developing an item bank is not simply just a large collection of items measuring the same concept; rather, the items typically have undergone rigorous qualitative and quantitative evaluation before they are ready for test administration (Reeve et al., 2007). Figure 8.8 provides an illustration of the major steps involved in developing an item bank that measures a single domain such as depressive symptomatology. To avoid redundancy, many item banking efforts collect items from existing instruments (A, B, and C in Fig. 8.8) and place them in an item pool. One may also write new items if existing questionnaires are not available or there are relevant attributes of the measured construct that are not covered by the current set of items. The item pool is then evaluated by content experts, potential respondents, and survey experts using methods such as focus groups and cognitive testing. When available, items from existing questionnaires can be evaluated for its psychometric properties through secondary data analysis. During the qualitative evaluation phase, items in the pool are often standardized to have a common response categories (e.g., 5-point Likert-type scale) and phrasing (e.g., similar reference period such as "In the past 7 days").

Then, the item pool is administered, via computer-based (preferably) or paper-and-pencil questionnaire, to a large representative sample of the target

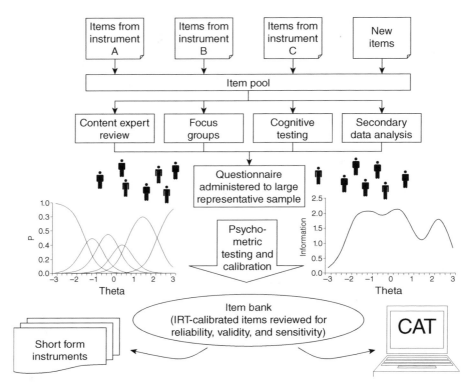

FIGURE 8.8. Figure demonstrates the general process for developing an item bank that is used as the foundation for administering computerized-adaptive tests (CATs) and static short forms. In the first two rows, items are either collected from existing questionnaires or written by a working group to form an item pool. Items then undergo a qualitative evaluation phase which includes expert and respondent input through such techniques as focus groups and cognitive interviews. Following qualitative review, the items are administered to a large representative sample and evaluated using a number of psychometric methods. Finally, the items are calibrated using IRT models to form the item bank.

population. Next, responses to the items are evaluated for validity, reliability, domain coverage, and DIF across measured groups. The items are calibrated with a set of statistical properties (e.g., the discrimination and threshold parameters), derived from IRT models, to form the item bank. With the item bank in place, developers can create fixed short-form instruments or can develop CATs.

Based on a person's response to an initial item, the CAT selects the next most informative item (based on the IRT item information curves) from the item bank to administer to the respondent, and this process is repeated until a minimal standard error or maximum number of administered items is achieved. The benefits of CAT technology include: decreased respondent burden because fewer items can be administered to achieve any given level of statistical precision; reduced "floor and ceiling" effects (often seen in fixed-length paper-and-pencil questionnaires); instant scoring; and widespread availability of this technology on many platforms (e.g., Internet, handheld devices, computer-assisted telephone interviewing).

One excellent example of use of qualitative and quantitative methodologies to develop item banks and CAT technology is the National Institutes of Health-sponsored initiative for developing the Patient-Reported Outcomes Measurement Information System (PROMIS®; Cella et al., 2007b). The goal of PROMIS is to provide researchers and clinicians access to efficient, precise, and valid measures of a person's health status. With questionnaires available for children or adults, the PROMIS measures such constructs as physical function, pain, fatigue, depression, anxiety, and anger. PROMIS provides both fixed questionnaires (i.e., short forms) or CAT-based assessment and can be used across a wide variety of chronic diseases and conditions and in the general population. IRT played a dominant role for designing the scales, assessing item performance, testing for DIF, and calibration of the item bank (Reeve et al., 2007). A demonstration of CAT and background information on the development of PROMIS measures is available at http://www.nihpromis.org/.

8.8 CONCLUSION

Applications of IRT models has spread across many research fields from educational testing, to psychological assessment, and to health outcomes research; however, there remain obstacles to widespread application. First, many researchers have been trained in CTT statistics, are comfortable interpreting these statistics, and can use readily available software to generate easily familiar summary statistics, such as Cronbach's coefficient α or item-total score correlations. In contrast, IRT modeling requires an advanced knowledge of measurement theory to understand the mathematical complexities of the models, to determine whether the assumptions of the IRT models are met, and to choose the model from within the large family of IRT models that best fits the data and the measurement task at hand. In addition, the supporting software and literature are not well adapted for researchers outside the field of educational testing. Another obstacle is that the algorithms in the IRT parameter-estimation process require large sample sizes to provide stable estimates. Sample size requirements vary depending on a number of factors, including choice of IRT model and data characteristics, from the simplest IRT model requiring at least 100 people to some models requiring 500 or more respondents to estimate precise parameter estimates for scoring individuals at all levels of the underlying construct.

Also, it must be reiterated that IRT models are probabilistic models that relate responses to individual questions to the latent variable being measured by the scale. Thus, IRT models are not applied to more observable types of variables such as sex, weight, dietary consumption, or employment status where typically single questions are appropriate or the questions are used simply for description and not summed together. IRT models are latent trait models and have been applied to assessing variables such as math ability, psychological distress, nicotine dependence, physical function, and extroversion. Each of these constructs typically includes multiple questions to capture a person's level on the domain.

Despite the conceptual and computational challenges, the many potential applications of IRT modeling for evaluating and refining questionnaires should not be ignored. Knowledge of IRT is spreading as more courses are taught in the academic disciplines of psychology, education, and public health, and at seminars and conferences throughout the world. An increased number of books and tutorials are being written on the subject and more user-friendly software is being developed. Research that applies IRT models appears more frequently in the literature, and many concluding comments are directed toward the benefits and limitations of using the IRT methodology in various fields. For all of these reasons, a better understanding of the models and applications of IRT will emerge in the questionnaire development and evaluation field.

As emphasized in the introduction of this chapter, the ability of an instrument developer to use both qualitative and quantitative methodologies in designing the questionnaire will improve the quality of the measure to capture the construct of interest in the target population. While each method provides unique information about questionnaire properties and function, the methods also can provide complementary information to design a high-quality instrument (Bjorner et al., 2003). In initial phases of questionnaire development, qualitative methods like focus groups, cognitive testing, and behavioral coding may be used. Once pilot data have been collected, quantitative methods like IRT modeling can then evaluate item and scale performance. Findings from quantitative testing (e.g., findings of DIF for an item) may suggest the need to go back to cognitive interviewing or focus groups to identify why a particular item or set of items perform in a way inconsistent with what was expected. Thus, a mixed-method approach for designing questionnaires will result in measures that are shorter, more precise, more valid, and better targeted toward the population of interest.

REFERENCES

Bjorner JB, Ware JE, Kosinski M (2003). The potential synergy between cognitive models and modern psychometric models. Quality of Life Research; 12(3): 261–274.

Bjorner JB, Chang C-H, Thissen D, Reeve, BB (2007). Developing tailored instruments: item banking, and computerized adaptive assessment. Quality of Life Research; 16(1):95–108.

Bond TG, Fox CM (2007). Applying the Rasch Model: Fundamental Measurement in the Human Sciences, 2nd ed. Mahwah, NJ: Lawrence Erlbaum.

Cella D, Gershon R, Lai JS, Choi S (2007a). The future of outcomes measurement: item banking, tailored short-forms, and computerized adaptive assessment. Quality of Life Research; 16(1):133–141.

Cella D, Yount S, Rothrock N, Gershon R, Cook K, Reeve B, Ader D, Fries JF, Bruce B, Rose M (2007b). The Patient-Reported Outcomes Measurement Information System (PROMIS): progress of an NIH roadmap cooperative group during its first two years. Medical Care; 45(5 Suppl. 1):S3–S11.

Dorans NJ (2007). Linking scores from multiple health outcome instruments. Quality of Life Research.; 16(Suppl. 1):85–94.

Edelen MO, Reeve BB (2007). Applying item response theory (IRT) modeling to questionnaire development, evaluation, and refinement. Quality of Life Research; 16(Suppl. 1):5–18.

Embretson SE, Reise SP (2000). Item Response Theory for Psychologists. Mahwah, NJ: Lawrence Erlbaum.

Hambleton RK, Han N (2005). Assessing the fit of IRT models to educational and psychological test data: a five-step plan and several graphical displays. In: Lenderking WR, Revicki DA, editors. Advancing Health Outcomes Research Methods and Clinical Applications. McLean, VA: Degnon; pp. 57–78.

Hambleton RK, Swaminathan H, Rogers H (1991). Fundamentals of Item Response Theory. Newbury Park, CA: Sage.

Lord FM (1980). Applications of Item Response Theory to Practical Testing Problems. Hillsdale, NJ: Lawrence Erlbaum.

McHorney CA, Cohen AS (2000). Equating health status measures with item response theory: illustrations with functional status items. Medical Care; 38(9 Suppl. II): 43–59.

Meredith W, Teresi JA (2006). An essay on measurement and factorial invariance. Medical Care; 44(11 Suppl. 3):S69–S77.

Reeve BB, Hays RD, Bjorner JB, Cook KF, Crane PK, Teresi JA, Thissen D, Revicki DA, Weiss DJ, Hambleton RK, Liu H, Gershon R, Reise SP, Lai JS, Cella D (2007). Psychometric evaluation and calibration of health-related quality of life item banks: plans for the Patient-Reported Outcomes Measurement Information System (PROMIS). Medical Care; 45(5):S22–S31.

Teresi JA, Fleishman JA (2007). Differential item functioning and health assessment. Quality of Life Research; 16(Suppl. 1):33–42.

Thissen D, Wainer H, editors (2001). Test Scoring. Mahwah, NJ: Lawrence Erlbaum.

Thissen D, Steinberg L, Wainer H (1993). Detection of differential item functioning using the parameters of item response models. In: Holland PW, Wainer H, editors. Differential Item Functioning. Hillsdale, NJ: Lawrence Erlbaum; pp. 67–113.

Thissen D, Reeve BB, Bjorner JB, Chang C-H (2007). Methodological issues for building item banks and computerized adaptive scales. Quality of Life Research; 16(Suppl. 1):109–119.

van der Linden WJ, Hambleton RK (1997). Handbook of Modern Item Response Theory. New York: Springer.

9 Response 1 to Reeve's Chapter: Applying Item Response Theory for Questionnaire Evaluation

RON D. HAYS

University of California, Los Angeles

9.1 INTRODUCTION

The chapter by Reeve (this volume) presents a general overview of item response theory (IRT) with implications for questionnaire evaluation. He discusses IRT models and assumptions along with examples of how IRT can help in the evaluation of questionnaire items. This response focuses on the features of IRT mentioned by Reeve (this volume) that have the greatest implications for questionnaire evaluation and development: category response curves, differential item functioning, information, and computerized-adaptive testing. A couple of issues not covered in Reeve (this volume), person fit and the appropriate unit of analysis, are also touched upon here.

9.2 CATEGORY RESPONSE CURVES (CRCS)

CRCs help in the evaluation of the response options of each item. The CRCs display the relative position of the response categories along the underlying continuum. In the example shown in Figure 8.7 of the Reeve (this volume) chapter, the categories fall along the expected ordinal order from *no change* to a *very great* change. In addition, 2 of 6 response options (i.e., *very small change* and *small change*) are never most likely to be chosen by respondents

Question Evaluation Methods: *Contributing to the Science of Data Quality*, First Edition.
Edited by Jennifer Madans, Kristen Miller, Aaron Maitland, Gordon Willis.
© 2011 John Wiley & Sons, Inc. Published 2011 by John Wiley & Sons, Inc.

across all levels of the posttraumatic growth scale continuum. Similarly, examination of empirical response curves for the Positive and Negative Syndrome Scale revealed that a responses of "1" and "2" indicated similar locations on the underlying trait (Kay et al., 1987).

Reeve suggests that one or both of the response categories could be dropped or reworded to improve the response scale. Dropping unnecessary categories could increase the readability of the survey item, but it could also reduce respondent burden. A rule of thumb is that about three to five survey items can be administered per minute. But survey items with fewer response options are more quickly administered. For example, Hays et al. (1995) found that 832 clients enrolled in treatment programs for impaired (drinking) drivers completed about 4.5 items per minute for an alcohol screening scale with polytomous response options. However, they were able to complete about 8 items per minute for scales with dichotomous response options.

But it might be challenging to determine what it is about *very small* and *small* that makes them less likely to be endorsed than *no change* for respondents in the −2 to −1 theta range of posttraumatic growth. Additional information such as cognitive interviews may provide insights (Hays and Reeve, 2008). One might also consider eliciting perceptions of where the problematic response options and alternative possibilities lie along the underlying construct continuum using the method of equal-appearing intervals (Thurstone and Chave, 1929). In this method, a sample of raters is asked to rate the position of intermediate response choices using a 10-cm line anchored by the extreme (lowest and highest) response choices (Ware et al., 1996).

Karmaliani et al. (2006) analyzed measures of emotional distress that included *never* to *always* frequency response options. One measure used four categories (*never, sometimes, often, always*) while the other used five (adding *rarely* in-between *never* and *sometimes*). CRCs showed that the *rarely* and *often* response options were *never most likely* to be chosen for some of the items. They recommended for persons with lower literacy that the number of response options be reduced to three: *never, sometimes*, and *almost always*. Whether this is an optimal solution or not is debatable. Alternatively, one might consider replacing *often* with another descriptor such as *usually*. Or one might want to try using *never, sometimes, usually, almost always*, and *always*.

Analyses of Consumer Assessment of Healthcare Providers and Systems (CAHPS®) *0–10* global rating items reveal that most of the information in the 11 response categories is captured by about 3 levels on the scale, with variation in where the thresholds occur (Damiano et al., 2004). One leading alternative is to recode 10 as one category, 9 a second category, and *8–0* as a third category. Another alternative is to recode the data so that *10–9* is one category to compensate for the fact that less educational attainment is associated with use of both extremes of the *0–10* rating scale (Elliott et al., 2009). Despite the fact that there are only about 3 levels of information, the CAHPS® investigators recommend administering the *0–10* scale as is and then collapsing the categories in analyses.

9.3 DIFFERENTIAL ITEM FUNCTIONING

If response options are changed and it is important to know the impact of the revisions, one can evaluate differential item functioning (DIF). When the target construct is less well defined, questions that tap it are more sensitive to changes in response categories (Rockwood et al., 1997). DIF is also more likely to be present if response options are changed to a larger extent rather than a little bit. It has been shown, for example, that items whose response anchors changed the most were more likely to exhibit DIF across 2 years of administration of an attitude survey to Federal Aviation Administration employees (Farmer et al., 2001).

Coons et al. (2009) note that substantial changes to items (stems or response scales) may make equivalence of the old and new items irrelevant. However, estimating the comparability of the old and new versions may still be valuable for purposes such as bridging scores. For example, a few years ago, UCLA transitioned from the Picker Hospital Survey to the CAHPS® Hospital Survey. Differences in wording, response options, and cut-points for "problem scores" yielded large differences in problem score rates between the two survey instruments that required bridging formulas. Tetrachoric correlations for 5 of 6 item pairs indicated high correspondence (r's of 0.71–0.97) in the underlying constructs. Bridged scores contained less information than directly measured new scores, but with sufficient sample sizes they could be used to detect trends across the transition (Quigley et al., 2008).

Examination of DIF may need to be done for different modes of administration because the effect of changes in response options might vary by mode. For example, the CAHPS clinician/group survey can be administered using either a four-category (*never, sometimes, usually, always*) or six-category (*never, almost never, sometimes, usually, almost always, always*) response scale. There is likely to be greater DIF for telephone administration than self-administration on the 6-point scale because of the effect of memory limitations on responses over the phone.

9.4 INFORMATION

Another advantage of IRT mentioned by Reeve (this volume) is that information (precision) is estimated at different points along the construct continuum. Reliability in its basic formulation is equal to $1 - SE^2$, where the $SE = 1/(\text{information})^{1/2}$. The standard error (SE) and confidence interval around estimated scores for those with milder levels of depressive symptoms are larger than for those with moderate to severe depressive symptoms (see Fig. 8.4 of Reeve, this volume). Lowering the SE for those with mild symptoms requires adding or replacing existing items with more informative items at this range of the depressive symptoms continuum.

But this is easier said than done. As noted by Reise and Waller (2009), "the low end of depression is not happiness but rather the lack of depression"

(p. 31). Depressive symptoms are a quasi-trait in the sense that it is unipolar or relevant in only one direction. There is also a limit on the number of ways to ask about a targeted range of the construct. One needs to avoid including essentially the same item multiple times. For example, "I'm generally sad about my life" and "My life is generally sad" are so similar that this could lead to violations of the IRT local independence assumption. That is, these depressive symptom items would be correlated above and beyond what the common depression construct would predict. One would then find significant residual correlations for the item pair after controlling for the common factor defining the depressive symptom items. Ignoring the local dependency would lead to inflated slope or discrimination parameters for the pair of items.

Candidate global physical health items administered in the Patient-Reported Outcome Measurement Information System (PROMIS) included the commonly used *excellent* to *poor* rating of health and a parallel rating of physical health (Hays et al., 2009). A single-factor categorical confirmatory factor analytic model for five global physical health items including this pair showed less than adequate fit. Adding a residual correlation ($r = 0.29$) between the item pair lead to noteworthy improvement in model fit. When the graded response model was fit to the five items, the discrimination parameters for the locally dependent items were 7.37 and 7.65; the next largest value for an item was 1.86. Dropping the "In general, how would you rate your health" item resulted in a discrimination parameter of 2.31 for the rating of your physical health item, and this was no longer the largest discrimination parameter of the remaining four physical health items (it was second largest) (Table 9.1).

With IRT-based SEs and precision, users have access to a more appropriate confidence interval around an individual's score for clinical applications. This is very important because use of patient-reported outcomes (PROs) at the individual level requires very high reliability (e.g., 0.90 or higher reliability was

TABLE 9.1. Item Parameters (Graded Response Model) for Global Physical Health Items in Patient-Reported Outcomes Measurement Information System

Item	a	b1	b2	b3	b4
Global01	7.37 (na)	−1.98 (na)	−0.97 (na)	0.03 (na)	1.13 (na)
Global03	7.65 (2.31)	−1.89 (−2.11)	−0.86 (−0.89)	0.15 (0.29)	1.20 (1.54)
Global06	1.86 (2.99)	−3.57 (−2.80)	−2.24 (−1.78)	−1.35 (−1.04)	−0.58 (−0.40)
Global07	1.13 (1.74)	−5.39 (−3.87)	−2.45 (−1.81)	−0.98 (−0.67)	1.18 (1.00)
Global08	1.35 (1.90)	−4.16 (−3.24)	−2.39 (−1.88)	−0.54 (−0.36)	1.31 (1.17)

Note: Parameter estimates for 5-item scale are shown first, followed by estimates for 4-item scale (in parentheses).
Global01: In general, would you say your health is . . . ? Global03: In general, how would you rate your physical health? Global06: To what extent are you able to carry out your everyday physical activities? Global07: How would you rate your pain on average? Global08: How would you rate your fatigue on average?
a = discrimination parameter; b1 = 1st threshold; b2 = 2nd threshold; b3 = 3rd threshold; b4 = 4th threshold; na = not applicable.

advocated by Nunnally, 1978). Having the best possible information is especially important as PROs are used in the future to monitor effects of different treatment options and as input to clinical decisions (Fung and Hays, 2008). Accurate information about individual SEs is critical to determine who "responds" to treatment because the statistical significance of individual change is driven by the SE of measurement (Hays et al., 2005).

9.5 COMPUTERIZED-ADAPTIVE TESTING (CAT)

The capacity to include an unlimited number of items (item bank) and rely upon CAT to tailor the number and kinds of items administered to different respondents is one of the most exciting aspects of IRT for questionnaire developers. By tailoring items based on iterative theta estimates, the fewest number of items can be administered to achieve a target level of precision for each individual. The items chosen to be administered to an individual are those that provide the most information at the person's estimated theta level. As noted by Reeve (this volume), information is driven by the slope or discrimination parameter and revealed in the item information curve or function.

Because different items and sequence of items are administered in a CAT, attention needs to be given to the potential for context effects. Lee and Grant (2009) randomly assigned 1191 English-language and 824 Spanish-language participants in the 2007 California Health Interview Survey to different orders of administration of a self-rated health item and a list of chronic conditions. They found no order effect for English-language respondents, but for Spanish-language respondents, self-rated health was reported as worse when it was asked before compared to after the chronic conditions. The National Health Measurement Study of 3844 individuals found that the responses to the EQ-5D (Visual Analog Scale and U.S. preference-weighted score) and the short form (SF)-36 (SF6D and Mental Health Component Summary Score) were significantly more positive when administered later in a telephone interview (Hays, 2009). But the magnitude of the order effects was small.

There is increasing interest in the use of multidimensional CAT so that scores on one domain can be estimated using information from other (correlated) domains as well as items on the target domain. Wang et al. (2004) noted savings in number of items administered by multidimensional CAT of about 50% compared to unidimensional CAT.

9.6 CHOICE OF IRT MODEL

The Rasch model is the simplest IRT model in the sense that it estimates only the difficulty or threshold parameter. The PROMIS project preferred the graded response model over the Rasch model because it is a flexible model from the parametric, unidimensional, polytomous-response IRT family of models. Because the graded response model allows discrimination to vary item by item, it typically fits response data better than the Rasch model. In addition, compared with alternative 2-parameter models such as the generalized partial credit model, the model is relatively easy to understand and illustrate and retains its functional form when response categories are merged (Reeve et al., 2007). However, Reise and Waller (2009) noted that inspection of item response curves where item endorsement rates are plotted by corrected total scores shows that the assumption of 0.00 lower and upper asymptotes for the graded response model is often violated. Hence, a 4-parameter model is needed for inclusion of upper and lower asymptote parameters. Failure to do so can lead to overly optimistic evaluation of the precision of the scale.

More generally, the need to consider alternative IRT models is evident by the fact that if data fit one IRT model, it also will fit other IRT models (Lord, 1975). The recognition of the importance of considering alternative models in the parallel field of structural equation modeling has been recognized for a number of years (Hays and White, 1987).

9.7   PERSON FIT

An issue not addressed by Reeve (this volume) that has important implications for questionnaire evaluation and interpretation is person fit. Person fit evaluates the extent to which a person's pattern of item responses is consistent with the IRT model (Reise, 1990). It is in some sense a microlevel evaluation of DIF. The standardized Z_L Fit Index (Drasgow et al., 1985) is an example person fit index. Large negative Z_L values indicate misfit. Large positive Z_L values indicate response patterns that are higher in likelihood than the model predicts. Person misfit can be suggestive of response carelessness or cognitive errors. These may occur, for example, if the readability of the survey exceeds the literacy of the respondent (Paz et al., 2009). One study interviewed children who displayed significant misfit on a widely used measure of self-concept and concluded that they had problems understanding the meaning of the questions (Meijer et al., 2008; cited by Reise and Waller, 2009). Cases with significant person misfit can be excluded or at least flagged to determine impact on conclusions from analyses.

The Z_L person fit index was computed for the PROMIS physical functioning item bank using a program written in R (Choi, 2007). A scatter plot of the Z_L index by theta from answers to the physical functioning item bank is provided in Figure 9.1. Seventy-five cases had person misfit that was significant at $P < 0.01$. One of these cases had a Z_L value of -3.13. This person responded to 14 of the items in the bank. The expected a posteriori (EAP) estimated theta was 1.18 and estimated SE of theta was 0.55 for this person. These 14 items were administered using a five-category difficulty response scale: (1) *unable to do*; (2) *with much difficulty*; (3) *with some difficulty*; (4) *with a little difficulty*; and (5) *without any difficulty*. For 13 of the items, this person responded that they were able to do it without any difficulty. This included a

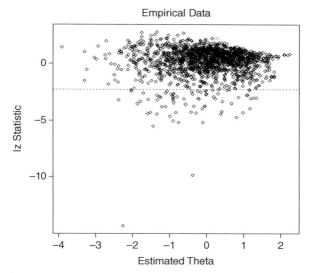

FIGURE 9.1. Scatterplot of Person Fit Index by theta for PROMIS physical functioning bank.

response to a question about being able to run 5 miles. The threshold between *with a little difficulty* and *without any difficulty* for this item is 1.91, indicating a very high level of functioning is needed to have a 50% probability of responding *without any difficulty*. However, this same person reported *a little difficulty* in being out of bed for most of the day. The threshold between *a little difficulty* and *without any difficulty* for this item is −2.02, indicating that a very low level of functioning is needed to report *without any difficulty*. Thus, the person's answer to this question was inconsistent with their other responses (Fig. 9.1).

Another case had a Z_L value of −2.49. This person only responded to 7 of the items in the bank, as indicated below and had an EAP estimated theta of −1.07 and SE of theta of 0.42.

Are you able to walk up and down two steps? Response: *Without any difficulty.*

Are you able to run ten miles? Response: *With much difficulty.*

Are you able to stand without losing your balance for several minutes? Response: *With some difficulty.*

Are you able to sit down in and stand up from a low, soft couch? Response: *With some difficulty.*

Are you able to be out of bed most of the day? Response: *Without any difficulty.*

Are you able to turn from side to side in bed? Response: *Without any difficulty.*

Are you able to get in and out of bed? Response: *With a little difficulty.*

The reason this person does not fit the IRT model is because he/she reported *some difficulty* with low levels of physical functioning (standing without losing balance, sit down and stand up from a soft couch, getting in and out of bed) while having *no difficulty* with a higher level of physical functioning (walking up and down two steps). Nonetheless, the inconsistency for this case is not as problematic as the inconsistency seen in the first example.

We predicted person misfit on the physical functioning measure in PROMIS from the following individual characteristics: age, gender, race, education, marital status, and count of chronic conditions (hypertension, angina, coronary artery disease, heart failure, heart attack, asthma, Chronic Obstructive Pulmonary Disease, stroke, spinal cord injury, multiple sclerosis, Parkinson's disease, epilepsy, amyotrophic lateral sclerosis, depression, anxiety, drug/alcohol problems, cancer, diabetes, arthritis, liver disease, kidney disease, migraines, HIV, sleep disorder). We began with all independent variables in the model and then trimmed the model by removing the one variable with the largest p-value iteratively until only significant ($P < 0.05$) variables remained in the model. Three variables were significantly related to misfit: non-white race (white race OR = 0.55), less than high school education (OR = 3.59), and more chronic conditions (OR = 1.13). The significant association of less education with person misfit provides support for cognitive limitations as a cause. We then looked to see if response time predicted person misfit. Based on preliminary analyses, response time was coded in quartiles, and the third and fourth (longer) quartiles were combined because they were associated with similar levels of misfit. We included the same other variables as above but added response time to the multivariate model. The resulting trimmed model had three variables significantly associated with misfit: younger age (OR = 0.98), more chronic conditions (OR = 1.18), and longer response times (1st quartile of response time OR = 0.31; 2nd quartile of response time OR = 0.49). The significant association of longer response time with person misfit again suggests cognitive limitations or difficulty responding to the survey items as a potential cause.

9.8   UNIT OF ANALYSIS

Most applications of IRT in the patient-reported outcomes field analyze the data at the individual level. But the appropriate unit of analysis for some PRO applications is a higher-level aggregate. For example, the CAHPS health plan surveys focus on the health plan rather than the individual completing the survey. Hence, psychometric analyses need to be conducted at the health plan level. Group-level IRT analyses of CAHPS health plan survey data from a sample of 35,572 Medicaid recipients nested within 131 health plans confirmed that within-plan variation dominates between-plan variation for the items (Reise et al., 2006). Hence, large sample sizes are needed to reliably differentiate among plans than individuals. A plan-level 3-parameter IRT model showed that CAHPS items had small item discrimination parameter estimates relative to the person-level estimates. In addition, CAHPS items had large lower asymptote parameters at the health plan level. While these results are not surprising in and of themselves, the performance of the CAHPS survey at the unit of analysis for which it is being used is important.

9.9 CONCLUSION

There are multiple features of IRT analyses that can be helpful in questionnaire evaluation. This response attended to a number of issues in IRT analyses including CRCs, DIF, information, CAT, choice of IRT model, person fit, and unit of analysis. Increased application of IRT will enhance knowledge of how surveys perform and lead to better survey items in the future. It is important to emphasize that the quantitative data from IRT provide clues about survey item performance, but that good surveys are produced by using a combination of quantitative and qualitative methods (e.g., focus groups, cognitive interviews).

ACKNOWLEDGEMENT

Ron Hays was supported in part by NIH/NIA Grants P30-AG028748 and P30-AG021684 and by NCMHD Grant 2P20MD000182.

REFERENCES

Choi SW (2007). Person$_z$: Person Misfit Detection Using the lz Statistic. Evanston, IL: Evanston Northwestern Healthcare Research Institute.

Coons SJ, Gwaltney CJ, Hays RD, Lundy JJ, Sloan JA, Revicki DA, Lenderking WR, Cella D, Basch E (2009). Recommendations on evidence needed to support measurement equivalence between electronic and paper-based Patient-Reported Outcome (PRO) Measures: ISPOR ePRO good research practices task force report. Value in Health; 12:419–429.

Damiano PC, Elliott M, Tyler MC, Hays RD (2004). Differential use of the CAHPS 0–10 global rating scale by Medicaid and commercial populations. Health Services and Outcomes Research Methodology; 5:193–205.

Drasgow F, Levine MV, Williams EA (1985). Appropriateness measurement with polytomous item response models and standardized indices. The British Journal of Mathematical and Statistical Psychology; 38:67–68.

Elliott MN, Haviland AM, Kanouse DE, Hambarsoomian K, Hays RD (2009). Adjusting for subgroup differences in extreme response tendency in ratings of health care: impact on disparity estimates. Health Services Research; 44:542–561.

Farmer WL, Thompson RC, Heil SKR, Heil MC (2001). Latent trait theory analysis of changes in item response anchors. U.S. Department of Transportation, Federal

Aviation Administration, National Technical Information Service, Springfield, Virginia 22161.

Fung CH, Hays RD (2008). Prospects and challenges in using patient-reported outcomes in clinical practice. Quality of Life Research; 17:1297–1302.

Hays RD (2009). Do Generic Health-Related Quality of Life Scores Vary by Order of Administration? Los Angeles, CA: UCLA Department of Medicine Research Day.

Hays RD, Reeve BB (2008). Measurement and modeling of health-related quality of life. In: Heggenhougen K, Quah S, editors. International Encyclopedia of Public Health. San Diego: Academic Press; pp. 241–251.

Hays RD, White K (1987). The importance of considering alternative structural equation models in evaluation research. Evaluation & the Health Professions; 10:90–100.

Hays RD, Merz JF, Nicholas R (1995). Response burden, reliability, and validity of the CAGE, Short-MAST, and AUDIT alcohol screening measures. Behavior Research Methods, Instruments, & Computers; 27:277–280.

Hays RD, Brodsky M, Johnston MF, Spritzer KL, Hui K (2005). Evaluating the statistical significance of health-related quality of life change in individual patients. Evaluation and the Health Professions; 28:160–171.

Hays RD, Bjorner J, Revicki DA, Spritzer K, Cella D (2009). Development of physical and mental health summary scores from the Patient-Reported Outcomes Measurement Information System (PROMIS) global items. Quality of Life Research; 18:873–880.

Karmaliani R, Bann CM, Mahmood MA, Harris HS, Akhtar S, Goldenberg RL, Moss N (2006). Measuring antenatal depression and anxiety: findings from a community-based study of women in Hyderabad, Pakistan. Women & Health; 44:79–103.

Kay SR, Fiszbein A, Opler LA (1987). The Positive and Negative Syndrome Scale (PANSS) for schizophrenia. Schizophrenia Bulletin; 2:261–276.

Lee S, Grant D (2009). The effect of question order on self-rated general health status in a multilingual survey context. American Journal of Epidemiology; 169: 1525–1530.

Lord FM (1975). The "ability" scale in item characteristic curve theory. Psychometrika; 40:205–217.

Meijer RR, Egberink IJL, Emons WHM, Sijtsma K (2008). Detection and validation of unscalable item score patterns using item response theory: an illustration with Harter's self-perception profile for children. Journal of Personality Assessment; 90:227–238.

Nunnally J (1978). Psychometric Theory, 2nd ed. New York: McGraw-Hill.

Paz SH, Liu H, Fongwa MN, Morales LS, Hays RD (2009). Readability estimates for commonly used health-related quality of life surveys. Quality of Life Research; 18:889–900.

Quigley D, Elliott MN, Hays RD, Klein D, Farley D (2008). Bridging from the picker hospital survey to the CAHPS® hospital survey. Medical Care; 46:654–661.

Reeve BB, Hays RD, Bjorner JB, Cook KF, Crane PK, Teresi JA, Thissen D, Revicki DA, Weiss DJ, Hambleton RK, Liu H, Gershon R, Reise SP, Cella D (2007). Psychometric evaluation and calibration of health-related quality of life item banks:

plans for the Patient-Reported Outcome Measurement Information System (PROMIS). Medical Care; 45:22–31.

Reise SP (1990). A comparison of item- and person-fit methods of assessing model-data fit in IRT. Applied Psychological Measurement; 14:127–137.

Reise SP, Waller NG (2009). Item response theory and clinical measurement. Annual Review of Clinical Psychology; 5:27–48.

Reise SP, Meijer RR, Ainsworth AT, Morales LS, Hays RD (2006). Application of group-level item response models in the evaluation of consumer reports about health plan quality. Multivariate Behavioral Research; 41:85–102.

Rockwood TH, Sangster RL, Dillman DA (1997). The effect of response categories on questionnaire answers. Sociological Methods & Research; 26:118–140.

Thurstone LL, Chave EJ (1929). The Measurement of Attitude. Chicago: University of Chicago Press.

Wang WC, Chen PH, Cheng YY (2004). Improving measurement precision of test batteries using multidimensional item response models. Psychological Methods; 9:116–136.

Ware JE, Gandek BL, Keller SD; The IQOLA Project Group (1996). Evaluating instruments used cross-nationally: methods from the IQOLA project. In: Spilker B, editor. Quality of life and Pharmacoeconomics in Clinical Trials, 2nd ed. Philadelphia, PA: Lippincott-Raven; pp. 681–692.

10 Response 2 to Reeve's Chapter: Applying Item Response Theory for Questionnaire Evaluation

CLYDE TUCKER, BRIAN MEEKINS,
and JENNIFER EDGAR

Bureau of Labor Statistics

PAUL P. BIEMER

RTI International and University of North Carolina at Chapel Hill

10.1 BACKGROUND

First, let me congratulate Dr. Reeve for a well-written chapter. Rarely does a researcher write, as in this case, a work that is accessible to a wide audience. The highest compliment that can be made to an author is that it got me thinking.

When I was in graduate school in political science and statistics 35 years ago, the three most influential books I read were neither substantive political science nor the standard statistical textbooks. They were books on deeper explanations of quantitative methodology in social science. The first was Clyde Coombs' (1964) *A Theory of Data*, the second was by Stouffer et al.'s (1950) *Measurement and Prediction*, the fourth volume of *The American Soldier: Studies in Social Psychology in World War II*. The last was John Tukey's (1977) *Exploratory Data Analysis*. They were so important because they opened up the world of a quantitative methodology that could be applied to the social sciences. Up until that point, I had training largely in traditional political

Question Evaluation Methods: *Contributing to the Science of Data Quality,* First Edition.
Edited by Jennifer Madans, Kristen Miller, Aaron Maitland, Gordon Willis.
© 2011 John Wiley & Sons, Inc. Published 2011 by John Wiley & Sons, Inc.

science; and, after reading these books, I knew there could be quantification in social science—just as in the so-called "hard" sciences. In defining the levels of measurement (nominal, ordinal, interval, and ratio), psychophysicist S.S. Stevens (1946) presented a challenge to social scientists to go beyond names to numbers. Social scientists are still grappling with that challenge. The item response theory (IRT) Model, like latent class analysis (LCA), attempts to do just that. In the case of LCA, both Goodman and Clogg have developed methods for moving not only to the ordinal but the interval level (Goodman, 1984; Clogg and Shihadeh, 1994). Of course, both IRT and LCA rely on probability and are not deterministic, but so are quantum mechanics, the photoelectric effect, and Brownian motion (Heisenberg, 1927; Einstein 1905a, b).

10.2   IRT: LIMITATIONS AND POTENTIAL USES IN SURVEY RESEARCH

IRT has the advantage of being able to handle both dichotomous and polytomous variables, but it has two critical assumptions. One is unidimensionality, and the other is local independence. Without unidimensionality, IRT loses it value as a scaling technique, but this is true for other techniques such as LCA and Guttman scaling. Multidimensionality might be overcome by selecting items on the basis of a particular factor identified using factor analysis. Local independence, on the other hand, is more problematic. The more we learn about question ordering (not to mention wording), the less we can assume local independence. In addition, the topics covered in the survey can affect scale performance (see Krosnick and Presser, 2010).

The bigger problem for government surveys is that the questions that are asked are generally factual. This would not be a problem if the true answers were known (i.e., educational testing), but they are not (e.g., consumer expenditures, disability and substance abuse, job search, time use, crime victimization, and use of mass transit). Nevertheless, IRT can be very useful to survey methodologists when it comes to measuring data quality when quality is based on observations of actual behavior and not just reports, in particular indicators of the respondent's level of effort.

In the case of level of effort, the measures can be scales or dichotomies, and the IRT model allows the methodologist to evaluate the power of each item (i.e., differential item functioning) more rigorously than just subjectively judging the results of univariate or bivariate tables of indicator distributions. If the IRT parameters are invariant except for linear transformations, IRT also provides the opportunity to compare item performance and scale scores over subgroups and hopefully across time, as the population as well as the survey design changes (van der Linden and Hambleton, 1997). The time dimension is crucial for monitoring respondent performance by establishing a baseline and using time series analysis to gauge changes in the level of effort.

IRT could prove valuable not only for making comparisons over time but also for deepening our understanding of the characteristics of the indicators, including their power of discrimination. The model can provide functions describing the indicators' characteristics in both the population and various subpopulations. These functions generally will not require the guessing parameter.

IRT would gain immediate utility when paired with LCA. In recent years, LCA has been used to create a unidimensional measure of data quality in surveys (Van de Pol and de Leeuw, 1986; Tucker, 1992; Van de Pol and Langeheine, 1997; Bassi et al., 2000; Biemer and Bushery, 2001;Biemer and Wiesen, 2002 Tucker et al., 2003, 2004, 2005; Biemer, 2004). LCA, a theory for detecting unobserved variables, was developed by Paul Lazarsfeld (1950). According to Lazarsfeld, an unobserved variable (measurement error in this case) could be constructed by taking into account the interrelationships among observed or "manifest" variables. The mathematics underlying this theory was extended by Lazarsfeld and Henry (1968) and Goodman (1974).

The concept underlying LCA is relatively straightforward. The idea is to find a latent variable that explains the relationships between the observed variables. Thus, statistically speaking, the relationships between the observed variables disappear after conditioning on the latent variable. In the case of measurement error, the observed variables are often repeated measures or a set of indicators that measure various aspects of respondent performance. Maximum likelihood estimation, using an EM algorithm, is used to identify the latent model. A chi-square test is used to measure the goodness of fit.

The problem arises when trying to interpret the results. Although the mathematics is understandable, the estimation procedure does operate something like a "black box." While the fit may be good from a statistical standpoint, the question remains as to whether valid conclusions can be drawn about the structure of measurement error in the data. Much of the answer to this question lies in having strong theoretical reasons for the choice of observed variables, understanding what their interrelationships say about measurement error, and being able to see how this information is captured in the latent variable. In the latter case, substantive, and not statistical diagnostics, are essential. This is where IRT could complement LCA by providing information about the power of the individual performance indicators and guide the creation of the latent variable. At that point, the tabular displays could serve a more confirmatory role.

10.3   AN EXAMPLE EVALUATING LATENT CLASS MODELS

A recent analysis of measurement error in interviews collected over 6 years of the Bureau of Labor Statistics Consumer Expenditure Interview Survey (CE 1996 through 2001) will illustrate the current substantive diagnostics used to validate an LCA model that measures data quality (see Tucker et al., 2004,

2005). Given that many models involving various combinations of the manifest indicators were done for respondent reports for several separate expenditure categories, the evaluation of these diagnostics was labor intensive. IRT might produce a superior product that requires less time once model construction begins.

10.3.1   Indicators from CE

Although respondents complete five quarterly waves of the survey, only second-wave data (the first-quarter data are only for bounding) from those respondents actually reporting expenditures in certain commodity classes were included (over 40,000 interviews). Thus, measures of response error only were considered for those respondents reporting expenditures and not for those who said they had no expenditures in these categories. A number of manifest indicators of response error were created; however, only one of them was directly related to expenditure information. These indicators are listed below, with the coding scheme used for each:

- Number of contacts the interviewer made to complete the interview (1 = 0–2; 2 = 3–5; 3 = 6+)
- The ratio of respondents to total number of household members (1 ≤ 0.5; 2 > 0.5)
- Whether the household was missing a response on the income question (1 = present; 2 = missing)
- The type and frequency of records used. This variable indicates whether respondents used bills or their checkbook to answer questions, and how often they did so. (1 = never; 2 = single type or sometimes; 3 = multiple types or always)
- The length of the interview (1 < 45 minutes; 2 = 45–90; 3 > 90)
- A ratio of expenditures reported for the last month of the 3-month reporting period to the total expenditures for the 3 months (1 < 0.25; 2 = 0.25–0.5; 3 ≥ 0.5)
- A combination of type of record used and the length of the interview. (1 = poor; 2 = fair; 3 = good)

Each quarter these survey process variables are collected to gain some understanding of the respondent's performance. Difficulty contacting a respondent could indicate some reluctance to participate in the survey or, at least, a limited amount of time to devote to it. The more household members that actually participate directly in the interview, the better the reporting should be. Missing data on the income question is one indication that a respondent is not fully cooperative, and this might extend to the reporting of expenditures. Extensive record use indicates a respondent's willingness to provide as complete a report as possible and limits the need for recall. Longer interviews are usually associated with providing more detailed information. One could argue that interview length is directly related to the amount of expenditures reported, at least overall, but it is no doubt also a measure of respondent performance. Previous research by Silberstein (1989) demonstrated that higher expenditure reports in the most recent month was associated with forgetting expenditures made during the first two months of the reporting period. Finally, we believed that a variable combining both record use and interview length might be at least as powerful as the two alone.

10.3.2   Model Selection

Because only respondents reporting expenditures in a particular category were included in the analysis, for each of the seven expenditure categories studied, separate models were estimated using the lEM LCA software developed by Vermunt (1997). We ran all possible combinations of only three and four indicators for each expenditure category in order to maintain adequate cell sizes in the manifest tables. Restricted models forced the ordering of the manifest indicators based on the hypothesized relationship of each indicator to response error (ordering the latent classes in what we believed to be an interpretable manner). Every model was run with several different sets of starting values to avoid reaching only a local solution. The models selected and displayed in Table 10.1 were three-class LCA models involving either three or four manifest indicators. Their selection was based on the statistical diagnostics in the lEM package. They are the Bayesian Information Criteria (BIC) and the Dissimilarity Index. A negative BIC indicates a good fit, although its size depends upon the degrees of freedom in the model. Essentially, it is a log-likelihood ratio chi-square adjusted for degrees of freedom. The Dissimilarity Index specifies the proportion of cases that would be incorrectly assigned to a manifest cell based on the latent model. A Dissimilarity index less than 0.05 is desirable. The same set of manifest indicators were not used for the best model in each case, but the statistical diagnostics confirm a good fit for all final models.

TABLE 10.1. Results from Latent Class Analysis Creating Response Error Measures for Purchasers of Selected Commodities

Commodity	BIC	Dissimilarity Index
Kid's clothing	−7.4029	0.0040
Women's clothing	−23.1575	0.0221
Men's clothing	−234.0034	0.0258
Furniture	−239.6923	0.0327
Electricity	−8.6450	0.0021
Minor vehicle repairs	−221.5239	0.0306
Kitchen accessories	−119.7008	0.0288

The substantive diagnostics used in final model selection for three of the commodity categories are provided in Tables 10.2–10.9 and correspond to the selected models in Table 10.1. Remember, however, these tables were produced for each model that had acceptable values on the statistical diagnostics. For each commodity's model (Tables 10.2, 10.4, and 10.6), we examined both conditional probabilities of the latent variable given each value of each

TABLE 10.2. Conditional Probabilities Associated with Each Manifest Variable in the LCA Model for Minor Vehicle Repairs

		Latent Class					
		P(A|X)			P(X|A)		
Manifest Variable	Manifest Category	x = 1 (Poor)	x = 2 (Fair)	x = 3 (Good)	x = 1 (Poor)	x = 2 (Fair)	x = 3 (Good)
Contacts	1 (0–2)	0.33	0.27	0.17	0.22	0.69	0.10
	2 (3–5)	0.40	0.46	0.57	0.15	0.66	0.19
	3 (6+)	0.26	0.26	0.26	0.17	0.67	0.15
Income	1 (present)	0.75	0.84	0.99	0.15	0.66	0.18
missing	2 (missing)	0.25	0.16	0.01	0.28	0.71	0.01
Interview	1 (45 minutes)	0.82	0.03	0.03	0.87	0.11	0.03
length	2 (45–90 minutes)	0.00	0.77	0.00	0.00	1.0	0.00
	3 (>90 minutes)	0.18	0.20	0.97	0.10	0.42	0.48
Third	1 (<0.25)	0.38	0.36	0.38	0.18	0.66	0.16
month	2 (0.25)	0.32	0.33	0.37	0.17	0.66	0.17
ratio	3 (≥0.5)	0.30	0.31	0.26	0.17	0.69	0.13

TABLE 10.3. Conditional Probabilities Associated with Cells of the Manifest Table

Manifest Variable Category				Latent Class					
(A)	(B)	(C)	(D)	P(ABCD|X = x)			P(X = x|ABCD)		
Contacts	Income Missing	Interview Length	Third Month Ratio	x = 1 (Poor)	x = 2 (Fair)	x = 3 (Good)	x = 1 (Poor)	x = 2 (Fair)	x = 3 (Good)
1	1	1	1	0.08	0.00	0.00	0.89	0.10	0.02
1	1	1	2	0.06	0.00	0.00	0.88	0.10	0.02
1	1	1	3	0.00	0.06	0.00	0.88	0.10	0.01
3	2	3	1	0.00	0.00	0.00	0.26	0.69	0.05
3	2	3	2	0.00	0.00	0.00	0.25	0.70	0.05
3	2	3	3	0.00	0.00	0.00	0.25	0.71	0.04

Note: The complete table was too large for publication in the book. This excerpt provides an example of the information used in the analysis.

TABLE 10.4. Conditional Probabilities Associated with Each Manifest Variable in the LCA Model for Women's Clothing

		Latent Class					
		P(A|X)			P(X|A)		
Manifest Variable	Manifest Category	x = 1 (Poor)	x = 2 (Fair)	x = 3 (Good)	x = 1 (Poor)	x = 2 (Fair)	x = 3 (Good)
Contacts	1 (0–2)	0.27	0.27	0.27	0.37	0.35	0.28
	2 (3–5)	0.35	0.35	0.72	0.28	0.27	0.44
	3 (6+)	0.38	0.38	0.00	0.51	0.49	0.00
Third	1 (<0.25)	0.24	0.40	0.33	0.30	0.39	0.31
month	2 (0.25)	0.56	0.28	0.48	0.43	0.30	0.28
ratio	3 (≥0.5)	0.20	0.33	0.19	0.33	0.44	0.24
Record	1 (poor)	0.72	0.25	0.25	0.63	0.21	0.16
length	2 (fair)	0.21	0.41	0.37	0.23	0.45	0.32
	3 (good)	0.07	0.34	0.38	0.11	0.48	0.41

indicator and the conditional probabilities of each indicator given each value of the latent variable. We also provide, in Tables 10.3, 10.5, and 10.7 the actual probabilities of an individual respondent being in a particular latent class given the manifest cell location, as well as the proportion of cases assigned to each manifest cell by the latent class model.

10.3.3 Minor Vehicle Repairs Model

Turning to Table 10.2 for minor vehicle repairs, notice first that the best model included four indicators—number of contact, income missing, interview length, and the ratio of third month expenditures to the total. The variable labeled **X** is the latent variable and the values are "1" for "poor" reporting, "2" for "fair" reporting, and "3" for "good" reporting. Looking at the P(A\|X) column, the least differentiation across the latent classes occurs for the third month ratio indicator, and number of contacts is not particularly powerful either. The proportion of respondents missing income drops when moving from poor to good on the response error scale. Clearly, interview length has the most powerful relationship with the latent variable. The next column makes it apparent that most respondents will fall in the fair category, although interview length stands out even here. Table 10.3 gives the probabilities $P(X = x|ABCD)$ that respondents in a given cell on the manifest table will fall in a particular latent class. In this case, the model has worked quite well, in that, the modal probabilities for most cells are around 0.7 or above, with many above 0.8 or even 0.9. Note, however, that respondents in only a few cells will be assigned to the good category. In those cases, the interview was long, income was not missing, but the modal probability was not that large.

TABLE 10.5. Conditional Probabilities Associated with Cells of the Manifest Table

Manifest Variable Category			Latent Class					
(A)	(B)	(C)	P(X = x|ABCD)			P(X = x|ABCD)		
Contacts	Third Month Ratio	Record Length	x = 1 (Poor)	x = 2 (Fair)	x = 3 (Good)	x = 1 (Poor)	x = 2 (Fair)	x = 3 (Good)
1	1	1	0.05	0.02	0.02	0.56	0.25	0.20
1	1	2	0.01	0.04	0.03	0.19	0.47	0.34
1	1	3	0.00	0.03	0.03	0.08	0.49	0.43
1	2	1	0.11	0.03	0.03	0.68	0.16	0.15
1	2	2	0.03	0.04	0.05	0.28	0.39	0.33
1	2	3	0.01	0.04	0.05	0.13	0.42	0.44
1	3	1	0.04	0.02	0.01	0.58	0.27	0.15
1	3	2	0.01	0.03	0.02	0.20	0.53	0.26
1	3	3	0.00	0.03	0.02	0.09	0.57	0.34
2	1	1	0.06	0.03	0.06	0.46	0.20	0.33
2	1	2	0.02	0.05	0.09	0.14	0.35	0.52
2	1	3	0.01	0.04	0.09	0.06	0.34	0.61
2	2	1	0.14	0.03	0.09	0.59	0.14	0.27
2	2	2	0.04	0.06	0.13	0.21	0.29	0.51
2	2	3	0.01	0.05	0.13	0.09	0.29	0.62
2	3	1	0.05	0.02	0.03	0.50	0.23	0.26
2	3	2	0.01	0.04	0.05	0.16	0.42	0.42
2	3	3	0.01	0.03	0.05	0.07	0.42	0.51
3	1	1	0.07	0.03	0.00	0.69	0.31	0.00
3	1	2	0.02	0.05	0.00	0.28	0.71	0.01
3	1	3	0.01	0.04	0.00	0.14	0.85	0.01
3	2	1	0.15	0.04	0.00	0.80	0.19	0.00
3	2	2	0.04	0.06	0.00	0.42	0.57	0.01
3	2	3	0.02	0.05	0.00	0.24	0.75	0.01
3	3	1	0.05	0.03	0.00	0.68	0.32	0.00
3	3	2	0.02	0.04	0.00	0.27	0.72	0.00
3	3	3	0.00	0.04	0.00	0.14	0.86	0.01

10.3.4   Women's Clothing Model

The model for women's clothing (Tables 10.4 and 10.5) included three indicators—number of contacts, the ratio of third month expenditures to the total, and the combination of record use and interview length (RECLEN). The P(A|X) column shows that few respondents requiring a great many contacts are good reporters. By itself, the ratio variable does not provide that much discrimination. As for RECLEN, those with the shortest interviews and who rely least on records are more likely to fall in the poor reporter category. Table 10.5 provides a contrast to Table 10.3. In general, the modal values are not as large, and the three latent categories are not as well differentiated by

TABLE 10.6. Conditional Probabilities Associated with Each Manifest Variable in the LCA Model for Electricity

		Latent Class					
		P(A|X)			P(X|A)		
Manifest Variable	Manifest Category	x = 1 (Poor)	x = 2 (Fair)	x = 3 (Good)	x = 1 (Poor)	x = 2 (Fair)	x = 3 (Good)
Contacts	1 (0–2)	0.21	0.27	0.44	0.44	0.14	0.43
	2 (3–5)	0.35	0.57	0.56	0.46	0.19	0.35
	3 (6+)	0.43	0.17	0.00	0.91	0.09	0.00
Respondent to HH ratio	1 (≤0.5)	0.20	0.13	0.13	0.68	0.11	0.21
	2 (>0.5)	0.80	0.87	0.87	0.56	0.15	0.29
Third month ratio	1 (<0.25)	0.22	0.60	0.17	0.49	0.33	0.18
	2 (0.25)	0.59	0.07	0.70	0.63	0.02	0.36
	3 (≥0.5)	0.20	0.33	0.12	0.58	0.24	0.17

TABLE 10.7. Conditional Probabilities Associated with Cells of the Manifest Table

Manifest Variable Category			Latent Class					
(A)	(B)	(C)	P(ABCD|X = x)			P(X = x|ABCD)		
Contacts	Respondent to HH Ratio	Third Month Ratio	x = 1 (Poor)	x = 2 (Fair)	x = 3 (Good)	x = 1 (Poor)	x = 2 (Fair)	x = 3 (Good)
1	1	1	0.01	0.02	0.01	0.48	0.27	0.25
1	1	2	0.02	0.00	0.04	0.55	0.01	0.43
1	1	3	0.01	0.01	0.01	0.57	0.20	0.23
1	2	1	0.04	0.14	0.07	0.36	0.33	0.30
1	2	2	0.10	0.02	0.27	0.43	0.02	0.55
1	2	3	0.03	0.08	0.05	0.45	0.26	0.29
2	1	1	0.03	0.05	0.01	0.47	0.34	0.19
2	1	2	0.04	0.01	0.05	0.61	0.02	0.37
2	1	3	0.01	0.02	0.01	0.57	0.25	0.18
2	2	1	0.06	0.29	0.09	0.35	0.42	0.23
2	2	2	0.17	0.04	0.34	0.49	0.03	0.48
2	2	3	0.06	0.16	0.06	0.45	0.33	0.23
3	1	1	0.02	0.01	0.00	0.85	0.15	0.00
3	1	2	0.05	0.00	0.00	0.99	0.01	0.00
3	1	3	0.02	0.01	0.00	0.90	0.10	0.00
3	2	1	0.08	0.09	0.00	0.78	0.22	0.00
3	2	2	0.20	0.01	0.00	0.99	0.01	0.00
3	2	3	0.07	0.05	0.00	0.85	0.15	0.00

TABLE 10.8. Mean Expenditures for Selected Commodities by Latent Class (Purchasers Only)

| | | Value of Latent Class | | |
		1 = Poor	2 = Fair	3 = Good
Kid's clothing	Mean	192.65(a)	198.42(a)	220.74(b)
	N	5749.00	3606	2039
Women's clothing	Mean	212.70(a)	237.46(b)	231.30(b)
	N	9922	7035	6877.00
Men's clothing	Mean	194.90(a)	199.38(a)	188.57(a)
	N	14,621	452	3398
Furniture	Mean	724.09(a)	517.46(b)	884.05(c)
	N	3795	2088	979
Electricity	Mean	253.64(a)	218.59(b)	248.71(a)
	N	29,900	3982	5132
Minor vehicle	Mean	216.55(a)	219.04(a)	226.86(a)
expenditures	N	1798	6844	2483
Kitchen accessories	Mean	122.36(a)	153.30(b)	141.67(b)
	N	5056	1005	4526

TABLE 10.9. Mean Expenditures for Selected Commodities by Latent Class (Purchasers and Nonpurchasers)

| | | Value of Latent Class | | |
		1 = Poor	2 = Fair	3 = Good
Kid's clothing	Mean	44.90(a)	59.62(b)	71.09(c)
	N	24,666	12,001	6,331
Women's clothing	Mean	99.00(b)	148.08(a)	152.94(a)
	N	21,316	11,281	10,401.00
Men's clothing	Mean	78.98(b)	107.04(a)	105.46(a)
	n	36,080	842	6076
Furniture	Mean	117.25(a)	66.22(b)	266.63(c)
	n	23,437	16,315	3246
Electricity	Mean	230.47(a)	198.87(b)	223.30(c)
	n	32,905	4377	5716
Minor vehicle repairs	Mean	39.47(a)	57.03(b)	82.28(c)
	n	9864	26,288	6846
Kitchen accessories	Mean	23.27(b)	52.51(a)	47.58(a)
	n	26,589	2934	13,475

the P(X = x|ABCD). Also, for several of the cells, the probabilities for categories 2 and 3 are quite close to one another. The largest modal values tend to occur for respondents requiring a large number of contacts.

10.3.5   Electricity Expenditures Model

Table 10.6 and 10.7 give the latent model information for electricity expenditures. This is a three-indicator model including number of contacts, proportion of household members participating in the survey, and the ratio of third month expenditures to the total. Virtually none of the respondents who are difficult to contact are found in the good reporter category. The proportion of household members participating does not provide much discrimination on its own. The ratio variable does behave somewhat differently in this case. Very few respondents in the middle category of this variable are fair reporters. They tend to be either poor or good reporters. Table 10.7 presents a troubling picture. Although a few of the modal probabilities are not very large, a review of the P(X = x|ABCD) indicates that all but two cells would be assigned to the poor reporter category, and the ones that are not do not have large modal probabilities. These results demonstrate that, while the statistical diagnostics indicated a good fit, a closer look at the substantive workings of the model cause one to suspect its validity.

10.3.6   Analysis of Expenditure Means by Latent Class

To gain a further understanding of the models (given the variability in the substantive diagnostics), we decided to examine expenditure means for the three latent classes for respondents purchasing items in each of the seven commodity categories. These means are reported in Table 10.8. The results, while not completely disconfirming, are not that promising. Looking across all seven categories of expenditures, there are none where each of the three means can all be distinguished from one another and in the direction we would expect. However, for kid's clothing, women's clothing, and kitchen accessories, two separate reporting groups can be identified that meet our expectations. For men's clothing and minor vehicle repairs, none of the means differ from one another. Furniture has means that are all different, but the fair category mean is much smaller than the poor category one. In the case of electricity, the middle category mean is also the smallest. So, at most we can distinguish only two groups of reporters, and for minor vehicle repairs, which had the best diagnostics up to this point, we cannot even do that.

We explored including in our analysis those interviewed who reported no expenditures in these categories. Perhaps, by including them, we would get a clearer picture of the differences between categories of reporters. We assigned these respondents to latent classes based on their values on the indicators in the models created from the analysis of the purchasers. The new latent class means for the seven expenditure categories are shown in Table 10.9. Of course, the means are smaller, but there are notable changes in the pattern in some cases. Here we concentrate on the three categories that were analyzed in detail earlier. Recall that the assignment probabilities in Table 10.3 for minor vehicle repairs discriminated well between the three classes of the latent variable, and at least two of the indicators had a relationship with the latent variable that was clearly in the expected direction. The pattern of expenditure means now bears this out. The means increase monotonically moving across the scale, and the three means are significantly different from one another. In the case of women's clothing, two of the indicators had a relationship with the latent variable in the direction expected, but the assignment probabilities were not as good as for vehicle repair. Again, in some cases, the assignment probabilities for the fair and good categories were quite close. As in Table 10.8, the pattern of the means bears this out, but at least the differentiation is more pronounced. The poor category mean is significantly lower than the other two, but the fair and good category means are much the same but in the right direction. It is possible that the models really can only differentiate two categories of reporters in this case. For electricity, one of the most confusing cases along with maybe furniture, the pattern of the means still makes little sense. As expected, most respondents fall in the poor category, but the mean is highest here and is not meaningfully different from that of the good reporters. This is not that surprising because the assignment probability for the respondents in the good category is not much greater than it was for the poor category. Although the mean for the fair category is the lowest, given the magnitudes of the means for electricity, it is not that much different from the other two, and few respondents end up in that category anyway. We might conclude that the LCA model failed in this case. In fact, we think it did, and it may be because almost everyone provides an expenditure amount. They all have it and probably have good records for it.

Even including the nonpurchasers does not show that the LCA models are accomplishing what we would like. After all, they were created just based on purchasers and could not make distinctions between them in terms of size of expenditure. In the cases where a distinction could be made, it seems that a two-class model would have been more appropriate in some cases. Even if we can explain the problem with electricity, we still have the problem with furniture. On the other hand, the indicators we chose and the models we created tell us something when we include those who reported no purchases. Clearly, they are respondents who do not score well on the performance indicators. When they are included in the mix, providing a more varied set of respondents, the results from the models improve significantly for several of the expenditure categories. In either case, however, most of the commodities have at least half of the respondents assigned to the poor reporter class.

10.4 CONCLUSION

This exercise demonstrates the tedious process required for choosing between many different possible LCA models. The examination of the class means in

Tables 10.8 and 10.9 illustrate how fraught with difficulties this can be. In the future, as additional indicators become available, the sorting process could be even more complicated. At least with IRT, another, more rigorous layer of indicator selection is available. Using IRT could not only reduce the time necessary to identify the best indicators, but it is likely to improve the selection process. Plans also are to increase both the number of indicators in a model and the number of latent classes to the extent allowed by the sample size. IRT could have some value in this case, too.

REFERENCES

Bassi F, Hagenaars JA, Croon MA, Vermunt J (2000). Estimating true changes when categorical panel data are affected by uncorrelated and correlated classification errors: an application to unemployment data. Sociological Methods & Research; 29:230–268.

Biemer P (2004). Analysis of classification error for the revised current population survey employment questions. Survey Methodology; 30(2):127–140.

Biemer P, Bushery J (2001). Application of Markov latent class analysis to the CPS. Survey Methodology; 26(2):136–152.

Biemer P, Wiesen C (2002). Latent class analysis of embedded repeated measurements: an application to the national household survey on drug abuse. Journal of the Royal Statistical Society, Series A; 165(1):97–119.

Clogg CC, Shihadeh ES (1994). Statistical models for ordinal variables. SAGE Series on Advanced Quantitative Techniques. Thousand Oaks, CA: Sage Publications.

Coombs CH (1964). A Theory of Data. New York: Wiley.

Einstein A (1905a). On a heuristic viewpoint concerning the production and transformation of light. Annalen der Physics; 17:132–148.

Einstein A (1905b). On the motion required by the molecular kinetic theory of heat of small particles suspended in a stationary liquid. Annalen der Physics; 17:549–560.

Goodman LA (1974). Exploratory latent structure analysis using both identifiable and unidentifiable models. Biometrika; 61:215–231.

Goodman LA (1984). The Analysis of Cross-Classified Data Having Ordered Categories. Cambridge: Harvard University Press.

Heisenberg W (1927). Uber den Anschulichen Inhalt der Quantentheoretischen Kinematik und Mechanik. Zeitschrift fur Physiotherapie; 43:172–198.

Krosnick JA, Presser S (2010). Question and questionnaire design. In: Marsden PV, Wright JD, editors. Handbook of Survey Research. Bingley, UK: Emerald; pp. 263–313.

Lazarsfeld PF (1950). The logical and mathematical foundation of latent structure analysis. In: Stouffer S, Guttman L, Suchman EA, Lazarsfeld PF, Starr SA, Clausen J, editors. Studies on Social Psychology in World War II, Vol. 4, Measurement and Prediction. Princeton, NJ: Princeton University Press.

Lazarsfeld PF, Henry NW (1968). Latent Structure Analysis. Boston, MA: Houghton-Mifflin.

Silberstein AR (1989). Recall effects in the U.S. consumer expenditure interview survey. Journal of Official Statistics; 5:125–142.

Stevens SS (1946). On the theory of scales of measurement. Science; 103:677–680.

Stouffer SA, Guttman L, Suchman EA, Lazarsfeld PF, Star SA, Clausen JA (1950). Studies on Social Psychology in World War II, Vol. 4, Measurement and Prediction. Princeton, NJ: Princeton University Press.

Tucker C (1992). The estimation of instrument effects on data quality in the consumer expenditure diary survey. Journal of Official Statistics; 8:41–61.

Tucker C, Biemer P, Meekins B (2003). Latent class modeling and estimation errors in consumer expenditure reports. In: American Statistical Association 2003 Proceedings of the Section on Survey Research Methods. Washington, DC: American Statistical Association.

Tucker C, Biemer P, Meekins B, Shields J (2004). Estimating the level of underreporting of expenditures among expenditure reporters: a micro-level latent class analysis. In: American Statistical Association 2004 Proceedings of the Section on Survey Research Methods. Washington, DC: American Statistical Association.

Tucker C, Biemer P, Meekins B (2005). Estimating the level of underreporting of expenditures among expenditure reporters: a further micro-level latent class analysis. In: American Statistical Association 2005 Proceedings of the Section on Survey Research Methods. Washington, DC: American Statistical Association.

Tukey JW (1977). Exploratory Data Analysis. Boston, MA: Addison-Wesley.

Van de Pol F, de Leeuw J (1986). A latent Markov model to correct for measurement error. Sociological Methods & Research; 15(1–2):118–141.

Van de Pol F, Langeheine R (1997). Separating change and measurement error in panel surveys with an application to labor market data. In: Lyberg LE, Biemer P, Collins M, de Leeuw ED, Dippos C, Schwarz N, Trewin D, editors. Survey Measurement and Process Quality. New York: Wiley.

van der Linden WJ, Hambleton RK (1997). Handbook of Modern Item Response Theory. New York: Springer.

Vermunt J (1997). lEM: A General Program for the Analysis of Categorical Data. Tilburg, The Netherlands: Tilburg University.

PART IV
Latent Class Analysis

11 Some Issues in the Application of Latent Class Models for Questionnaire Design

PAUL P. BIEMER

RTI International, University of North Carolina at Chapel Hill

MARCUS BERZOFSKY

RTI International

11.1 INTRODUCTION

A key objective of questionnaire evaluation and improvement is the reduction of measurement error. There may be hundreds of questions in a questionnaire, each designed with a specific purpose. Without information on measurement error, determining which questions are flawed and need improvement is speculative. Questionnaire improvement can be accomplished best with some quantification of measurement error. In addition, measurement error data can provide clues regarding the sources of error and its causes. Thus, evaluation of measurement error is critical to the questionnaire improvement process.

Many survey questions classify respondents into two or more categories; examples include demographic groups, income classes, opinion-, attitude-, or behavior-types, and so on. For categorical responses, measurement errors result in misclassifications or classification error. Thus, methods for evaluating classification errors are an important area of survey research and are the focus of this chapter.

Evaluation of measurement error requires at least two measurements of the underlying construct or latent variable. If just two measurements are

---

Question Evaluation Methods: Contributing to the Science of Data Quality, First Edition.
Edited by Jennifer Madans, Kristen Miller, Aaron Maitland, Gordon Willis.
© 2011 John Wiley & Sons, Inc. Published 2011 by John Wiley & Sons, Inc.

available, (e.g., the original question and an exact repeat of the question) test/retest reliability can be estimated under the assumption that the two measurements are parallel (i.e., their errors are independent and identically distributed). However, reliability analysis is quite limiting (see Biemer, 2009). Biemer (2009) demonstrates that an alternative analysis based upon *latent class analysis (LCA)* can provide more information regarding the classification error structure under constraints proposed by Hui and Walter (1980). Rather than requiring that the two measurements be parallel, Hui and Walter introduced a categorical grouping variable, G, with the properties that the true characteristic varies by level of G while the classification error parameters do not.

Sinclair (1994) and Kreuter et al. (2008) showed that the Hui–Walter restrictions can be problematic in analyses when the grouping variable only roughly satisfies the assumptions. In such cases, the estimates of the model parameters (prevalence and classification error probabilities) can be quite biased. Some authors (e.g., Biemer, 2001) have substituted alternate assumptions in such cases; however, these two may be too restrictive for general use. The only other alternative is to introduce a third repeated measurement. With three or more measurements (or *indicators*) of the same construct, the Hui–Walter assumptions can be relaxed and, in fact, a grouping variable is not needed. This paper is concerned with LCA when at least three indicators are available, referred to as the standard latent class (LC) model (Lazarsfeld and Henry, 1968).

This chapter documents key issues for LCA and provides guidance from the literature as well as new research on approaches to dealing with them. We begin with a brief overview of standard LCA with three indicators. Section 11.3 discusses the issues of unidentifiability, sparseness, boundary values, and flippage. Section 11.4 is devoted to the important issue of local independence. The major points of the chapter are summarized in Section 11.5.

11.2   BASIC LC MODEL FOR THREE INDICATORS

The standard LC model has three key assumptions: (a) homogeneity, (b) independent classification errors, and (c) univocality. Homogeneity means that the classification probabilities have essentially zero variance within the population subdomains specified by the model. Independent classification errors means that misclassification by one indicator does not influence classification error of other indicators. Univocality means that all the indicator variables are indicators of the same underlying latent variable. When (a), (b), and (c) are satisfied, the indicators satisfy a fourth condition referred to as *local independence*. Local independence means that the joint conditional probability of the indicators given the latent variable can be factored into the product of their respective conditional marginal probabilities. When one or more of these assumptions do not hold, the model parameter estimates will be biased.

There are a number of additional issues that can cause problems in LCA. These include unidentifiability, data sparseness, boundary values/improper solutions, local maxima, and LC flippage. Each of these issues creates challenges for the LC modeler. In addition, the current literature provides no systematic approach for addressing these issues, and current guidance is scattered and inconsistent.

11.2.1 Notation and Model Assumptions

Let X denote an individual's true classification for some categorical characteristic assumed to be observed with error. X is assumed to be unobservable (i.e., a *latent variable*). Let A, B, and C denote three indicators of X corresponding to the three survey items. Throughout this paper, we assume that the latent variable and the indicators have the same number of categories or classes denoted by K. However, in general, this assumption is not required for LCA. As an example, we may wish to classify persons into $K = 3$ labor force categories: employed (EMP), unemployed (UNE), and not in the labor force (NLF). Denote an individual's true labor force status by X, and denote by A, B, and C three questions or methods for measuring X. These three indicators are assumed to classify an individual (fallibly) into the same three labor force categories as X.

The notation will generally conform to conventions established by the LCA literature. Let $\pi_x^X = \Pr(X = x)$ or the prevalence of class x in the population. Let $\pi_{a|x}^{A|X} = \Pr(A = a \mid X = x)$; in other words $\pi_{a|x}^{A|X}$ is the conditional probability that a person whose true characteristic is x responds with a to question A with corresponding definitions for $\pi_{b|x}^{B|X}$ and $\pi_{c|x}^{C|X}$. If $X = x$ and $A = x$, then $\pi_{x|x}^{A|X}$ is the probability of a correct response to question A for persons truly in class x. Likewise, $\pi_{a|x}^{A|X}$ for $a \neq x$ are error probabilities for persons truly in class x. Thus, for the labor force example, $\pi_{2|3}^{A|X}$, $\pi_{2|3}^{B|X}$, and $\pi_{2|3}^{C|X}$ are the probabilities that persons who are truly UNE are classified as NLF by A, B, and C, respectively. Note that, for $K = 3$, there are nine conditional probabilities for each indicator. However, since the probabilities must sum to one within each LC, there are only six independent parameters per indicator. Therefore, the total number of independent conditional probabilities is $3 \times 6 = 18$. Adding this to the two independent prevalence probabilities yields a total 20 independent model parameters to be estimated. In general, for L indicators, the number of independent parameters is $(K-1)(LK+1)$.

Under the assumptions of LCA, these parameters can be estimated directly from this K^L table formed by cross-classifying the J indicators. For $J = 3$, we denote the cross-classification of A, B, and C by ABC. An arbitrary cell in ABC is denoted by abc corresponding to $A = a$, $B = b$, and $C = c$, for $a,b,c = 1, \ldots, K$. Maximum likelihood estimation (MLE) is used to obtain the model estimates. The probability of being classified in cell abc under the model is

$$\pi_{abc}^{ABC} = \sum_x \pi_x^X \, \pi_{a|x}^{A|X} \, \pi_{b|x}^{B|X} \, \pi_{c|x}^{C|X} \tag{11.1}$$

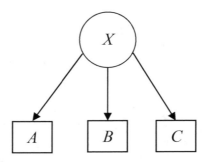

FIGURE 11.1. Path diagram for the standard three-indicator LC model.

which is referred to as the *likelihood kernel*. This model can be represented by the path diagram in Figure 11.1.

Software packages for estimating the LC model parameters typically use the EM algorithm (Dempster et al., 1977) for maximizing the likelihood. Unless otherwise stated, estimates of model parameters will be denoted by the parameter symbol with a "hat" or circumflex. For example, $\hat{\pi}_{b|x}^{B|X}$ denotes an estimate of $\pi_{b|x}^{B|X}$. Let n_{abc} denote the number of observations in cell abc, $n = \sum_{abc} n_{abc}$, and let $\hat{m}_{abc} = n\hat{\pi}_{abc}^{ABC}$ denote the model estimate of the cell frequency where $\hat{\pi}_{abc}^{ABC}$ is obtained from replacing each parameter by its estimate.

As in standard categorical data analysis (e.g., Agresti, 2002), a measure of model fit is provided by the likelihood ratio chi-squared statistic given by

$$L^2 = -2\sum_{abc} n_{abc} \ln \frac{\hat{m}_{abc}}{n_{abc}} \tag{11.2}$$

which is distributed approximately as a chi-square with degrees of freedom equal to the number of cells in the observed table minus the number of independent parameters to be estimated minus 1. For example, for $K = J = 3$, the model has six degrees of freedom. The model is said to adequately fit the data if the value of L^2 is smaller than the 95th percentile of a chi-square distribution with the corresponding degrees of freedom or if L^2 is otherwise sufficiently low so that any observed lack of fit can be attributable to chance alone. Alternatively, one can use the Pearson chi-squared statistic given by

$$X^2 = \sum_{a,b,c} \frac{(n_{abc} - \hat{m}_{abc})^2}{\hat{m}_{abc}}. \tag{11.3}$$

11.2.2   Multiple Group Analysis

Multiple group analysis is important for comparing misclassification probabilities for different subpopulations and analytic domains. For example, we may wish to compare the accuracy of labor force status reports for males and

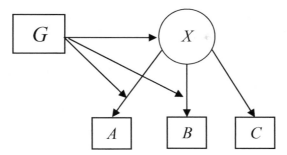

FIGURE 11.2. Path diagram for standard three-indicator LC model with grouping variable.

females. In that case, Equation 11.1 can be fit separately for males and females and the estimated parameters compared via t-tests, but this is an inefficient method. A more efficient approach is to fit a model simultaneously for males and females using the $GABC$ cross-classification table where $G = 1$ for males and $G = 2$ for females. The likelihood kernel for this model,

$$\pi_{gabc}^{GABC} = \pi_g^G \sum_x \pi_{x|g}^{X|G} \pi_{a|xg}^{A|XG} \pi_{b|xg}^{B|XG} \pi_{c|xg}^{C|XG} \tag{11.4}$$

is equivalent to fitting Equation 11.1 separately for each group. Figure 11.2 shows its path diagram.

One advantage of Equation 11.3 is that testing group equivalence is facilitated. For example, testing $H_0: \pi_{a|x1}^{A|XG} = \pi_{a|x2}^{A|XG}$ against $H_1: \pi_{a|x1}^{A|XG} \neq \pi_{a|x2}^{A|XG}$ for all x and a can be done with a single conditional chi-square test. The conditional chi-square test compares the likelihood ratio chi-square (L^2) for the *unrestricted* model Equation 11.4 and the *restricted* model, obtained by replacing $\pi_{a|xg}^{A|XG}$ by $\pi_{a|x}^{A|X}$ in Equation 11.4. The difference, $L^2(\text{restricted}) - L^2(\text{unrestricted})$, is distributed as a chi-square with degrees of freedom equal to $df(\text{restricted}) - df(\text{unrestricted})$. This test is also more powerful than the corresponding t-test discussed above (Agresti, 2002) and is valid even when the distribution of L^2 is not strictly chi-square.

As we will describe later in the paper, in addition to facilitating multigroup comparisons, multiple group LCA has at least three other purposes. First, incorporating multiple grouping variables helps achieve homogeneity (assumption (b) above) (see Section 11.4.1 for more details). Second, grouping variables help with model identifiability, but may introduce sparseness in the data (see Section 11.3.1 for more details). Third, grouping variables can lead to a better fitting model (see Section 11.4 for more details).

Adding grouping variables to a model is very much like adding explanatory variables to a regression equation with similar model building strategies. In the next section, the log-linear modeling framework for LCA that facilitates these strategies will be described.

11.2.3 Log-Linear Parameterization of the LC Model

The likelihood kernel in Equation 11.4 is specified in a *probabilistic model* framework. In that framework, cell probabilities or expected frequencies are expressed as products of conditional and marginal probabilities. How the likelihood is factored depends on the model assumptions; that is, local independence, assumed relationships between the grouping variables, the latent variable and the indicators, and so on. This formulation is somewhat restrictive in that certain higher order interactions that may not be significant are still required in the model in order to keep antecedent lower-order interactions in the model. A more flexible approach is the *logistic regression* framework.

The logistic regression framework takes advantage of the equivalent representation of conditional probabilities as log-linear models (see Haberman, 1979; Formann, 1992; Vermunt, 1997). For example, the conditional probability $\pi_{a|xg}^{A|XG}$ can be expressed as

$$\pi_{a|xg}^{A|XG} = \frac{\exp(u_a^A + u_{ga}^{GA} + u_{xa}^{XA} + u_{xga}^{XGA})}{\sum_a \exp(u_a^A + u_{ga}^{GA} + u_{xa}^{XA} + u_{xga}^{XGA})} \tag{11.5}$$

(see, e.g., Vermunt, 1997) where the u-parameters are obtained from the log-linear model for the expected frequencies of the XGA marginal table (collapsed from the full XGABC table); viz.,

$$\log(m_{xag}) = u_a^A + u_{ga}^{GA} + u_{xa}^{XA} + u_{xga}^{XGA}. \tag{11.6}$$

Because X is unknown, special algorithms such as the EM algorithm, Newton–Raphson, the method of scoring, and so on may be used to estimate the u-parameters in Equation 11.6. Vermunt (1997) discusses these methods in more detail. Now Equation 11.4 can be reformulated by substituting Equation 11.5 for $\pi_{a|xg}^{A|XG}$ and the remaining probabilities by

$$\pi_{x|g}^{X|G} = \frac{\exp(u_x^X + u_{xg}^{XG})}{\sum_x \exp(u_x^X + u_{xg}^{XG})}$$

$$\pi_{b|xg}^{B|XG} = \frac{\exp(u_b^B + u_{gb}^{GB} + u_{xb}^{XB} + u_{xgb}^{XGB})}{\sum_b \exp(u_b^B + u_{gb}^{GB} + u_{xb}^{XB} + u_{xgb}^{XGB})} \tag{11.7}$$

$$\pi_{c|xg}^{C|XG} = \frac{\exp(u_c^C + u_{gc}^{GC} + u_{xc}^{XC} + u_{xgc}^{XGC})}{\sum_c \exp(u_c^C + u_{gc}^{GC} + u_{xc}^{XC} + u_{xgc}^{XGC})}$$

Under this formulation, the model is sometimes called a *modified path model* (Goodman, 1973) since causal relationships among the variables is represented in the ordering of the conditional probabilities. As an example, if A is observed first in an interview, the response to B and C can be influenced or conditioned by A but not the converse. Likewise, if B is observed before C, then B can have a causal influence on C but not the converse. Thus, model building strategies need only consider conditional probabilities $\pi_{a|x}^{A|X}$, $\pi_{b|xa}^{B|XA}$, and $\pi_{c|xab}^{C|XAB}$ and their antecedents since terms such as $\pi_{a|xb}^{A|XB}$ and $\pi_{b|xc}^{B|XC}$ are theoretically implausible.

A shorthand notation for representing modified path models has been developed that takes advantage of the hierarchical structure of the models. For hierarchical modeling, Equation 11.5 can be represented unambiguously by $\{XGA\}$ since it is understood that u_{xga}^{XGA} as well as all its antecedents are also in the model. Using this notation, only the highest order interactions in the model need be specified. The combined model in Equation 11.4 can be specified as the union of the submodels for each conditional probability as $\{XGA\ XGB\ XGC\}$.

One of the advantages of the modified path model parameterization of LC models is that it can neatly and concisely represent many models such as $\{XGA\ XGB\ XC\ AB\}$ or $\{XG\ XA\ XB\ XC\ BC\}$ that cannot be simply represented in the probabilistic framework. In addition, the log-linear parameterization better lends itself to the goal of fitting parsimonious models that fit the data well.

The following example, taken from Biemer and Wiesen (2002), demonstrates some of the concepts covered thus far.

11.2.4   Example: Past Year Marijuana Use

Biemer and Wiesen (2002) consider the case of three measurements obtained in a single interview in an application to the National Household Survey on Drug Abuse (NHSDA). They defined three indicators of past year marijuana use from three questions asked at various points during the interview. Indicator A is the response to the recency of use question which asks about the length of time since marijuana or hashish was last used. Indicator B is the response to the frequency of use question which asks how frequently, if ever, the respondent has used marijuana or hashish in the past year. Indicator C is a composite of items referencing marijuana use from a drug answer sheet. An affirmative response to any one of the items is coded as "yes" for C and otherwise C is coded as "no." Their research was primarily focused on estimating the false positive and false negative probabilities separately for A, B, and C in order to determine the accuracy of each method, but especially A and B.

In preliminary analysis, Biemer and Wiesen (2002) noted a high degree of inconsistency among the indicators, particularly for C. But, because C is a composite of a number of questions, they hypothesized that C might be the more accurate indicator. In support of this conjecture, estimates of past year marijuana use were the highest for C, which could suggest greater accuracy since marijuana use tends to be underreported in the NHSDA (Mieczkowski, 1991; Turner et al., 1992).

TABLE 11.1. Comparison of Estimated Percent Classification Error by Indicator[1]

True Classification	Indicator of Past Year Use	Original	Revised
Yes ($X = 1$)	Recency = No ($A = 2$)	7.29 (0.75)	6.93 (0.72)
	Direct = No ($B = 2$)	1.17 (0.31)	1.18 (0.31)
	Composite = No ($C = 2$)	6.60 (0.70)	7.18 (0.72)
No ($X = 2$)	Recency = Yes ($A = 1$)	0.03 (0.02)	0.03 (0.02)
	Direct = Yes ($B = 1$)	0.73 (0.07)	0.76 (0.07)
	Composite = Yes ($C = 1$)	4.07 (0.15)	1.23 (0.09)

[1]Standard errors are shown in parentheses.

Biemer and Wiesen (2002) considered a number of models that were extensions of the basic LC model for three measurements including grouping variables defined by age (G), race (R), and sex (S). That simplest model allowed the prevalence of past year marijuana use, π_1^X, to vary by age, race, and sex while the error probabilities—$\pi_{a|x}^{A|X}$, $\pi_{b|x}^{B|X}$, and $\pi_{c|x}^{C|X}$—remained constant by group. The most complex model allowed error probabilities to vary by grouping variable. Since A, B, and C were collected in the same interview, the possibility of locally dependent errors was also considered in the analysis. The best model identified in their analysis was a locally independent model that incorporated simple two-way interaction terms between each grouping variable and indicator variable.

Estimates of the classification error rates for all three indicators of past year drug use were derived from the best model. Table 11.1 shows the estimated classification error rates (expressed as percentages) for the total population, for all 3 years including a revised data set which is identical to the 1994 data set for indicators A and B, but differ importantly for indicator C.

The large false positive rate for C suggest that the high inconsistency rate between C and the other two indicators is due to classification error in C and not classification error in the other two indicators as conjectured. They discovered that one of the questions used for constructing C seemed confusing and not closely aligned with marijuana use—a condition known as bivocality. They created a new indicator by deleting the suspicious question. The new false positive rate for C using the revised data set dropped to 1.23% from 4.07% which is closer to the other indicators.

11.3 SOME ISSUES IN ESTIMATION

Many analysts find the application of LC models difficult because a number of problems can produce unexpected results in the estimation of model parameters. These problems include model unidentifiability, boundary values, local maxima, and failure of model assumptions.

11.3.1 Model Unidentifiability

In order for the parameters of an LC model to be uniquely estimated, the model must be *identifiable*. The likelihood for an identifiable model has only one maximum. Like traditional log-linear models, an LC model is not identifiable if the number of parameters to be estimated exceeds the number of observed frequencies. Usually, this problem can be corrected by simply deleting some terms from the model or imposing parameter restrictions and constraints. However, with LC models, even models having non-negative degrees of freedom may still be unidentifiable. For example, the binary two-indicator model $\{XA\ XB\}$ with the constraint $XA = XB$ is unidentifiable even though the AB table contains four frequencies and the model has only three parameters.

Non-identified models are more common when the dimension of X is greater than the dimension of the indicators. For example, the model with four binary indicators (A,B,C,D) is not identified when X has three classes even though the model degrees of freedom is positive (Goodman, 1974b). If a model is not identifiable, the model estimates are not unique and are invalid. For example, repeated maximizations of the model likelihood will produce different solutions. Therefore, verifying that a model is identifiable is an important step in an LCA. Figure 11.3 provides a visual for the concept of identifiability. In this figure, $\mathsf{L}(\pi\,|\,\mathbf{n})$ denotes the likelihood function with maximum L_{max}.

Identifiability is not likely to be a problem for models that can be estimated with reasonable precision. Bartholomew and Knott (1999) suggest performing LCA only when the sample size is sufficiently large and the number of LCs to be estimated is four or fewer. However, such advice may be unrealistic in many applications. In addition, weak identifiability can usually be avoided if the indicators are highly correlated with X. Berzofsky (2009) showed that, for the three-indicator model, a single poor indicator can result in convergence to a local maximum with very high probability rendering the standard approach for checking identifiability misleading.

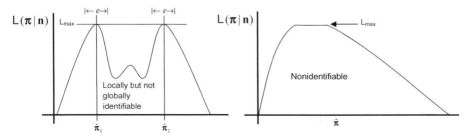

FIGURE 11.3. Locally identifiable and not identified likelihoods.

11.3.1.1   Checking Model Identifiability   In most situations, the best that can be done is to verify that the model is locally identifiable. According to Goodman (1974a), a model is *locally identifiable* if the maximum likelihood solution is unique within some ε-neighborhood of the solution. This is less restrictive than *global identifiability* in which the MLE must be unique throughout the entire parameter space. He goes on to show that a necessary and sufficient condition for *local identifiability* is that the information matrix, I, is positive definite (see also, Dayton and Macready, 1980). Formann (1985) provided a proof that the likelihood is locally identifiable if the eigenvalues of the information matrix are positive. This condition is easily checked in the estimation process when using methods that rely on second partial derivatives such as the Fisher's scoring method or Newton–Raphson. If the EM algorithm is used, the check is not done automatically because second derivatives are needed for parameter estimation. Nevertheless, software packages employing EM such as *l*EM (Vermunt, 1997), Mplus (Muthén and Muthén, 1998–2005), and Latent Gold (Vermunt and Magidson, 2005) still provide this check.

Tests for identifiability based on the rank of the information matrix typically do not include boundary parameters (e.g., probabilities estimated to 0 or 1). Therefore, this identification test does not check whether boundary parameters are identified. Identifiability of these parameters can be tested by other means. For example, parameters that are estimated zero because of *structural zeros* in the data as well as boundary parameters with large first-order derivatives may be considered identified. However, parameters that are estimated 0 or 1, with first-order derivatives about 0, may not be identified. The method of multiple starting values should be used to test whether boundary parameters are identified.

A common way of checking local identifiability when using the EM algorithm is to estimate the model multiple times using different starting values. If the solutions from these multiple maximizations are identical for the same maximum value of the likelihood, the model is probably identifiable. But should multiple runs with different starting values produce quite different parameter estimates for the same maximum likelihood value, the model is definitely not identifiable (Hagenaars, 1990, p. 112).

Bartholomew and Knott (1999, p. 151) discuss the relationship between identifiability and precision. If the likelihood is fairly flat in the neighborhood of an estimate, then the estimate will have a very high variance even though the likelihood is identifiable. As the likelihood approaches absolute flatness (as in Fig. 11.3), the estimate becomes unidentifiable. Therefore, estimates having very large variances could be *nearly* or *weakly unidentifiable*. One measure of the strength of model identification is the condition number defined by the ratio of the largest to the smallest eigenvalues of the information matrix. As a general rule of thumb, a ratio that exceeds 5000 is considered an indication of weak identification. In that case, standard errors of parameters will be large, and some parameters may be highly correlated. To achieve stronger identification, one or more of these parameters should be fixed or set equal to other parameters.

11.3.2   Data Sparseness

In fitting LC models with multiple grouping variables and interactions or that involve variables with many categories, the data table can become *sparse*. Sparseness is also a problem in less complex models when n is small. One general measure of data sparseness used throughout this literature is the ratio n/k where k is the number of cells in the table. A sparse data table is one with a low value of this ratio, say, less than 5. Sparse tables can cause a number of problems in an LCA as will be discussed in this section.

11.3.2.1   Sampling Zeros versus Structural Zeros   Zero count cells are particularly problematic since they can lead to inestimable parameters and biased tests of model fit. Zero count cells can be divided into two types: (1) *sampling zero* and (2) *structural* zero cells. *Sampling zero* cells have small, but nonzero, probabilities of being observed. When the sample size is small, such cells will contain no observations by chance. However, as the sample size increases, expected frequencies increase, and these zero cells can become populated. For example, the expected frequency of recent college graduates addicted to heroin may be quite small. Although such persons may exist in the population, they are unlikely to be sampled unless the sample size is large.

Structural zeros are cells that are theoretically impossible to observe. They either do not exist in the population or they exist but have a 0 probability of selection by the sample method. For example, consider the cross-classification of persons by gender and reason for a hospitalization. Some reasons such as pregnancy are gender-specific, causing certain cells to have 0 probability of being observed.

Sampling zero cells are considered as part of the observed data requiring no special constraints in the modeling. By contrast, structural zero cells are not part of the data. Proper modeling of these cells requires constraining the corresponding cell probabilities to 0. Otherwise, the model will treat these cells as sampling zeros with possibly important consequences for parameter estimation, model fitting, and so on. In addition, the degrees of freedom for the model should be adjusted to reflect the structural zero constraints imposed on the model. A general formula for adjusting the degrees of freedom is the following:

$$df = (k-z)-(npar-npar_0) \tag{11.8}$$

where z is the number of structural zero cells, *npar* is the number parameters in the model, $npar_0$ is the number of parameters that cannot be estimated because of zero cells. Note that when $npar_0 > 1$, the model is not identifiable. The methods described in the last section for obtaining an identifiable model (e.g., imposing constraints, reducing model complexity, and collapsing cells) should be applied in this situation.

11.3.2.2   Test Statistics and Data Sparseness   An important issue in dealing with sampling zero cells or, more generally, sparse tables is the validity of the chi-square test statistics. Particularly for LC models, sparseness can invalidate tests of model fit using L^2 and X^2 because they are no longer chi-square distributed. Haberman (1977) and Agresti and Yang (1987) have shown that for testing of nested models in log-linear analysis, the conditional likelihood test using L^2 is much less affected by data sparseness. Although it may be difficult to ascertain model fit, model selection can still proceed normally. For example, when determining whether XGA can be removed from the model, comparing the L^2 statistics from the models containing XGA versus the same model replacing XGA with GA XG XA only can still be considered a valid test when n/k is quite small.

Because both X^2 and L^2 provide a chi-square test of model fit, the question of which statistic is better arises. In most situations, the two measures will provide very nearly the same value so the question is moot. However, in sparse tables, they will differ as they are no longer chi-square or even identically distributed. If they differ considerably, then the validity of either chi-square goodness of fit test is questionable. Indeed, an important reason to compute both X^2 and L^2 is so that this comparison can be made. Studies have shown that X^2 is often closer to chi-square than L^2 in sparseness situations (Agresti, 1990, p. 247). However, Collins et al. (1993) warn that the variation in X^2 can still be quite large, and the test can be unreliable. Still, the literature seems to favor X^2 as long as $n/k > 1$.

One method for dealing with model testing for sparse tables is to reduce the number of grouping variables in the model or the categories of the manifest variables to reduce the number of cells, k. Reducing k can also restore identifiability if the reason for unidentifiability is empty cells or table sparseness. Another approach is to empirically estimate the exact distribution of either X^2 or L^2 for a particular data set and model using bootstrapping methods (Efron, 1979). Both the Latent Gold and Panmark (van de Pol et al., 1998) software have implemented this approach for LC models following on the work of Aitkin et al. (1981), Noreen (1989), and Collins et al. (1993). Alternatively, lower-order marginal distributions of the items can be used instead of the full table. Feiser and Lin (1999) have a test based upon first- and second-order marginal distributions of the items. Even when the full table is sparse, the first and second marginal distributions may not be, and their test is more valid.

Another option is to use a measure of model fit that is more robust to small cell sizes. One such measure is the *power divergence* statistic developed by Cressie and Read (1984), denoted by CR. As a measure of model deviance, CR is often less susceptible to data sparseness than either X^2 or L^2. It is implemented in most software packages with its power parameter (λ) equal to $2/3$ which then takes the form

$$CR = \frac{9}{5} \sum_{abc} n_{abc} \left[\left(\frac{n_{abc}}{\hat{m}_{abc}} \right)^{\frac{2}{3}} - 1 \right]. \tag{11.9}$$

When L^2 or X^2 and CR diverge, CR is likely to more closely approximate a chi-square random variable. However, CR will also be invalid if sparseness is severe. An alternative to model deviance measures is the dissimilarity index given by

$$D = \frac{\sum_{abc} |n_{abc} - \hat{m}_{abc}|}{2n} \tag{11.10}$$

can always be used to gauge the model lack of fit when chi-square statistics fail. Unfortunately, a statistical test of model fit using D is not available. In addition, D is not appropriate for testing and comparing nested models.

A useful approach for comparing alternative models, especially models that are not nested, is the Bayesian Information Criterion (BIC) defined as

$$\text{BIC} = -2 \log L_{max} + npar \times \log n \tag{11.11}$$

where L_{max} is the maximized likelihood value. The model having the lowest BIC is preferred under this criterion because it provides the best balance between two factors: model fit (reflected in the value of $\log L_{max}$) and model parsimony (represented by $npar \times \log n$). Note that as the number of parameters increases, $npar \times \log n$ increases as a sort of "penalty" for making the model more complex. However, $-2 \log L_{max}$ operates in the opposite direction, becoming more negative as more parameters are included and the likelihood increases. In this way information theoretic measures can be used to identify the model that strikes the best balance between model fit and the model parsimony. Unfortunately, the theory does not provide a statistical test for determining whether the BIC for one model is significantly lower than the corresponding measure for another model. Rather the measures are meant to be used as indexes of the comparative fit and somewhat of a rough guide for model selection.

If a problem occurs in the iterative algorithm for computing MLE estimates as a result of sampling zero cells or very small cells, the model parameters may not converge, and the software output will typically provide a warning such as "Information matrix is not of full rank" or "Boundary or non-identified parameters." In addition, one or more of the response probabilities will be at or very near their boundary values (i.e., 1 or 0) which corresponds to a log-linear parameter of $\pm\infty$. The estimated standard errors for these parameters will be quite large in comparison to the rest of the estimates. One method for solving this problem is to add a small constant (say, 0.001) to each cell of the observed frequency table. However, Agresti (1990, p. 250) advises caution in applying this method. He recommends rerunning the model with various additive constants (10^{-3}, 10^{-5}, 10^{-8}, etc.) to determine the sensitivity of the estimates to the choice. The smallest constant that will alleviate the computational problems without unduly influencing the key parameter estimates should be used.

11.3.3   Boundary Estimates and Improper Solutions

It is not uncommon for one or more model probability estimates to be either 0 or 1. Boundary estimates occur for a number of reasons.

(a) The probability may have been estimated to be very small, say 10^{-6}, which is essentially 0 at the level of precision of the printed output.
(b) The optimum value of the probability estimate is actually negative; however, the maximization algorithm stopped moving in the direction of the optimum once the estimate reached 0 and remained there until convergence.
(c) The data may contain zero cells which result in a u-parameter estimate of $\pm\infty$; that is, the parameter could not be estimated. Consequently, the corresponding error probability estimate is 0.
(d) The probability estimate may have been constrained to be 0.

The situation in (a) poses no problem for model testing because the estimate is optimal, and the standard error exists and is approximately 0 as well. The situation in (b) occurs when the sample size is too small or if the model is misspecified. If the model is misspecified, dropping one or more interaction terms from the model will alleviate the problem. Likewise, (c) may be a consequence of inadequate sample or model complexity. It can be handled using the methods described in the last section for data sparseness.

Finally, (d) is quite common when dealing with highly stigmatized questions such as drug use, abortions, and sexual perverseness. For these questions, it is highly unlikely that respondents will respond in error in socially undesirable ways. For example, if A is an indicator denoting the response to a question on whether or not the respondent has ever had an abortion, it may be reasonable to assume that the probability of a false positive is zero, so we constrain $\pi_{1|2}^{A|X}$ to be 0. Another situation that arises is testing for the number of classes for X. For example, in typology analysis, an analyst may wish to compare the fit of a two-class model and a three-class LC model. It can be shown that the two-class model can be obtained by setting $\pi_x^X = 0$ for either $x = 1, 2$, or 3 in the three-class model.

The boundary value estimates under (b) and (c) can pose problems for model testing since the information matrix is not of full rank and therefore cannot be inverted. This implies that the standard errors of the LC model parameters cannot be computed in the usual way. In addition, interval estimates or significance tests cannot be obtained by standard log-linear modeling procedures. One possible remedy is to constrain the LC model probabilities that are estimated on the boundary to either 1 or 0 and rerun the model. The degrees of freedom should also be adjusted to account for the fixed constraints. The standard errors of the unconstrained parameters are valid if one is willing to assume that these were a priori constrained based upon theoretical consid-erations as in (d). The resulting set of parameter estimates is referred to as a "terminal" solution. They are correct provided the true parameter value of the boundary estimate is actually 0 (Goodman, 1974a).

De Menezes (1999) proposed obtaining standard errors using the bootstrap variance estimation approach (see, e.g., Efron and Tibshirani, 1993). This is done by generating a large number of samples of size n using the estimated LC model of interest and then re-estimating the model for each synthetic sample. The square root of the variance of an estimate computed over these synthetic samples provides an estimator of the standard error of the estimate. Another approach is to impose a Bayesian prior distribution on the model parameters referred to as Bayesian posterior mode estimation by Galindo-Garre and Vermunt (2006). This latter approach has been implemented in the Latent Gold software (Vermunt and Magidson, 2005).

11.3.4   Local Maxima

The EM algorithm will sometimes converge to a local maxima rather than a global maximum. The usual strategy for avoiding this problem is to run the EM algorithm some number, say $R > 1$ times, each time using a different set of starting values for the parameters. Suppose the maximum likelihood value for the rth run is $\mathsf{L}_{\max,r}$. Then the global maximum value of the likelihood, say L_G will satisfy

$$\mathsf{L}_G \geq \max_r \{\mathsf{L}_{\max,r}\}. \tag{11.12}$$

A set of starting values that produces the global maximum is denoted by $\arg\max_r \{\mathsf{L}_{\max,r}\}$, and is called an *optimal set* of starting values. There may be many optimal sets; that is, $\arg\max_r \{\mathsf{L}_{\max,r}\}$ is not unique. If the parameter estimates associated with an optimal set are not the same across all optimal sets, then the model is unidentifiable. However, if the model is identifiable, then the unique set of estimates produce by each $\arg\max_r \{\mathsf{L}_{\max,r}\}$ is the model (local) MLE. In general, using the starting values associated with $\max_r \{\mathsf{L}_{\max,r}\}$ is no guarantee of the global maximum; however, the probability that it is the optimal set increases as R increases. Depending upon the complexity of the model and the sample size, R between 20 and 100 is usually sufficient to produce good results. However, as mentioned earlier, if the indicators are not sufficiently correlated with X, even $R = 1000$ may not find the global maximum (Berzofsky, 2009).

Small values of R can produce good results if good starting values are selected. For example, constraining the starting values of the error parameters to the interval (0.1, 0.5) will usually require fewer runs than the range (0,1) provided the optimal set is in the restricted range. However, Wang and Zhang (2006) showed that the probability that a randomly generated set of starting values is optimal is relatively high. Therefore, in most cases, the EM algorithm may only have to be rerun a dozen times or so to identify the global maximum. Some programs (e.g., Panmark) do this automatically for a user-specified R. For these programs, it is a good idea to specify an R of 100 or more. In other programs (e.g., *EM), rerunning the algorithm is a manual process, but it is easy to write a customized macro in a general software package (such as SAS or SPSS) to run the LCA software the desired number of times. Almost all LCA programs will list the seed value used for each run which can be saved by the custom macro. However, if the LCA software is rerun manually, usually 5–10 times is sufficient if each set of starting values produces the same set of estimates. This solution can be accepted as the global MLE. Otherwise, the algorithm should be rerun another five or more times, again with different starting values. If new maxima arise, the program should be rerun until one has confidence that the largest of the maxima has been found.

For very complex models, rerunning the algorithm to full convergence can be time-consuming. It is well-known that the EM algorithm increases the likelihood quickly in its early stage and slows as it approaches convergence. Wang and Zhang (2006) suggest rerunning the EM with a few (50–100) iterations to identify an optimal set of starting values. The algorithm can be rerun one final time with the optimal set to full convergence. This will greatly reduce the time required to find a global maximum. Wang and Zhang (2006) also note that the probability of finding the global maximum increases, for any R, as the sample size increases and as the correlation between the latent variable and indicators increases. On the other hand, the probability of a global maximum decreases with the amount of boundary or near-boundary values and as the number of LCs increases. To avoid local maximum solutions, it is a good idea is to keep the number of LCs small (e.g., five or fewer) if possible.

11.3.5 LC Flippage

Another problem that arises in fitting LC models is a misalignment of the categories of the latent variable X and the categories of the indicators, A, B, and C, referred to here as *LC flippage*. Consider the artificial data in Table 11.2. The proportion of positives for A, B, and C are 0.27, 0.36, and 0.17, respectively. However, fitting the standard LCA in *EM yields an estimated value of π_1^X, the latent prevalence rate, of 0.9. This is implausible, particularly since these data were generated using the value of $\pi_1^X = 0.1$. It appears that

TABLE 11.2. Artificial Data for Three Indicators

Pattern	111	112	121	122	211	212	221	222
Count	67	64	23	116	28	196	52	454

TABLE 11.3. Two Sets of Parameters that Will Generate the Data in Table 11.2

	π_1^X	$\pi_{1|2}^{A|X}$	$\pi_{2|1}^{A|X}$	$\pi_{1|2}^{B|X}$	$\pi_{2|1}^{B|X}$	$\pi_{1|2}^{C|X}$	$\pi_{2|1}^{C|X}$
Data generating model	0.1	0.2	0.1	0.3	0.15	0.1	0.2
Equivalent model	0.9	0.9	0.8	0.85	0.7	0.8	0.9

the software is confusing the positive and negative categories of X. Another consequence of LC flippage is that the error probabilities for the indicators are also flipped. The model used to generate the data in the table used $\pi_{1|2}^{A|X} = 0.2$ and $\pi_{2|1}^{A|X} = 0.1$. However, the software returned the implausible estimates of 0.9 and 0.8, respectively, for these two parameters. The estimates of the false positive and false negative probabilities for B and C were also implausibly large. After rerunning the model, a new solution emerged with estimates appropriately aligned to the data generating model; that is, with $\pi_1^X = 0.1$, $\pi_{1|2}^{A|X} = 0.1$ and $\pi_{2|1}^{A|X} = 0.2$. Apparently, the category of X which is associated with the positive categories of A, B, and C depends upon the set of starting values used to generate the solution.

LC flippage occurs as a result of the symmetry in the likelihood function for an LC model. The data in Table 11.2 were generated by the parameters in first row of Table 11.3. However, the same cell frequencies can be produced using the estimates in the second row of Table 11.3. Both sets of parameters produce the same data and the same value of the likelihood function. Depending upon where the likelihood maximization algorithm starts, it may find either solution. Particularly if the starting values are generated at random, a given software package may be just as likely to return the estimates in the first row as the second. It is up to the data user to determine whether LC flippage has occurred and, if so, to realign the estimates in the output appropriately.

LC flippage can be very difficult to discern in complex models such as models with polytomous indicators that interact with multiple grouping variables. It is not uncommon for LCs to flip inconsistently from group to group. It is therefore important to examine the LC and response probability estimates for each group to ensure they are plausible since this is generally not checked by the LCA software. Comparing the estimates over many runs and starting values is one way to detect the problem. One way to avoid LC flippage is to simply supply a set of plausible starting values for at least one of the indicators in order to align that indicator with latent when the iterative algorithm is initiated. As an example, for the data in Table 11.2, supplying the starting values $\pi_{1|2}^{A|X} = 0.2$ and $\pi_{2|1}^{A|X} = 0.1$ consistently produced the solution in the first row of Table 11.3 over 100 runs.

Supplying plausible starting values is not an absolute safeguard against LC flippage since the iterative algorithm can still venture off into an implausible area of the parameter space during the likelihood maximization process. In some cases, it may be necessary to supply starting values for all indicators as well as their interactions with the grouping variables (e.g., starting values for $\pi_{a|xgh}^{A|XGH}$ for grouping variables G and H). Note, however, these starting values should be removed when checking the models for identifiability.

11.4   LOCAL DEPENDENCE

A key assumption of LCA is local independence. Recall that three indicators, A, B, and C are locally independent if and only if

$$\pi_{abc|x}^{ABC|X} = \pi_{a|x}^{A|X} \pi_{b|x}^{B|X} \pi_{c|x}^{C|X}. \tag{11.13}$$

Failure of Equation 11.13 to hold implies that the indicators are *locally dependent*. If local dependence is not appropriately considered in the modeling process, LC models will exhibit poor fit (i.e., L^2 will be large), the estimates of the model parameters will be biased, sometimes substantially so depending upon the severity of the violation, and their standard errors will be too large (Sepulveda et al., 2008). One measure of this local dependence severity is the LD index defined by

$$LD_{abc|x} = \frac{\pi_{abc|x}^{ABC|X}}{\pi_{a|x}^{A|X} \pi_{b|x}^{B|X} \pi_{c|x}^{C|X}} \tag{11.14}$$

provided the probabilities in the denominator are positive. A necessary and sufficient condition for local independence is that $LD = 1$. For dichotomous outcome measures, Equation 11.14 simplifies to

$$LD_{111|x} = \frac{\pi_{111|x}^{ABC|X}}{\pi_{1|x}^{A|X} \pi_{1|x}^{B|X} \pi_{1|x}^{C|X}} = 1 \tag{11.15}$$

for $X = 1,2$. Because local dependence usually results in greater agreement among the indicators than would be explained by X alone, $LD_{111|x}$ is usually at least 1. The more the LD index deviates from 1, the greater the violation of the local independence assumption.

To illustrate the potential effects of local dependence, consider the data in Table 11.4 which were generated using the parameters in row 1 with $n = 5000$. In these data, the pairs A,C and B,C are locally independent while the pair A,B is not. Thus, $LD_{111|x}$ for $x = 1, 2$ is

TABLE 11.4. Illustration of Local Dependence between A and B

Parameter	π_1^X	$\pi_{1|2}^{A|X}$	$\pi_{2|1}^{A|X}$	$\pi_{1|2}^{B|X}$	$\pi_{2|1}^{B|X}$	$\pi_{1|2}^{C|X}$	$\pi_{2|1}^{C|X}$	$\pi_{2|11}^{B|AX}$	$\pi_{2|21}^{B|AX}$	$\pi_{1|12}^{B|AX}$	$\pi_{1|22}^{B|AX}$
True	0.10	0.050	0.100	0.104	0.150	0.100	0.200	0.130	0.280	0.360	0.090
Estimated	0.12	0.082	0.033	0.129	0.089	0.304	0.102	0.089	0.089	0.129	0.129

$$LD_{1|1} = \frac{\pi_{1|1|1}^{AB|X}}{\pi_{1|1}^{A|X}\pi_{1|1}^{B|X}} = \frac{\pi_{1|1}^{A|X}\pi_{1|1|1}^{B|XA}}{\pi_{1|1}^{A|X}\pi_{1|1}^{B|X}} = \frac{\pi_{1|1|1}^{B|XA}}{\pi_{1|1}^{B|X}} = \frac{(1-0.130)}{(1-0.150)} = 1.02$$

$$LD_{1|2} = \frac{\pi_{1|1|2}^{AB|X}}{\pi_{1|2}^{A|X}\pi_{1|2}^{B|X}} = \frac{\pi_{1|2}^{A|X}\pi_{1|2|1}^{B|XA}}{\pi_{1|2}^{A|X}\pi_{1|2}^{B|X}} = \frac{\pi_{1|2|1}^{B|XA}}{\pi_{1|2}^{B|X}} = \frac{(1-0.280)}{0.104} = 6.92.$$

$$(11.16)$$

The value of $LD_{1|2}$ suggests a high level of local dependence. The estimates from the model $\{XA\ XB\ XC\}$ appear in the last row of the table. Note that all the estimates are biased, not just the estimates for parameters involving A and B.

There are essentially three possible causes of local dependence including (a) unexplained heterogeneity, (b) dependent (or correlated) classification errors, and (c) departures from univocality (in particular, bivocality). Determining the cause of the local dependence is important, because it can provide clues about how to repair model failure. The next sections discuss these three causes and their remedies.

11.4.1 Unexplained Heterogeneity

As noted in Section 11.1, a key assumption of the LC model is that the error parameters $\pi_{a|x}^{A|X}$, $\pi_{b|x}^{B|X}$, ... are homogeneous in the population. However, this is seldom true for survey data analysis since the probabilities of correctly classifying individuals often depend upon individual characteristics such as age, race, gender, education, interest in the survey, and so on. If these variables are known for each respondent, this *unexplained* heterogeneity can be modeled by including grouping variables as suggested in Section 11.3. For example, if the error probability $\pi_{a|x}^{A|X}$ depends upon some grouping variable H, adding the term $\pi_{a|xh}^{A|XH}$ to the model will correct for this heterogeneity. If the probabilities $\pi_{a|xh}^{A|XH}$ are the same for all individuals in group h for all h, then H is said to explain the heterogeneity, and the assumption of homogeneous error probabilities will hold within each level of H.

In some cases, the heterogeneity cannot be explained exactly by the grouping variables available for modeling. This situation is often referred to as *unobserved* heterogeneity. It still may be possible to explain most of the heterogeneity using known grouping variables. To illustrate, suppose H is unobserved but another variable G that is observed is highly correlated with H. Then, adding the term $\pi_{a|xg}^{A|XG}$ to the model will approximately account for $\pi_{a|xh}^{A|XH}$; that is, the variation in $\pi_{a|x}^{A|X}$ will be small within the levels of G. If the correlation between G and H is weak, this approach will not be effective (Berzofsky, 2009). In some cases, several grouping variables may be required to explain the heterogeneity in the error probabilities.

A consequence of unexplained heterogeneity is the violation of Equation 11.13. To see this, suppose A and B satisfy conditional local independence given H; that is,

$$\pi_{ab|xh}^{AB|XH} = \pi_{a|xh}^{A|XH}\pi_{b|xh}^{B|XH}$$

$$(11.17)$$

or, in words, the classifications errors for A and B are independent within the groups defined by H. In order for unconditional local independence to hold, we must have

$$LD_{abx} = \frac{\pi_{abx}^{AB|X}}{\pi_{a|x}^{A|X} \pi_{b|x}^{B|X}} = \sum_h \pi_h \frac{\pi_{a|xh}^{A|XH} \pi_{b|xh}^{B|XH}}{\pi_{a|x}^{A|X} \pi_{b|x}^{B|X}} = 1 \qquad (11.18)$$

which, in general, is not true unless $\pi_{a|x}^{A|X} = \pi_{a|xh}^{A|XH}$ and $\pi_{b|x}^{B|X} = \pi_{b|xh}^{B|XH}$ for all h; that is, the error probabilities are homogeneous across the levels of H.

For example, let $G = 1$ for males and $G = 2$ for females and assume that males and females are equally divided in the population (i.e., $\pi_1^G = \pi_2^G = 0.5$). Suppose for measuring cocaine prevalence, the false negative probabilities for males and females is 0.4 and 0.05, respectively, for both A and B (equal error probabilities for both indicators); that is, $\pi_{2|11}^{A|XG} = 0.4$ and $\pi_{2|12}^{A|XG} = 0.05$. Further assume that the false positive probabilities are zero for both A and B. The marginal accuracy probabilities are then $\pi_{1|1}^{A|X} = \pi_{1|1}^{B|X} = 0.5(1-0.4) + 0.5(1-0.05) = 0.775$. Since A and B are locally independent conditional on G we can write,

$$\begin{aligned} LD_{1|1} &= \frac{\pi_{1|11}^{AB|X}}{\pi_{1|1}^{A|X} \pi_{1|1}^{B|X}} \\ &= \frac{\pi_1^G \pi_{1|11}^{A|XG} \pi_{1|11}^{B|XG} + \pi_2^G \pi_{1|12}^{A|XG} \pi_{1|12}^{B|XG}}{\pi_{1|1}^{A|X} \pi_{1|1}^{B|X}} \\ &= \frac{0.5(1-0.4)^2 + 0.5(1-0.05)^2}{0.775^2} = 1.05 \end{aligned} \qquad (11.19)$$

which, since it deviates from 1, indicates local dependence. If the variable sex (G) is added to the model via the terms $\pi_{a|xg}^{A|XG}$ and $\pi_{b|xg}^{B|XG}$, the model will be correctly specified, and this source of heterogeneity will be eliminated along with its consequential bias.

The best strategy for eliminating local dependence due to unexplained heterogeneity is to add grouping variables to the model that explain the variation in the error probabilities. Error probabilities usually vary by demographic variables such as age, race, sex, and education, which are also important for describing the variation in the structural components of the model. In addition, interviewer characteristics such as experience, age, sex, and, for some items, race, should also be considered for the measurement components. The usual model selection approaches (see Burnham and Anderson, 2002) can be applied to achieve model parsimony.

11.4.2   Correlated Errors

When the questions for assessing X are asked in the same interview, respondents may provide erroneous responses to each question either deliberately

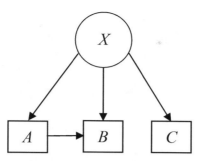

FIGURE 11.4. Figure path diagram of correlated errors.

or inadvertently, causing errors to be correlated across indicators. For example, one question may ask "Have you used marijuana in the past 30 days?" and another question later in the questionnaire may ask "In the past 30 days, have you smoked marijuana at least once?" If there is a tendency for respondents who falsely responded negatively to the first question to also respond negatively to the second question, a positive correlation in the false negative errors will result. Such correlation is sometimes called *behavioral correlation* (Biemer and Wiesen, 2002).

Even if the replicate measurements are separated by longer periods of time (e.g., days or weeks), correlated errors may still be a concern if respondents tend to use the same response process to arrive at their erroneous responses. Likewise, comprehension errors caused by the wordings of questions may cause errors that are correlated across the indicators. A path diagram depicting correlated errors is shown in Figure 11.4.

As an example, the analysis of Biemer and Wiesen (2002) from Section 11.2.4 found that respondents tended to underreport infrequent marijuana use for the question "Have you used marijuana in the past 30 days?" They speculated that the reason may be the word "use" which implies greater frequency of use as implied by the term "marijuana user." This effect is likely not to depend upon their memory of responses given in previous iterations of the same question. Therefore, if this question is asked in the same way in separate interviews occurring weeks apart, errors may still be correlated even though respondents may have no memory of their prior responses.

Errors will be correlated if respondents who provided a false negative response for indicator A have a greater probability of false negatively reporting on indicator B than respondents who correctly responded positively to A. Mathematically, this is written as $\pi_{2|12}^{B|XA} > \pi_{2|11}^{B|XA}$ or, equivalently,

$$\pi_{1|11}^{B|XA} > \pi_{1|12}^{B|XA} \qquad (11.20)$$

which indicates a tendency for respondents who answer A correctly to also answer B correctly. It follows that A and B are locally dependent because

$$
\begin{aligned}
LD_{1|1} &= \frac{\pi_{11|1}^{AB|X}}{\pi_{1|1}^{A|X}\,\pi_{1|1}^{B|X}} \\
&= \frac{\pi_{1|1}^{A|X}\,\pi_{1|11}^{B|XA}}{\pi_{1|1}^{A|X}\,(\pi_{1|1}^{A|X}\,\pi_{1|11}^{B|XA} + \pi_{2|1}^{A|X}\,\pi_{1|12}^{B|XA})} \\
&= \frac{\pi_{1|11}^{B|AX}}{\pi_{1|1}^{A|X}\,\pi_{1|11}^{B|XA} + \pi_{2|1}^{A|X}\,\pi_{1|12}^{B|XA}} \\
&> \frac{\pi_{1|11}^{B|AX}}{\pi_{1|1}^{A|X}\,\pi_{1|11}^{B|XA} + \pi_{2|1}^{A|X}\,\pi_{1|11}^{B|XA}} = 1.
\end{aligned}
\tag{11.21}
$$

Likewise, the correlated errors can occur for false positive responses as well; that is, $\pi_{1|21}^{B|XA} > \pi_{1|22}^{B|XA}$ with similar consequences for local dependence.

For modeling correlated errors between a pair of indicators, A and B, the *joint item method* can be used. For this method, the locally dependent pair of items is simply replaced in the model by a joint item, AB, having K^2 categories corresponding to the cells in the $K \times K$ cross-classification of A and B. In the model, the probabilities $\pi_{a|x}$ and $\pi_{b|x}$ are replaced by $\pi_{ab|x}$. For example, for dichotomous indicators $(K = 2)$, AB has four categories corresponding to $AB = 11, 12, 21, 22$, and the term $\pi_{ab|x}$ represents six parameters—two more than required for $\pi_{a|x}$ and $\pi_{b|x}$. With only three indicators, the unrestricted model with the joint item is unidentifiable since the model degrees of freedom are -2. However, in many situations, the model can be made identifiable through the addition of grouping variables with suitable parameter restrictions. This can be a trial-and-error modeling exercise where identifiability is checked for each restriction using the methods discussed in Section 11.3.1.

Note that the joint item method is equivalent to adding the term ABX to the submodel for $\pi_{b|xa}$ in the modified path model formulation. A more parsimonious solution which is just as effective for modeling behavioral local dependence in many applications is the *direct effect* method. For this method, XAB is replaced by AB in the submodel. Removing X from the AB interaction tacitly assumes that the correlation between the errors for A and B is the same in each LC.

11.4.3   Bivocality

A third source of local dependence is *bivocality*, sometimes called *multidimensional* or *multiple factor* models. Suppose A is an indicator of X and B is an indicator of Y. If $\pi_{y|x} < 1$ for $x = y$ (i.e., X and Y can disagree), A and B are said to be *bivocal* (Alwin, 2007). Bivocal indicators jointly measure two different latent constructs rather than one. As an example, suppose A asks "Have you ever been raped?" and B asks "Have you ever been sexually assaulted?" Clearly, A and B are not indicators of the same latent variable since sexual assault (Y) is a broader concept than rape (X). Therefore, $\pi_{1|1}^{Y|X} < 1$, and a person who has been sexually assaulted but not raped may answer "no" to A

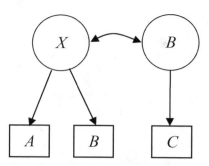

FIGURE 11.5. Bivocality with three indicators.

and "yes" to B without error. Two bivocal indicators may perfectly measure their respective constructs yet, a model that assumes univocality may attribute classification error to either or both indicators. An appropriate model for this situation specifies two latent variables which, depending on the correlation between X and Y and other indicators in the model, may or may not be identifiable.

Suppose there are three indicators A, B, and C where the first two are indicators of X and the third is an indicator of Y (see Fig. 11.5). In this situation, the model $\{XA\ XB\ YC\}$ is not identifiable so the analyst has little choice but to fit the model $\{XA\ XB\ XC\}$. In that case, the indicator C will exhibit very high error probabilities relative to A and B. In fact, it can be shown that the error probabilities for C satisfy

$$\pi_{c|x}^{C|X} = \sum_{y} \pi_{c|y}^{C|Y} \pi_{y|x}^{Y|X}.$$
(11.22)

Fortunately, A, B, and C are still locally independent, and hence, all the parameters of the model, including $\pi_{c|x}$ can still be estimated unbiasedly. To see this, note that

$$
\begin{aligned}
LD_{abc|x} &= \frac{\sum_{y} \pi_{abc|xy}^{ABC|XY} \pi_{y|x}^{Y|X}}{\pi_{a|x}^{A|X} \pi_{b|x}^{B|X} \pi_{c|x}^{C|X}} \\
&= \frac{\sum_{y} \pi_{a|x}^{A|X} \pi_{b|x}^{B|X} \pi_{c|y}^{C|Y} \pi_{y|x}^{Y|X}}{\pi_{a|x}^{A|X} \pi_{b|x}^{B|X} \pi_{c|x}^{C|X}} \\
&= \frac{\pi_{a|x}^{A|X} \pi_{b|x}^{B|X} \sum_{y} \pi_{c|y}^{C|Y} \pi_{y|x}^{Y|X}}{\pi_{a|x}^{A|X} \pi_{b|x}^{B|X} \pi_{c|x}^{C|X}} \\
&= \frac{\pi_{a|x}^{A|X} \pi_{b|x}^{B|X} \pi_{c|x}^{C|X}}{\pi_{a|x}^{A|X} \pi_{b|x}^{B|X} \pi_{c|x}^{C|X}} = 1.
\end{aligned}
$$
(11.23)

This situation is analogous to that considered by Biemer and Wiesen (2002) in Section 11.2.4. Indicator C in that analysis was found to be bivocal relative to A and B. However, replacing C by C' (a univocal indicator of X) did not change the estimates for A and B. This suggests that the local independence assumption was satisfied for both models.

Now suppose there are four indicators: A, B, C, and D. Further assume that items underlying C and D were poorly constructed and, rather than measuring X, they measure another (potentially closely related) latent variable, Y. Thus, A, B, C, and D are bivocal indicators of X. Bivocality is a common problem when multiple indicators of some latent construct are embedded in the same interview and the questionnaire designer tries to conceal the repetition from the survey respondent. If the same question wording is used to create the multiple indicators, respondents may become confused or irritated by the repetition if they become aware of it. Even if they do not perceive the repetition, using the same question wording may induce correlated errors. For this reason, questions are usually reworded and/or reformatted to obscure their repetitiveness. But this is very difficult to do for several indicators without changing the meaning of the questions and creating bivocal indicators.

For the situation in Figure 11.6, bivocality will induce local dependence sometimes referred to as *causal* local dependence, since, as indicators of X, C and D are locally dependent even though they may be locally independent indicators of Y. To see this, note that if we can assume that $\pi_{cd|y}^{CD|Y} = \pi_{c|y}^{C|Y} \pi_{d|y}^{D|Y}$, then

$$LD_{cd|x} = \frac{\pi_{cd|x}^{CD|X}}{\pi_{c|x}^{C|X} \pi_{d|x}^{D|X}} = \frac{\sum\limits_y \pi_{c|y}^{C|Y} \pi_{d|y}^{D|Y} \pi_{y|x}^{Y|X}}{\pi_{c|x}^{C|X} \pi_{d|x}^{D|X}} = \frac{\sum\limits_y \pi_{c|y}^{C|Y} \pi_{d|y}^{D|Y} \pi_{y|x}^{Y|X}}{\left(\sum\limits_y \pi_{c|y}^{C|Y} \pi_{y|x}^{Y|X}\right)\left(\sum\limits_y \pi_{d|y}^{D|Y} \pi_{y|x}^{Y|X}\right)} \qquad (11.24)$$

which in general is not equal to 1 unless $\pi_{y|x} = 1$ when $y = x$ (i.e., the indicators are univocal).

Determining whether the indicators in an analysis are bivocal is, to a large extent, subjective since a rigorous statistical test for bivocality does not exist.

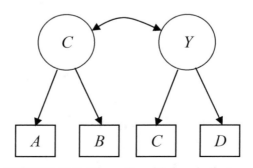

FIGURE 11.6. Bivocality with four indicators.

In order for a manifest variable, A, to be an indicator of the latent variable X, the correlation of A with X must exceed the correlation of A with any other potential latent variable. In other words, A is an indicator of X if it measures X better than any other latent variable *that could be* measured by A. Thus, the indicators in an LCA are univocal if and only if this is true for all the indicators. It is possible that $\pi_{a|x}^{A|X}$ for $a \neq x$ is very large yet A is still an indicator of X. For this reason, it may be difficult to distinguish between a poor indicator of X and a bivocal indicator other than subjectively using theoretical arguments.

As mentioned previously, a bivocal model can be used to address causal local dependence in situations where there are at least two indicators of each latent variable such as the model shown in Figure 11.6. The appropriate model is $\{XY\, AX\, BX\, CY\, DY\}$ which is identifiable as long as X and Y are sufficiently correlated. Using simulated data, Berzofsky (2009) showed that, as the correlation between X and Y decreases, the probability of finding a global maximum approaches 0—a condition known as *weak identifiability*.

11.4.4   A Strategy for Modeling Local Dependence

The possibility of local dependence from all three sources should be considered in any LCA. Fortunately, methods are available for modeling and remedying local dependence provided identifiable models can be found. The modeling fitting strategy is important since identifiability is a critical issue in modeling local dependence. A model building strategy we have used with some success is to consider the possibility of bivocality first. As noted previously, causal local dependence is not an issue when only one indicator is bivocal. When there are four or more indicators, bivocality should be considered by subjectively assessing whether each measure is an indicator of the same latent variable. If bivocality is suspected, a second latent variable should be added to the model.

The next step is to treat local dependency induced by unexplained heterogeneity by adding grouping variables to either the structural or measurement model components. When there are many possible explanatory variables from which to choose, the selection of grouping variables is best guided by a working theory regarding the drivers of the latent constructs as well as the errors of measurement. Age, race, sex, and education are typical correlates of latent constructs. However, for modeling measurement error, additional variables related to the measurement process should be considered. These might include interviewer characteristics, mode of interview, interview setting, interview task variables, as well as para-variables such as number of attempts to interview and respondent reluctance to be interviewed.

The third step is to determine whether one or more indicator by indicator interaction terms should be added to address any residual local dependence from the first two steps. Fortunately, some excellent diagnostic tools are available to guide this step of the modeling process. Most of these methods compare the observed and model-predicted cross-classification frequencies for all $\binom{I}{2}$ possible pairs of indicators. Association in the observed frequency table that is greater than the corresponding association in the predicted frequency table indicates unexplained local dependence in the current model. Various approaches incorporating this idea have been developed by Espeland and Handelman (1989), Garrett and Zeger (2000), Hagenaars (1988), Qu et al. (1996), and Magidson and Vermunt 2004). Sepulveda et al. (2008) provide a discussion of these approaches and proposes a new method based upon biplots. Their method is useful to distinguish between simple direct effects and second-order interactions (i.e., between AB and XAB).

Our preferred approach is the method proposed by Uebersax (2000) which is a variation of the log-odds ratio check (LORC) method of Garrett and Zeger (2000). The Uebersax method for two dichotomous indicators, I_1 and I_2 proceeds as follows:

1. Construct the observed and model predicted two-way cross-classification frequency table for a pair of items as in Table 11.5.
2. Calculate the log-odds ratio (LOR) in both the observed and predicted two-way tables where $LOR(\text{obs}) = \log(n_{11}n_{22} / n_{12}n_{21})$ with analogous definition for $LOR(\text{pred})$
3. Calculate the standard error of $LOR(\text{pred})$ using the approximation

$$s.e.[LOR(\text{pred})] = \sqrt{\hat{m}_{11}^{-1} + \hat{m}_{12}^{-1} + \hat{m}_{21}^{-1} + \hat{m}_{22}^{-1}}. \qquad (11.25)$$

4. If

$$|z| = \left| \frac{LOR(\text{obs}) - LOR(\text{pred})}{s.e.[LOR(\text{pred})]} \right| > 1.96 \qquad (11.26)$$

then I_1 and I_2 are significantly locally dependent.

TABLE 11.5. Observed and Predicted Frequencies for $I_1 \times I_2$

	$I_2 = 1$	$I_2 = 2$
Observed Frequencies		
$I_1 = 1$	n_{11}	n_{12}
$I_1 = 2$	n_{21}	n_{22}
Predicted Frequencies		
$I_1 = 1$	$\hat{m}_{11}$	$\hat{m}_{12}$
$I_1 = 2$	$\hat{m}_{21}$	$\hat{m}_{22}$

11.4.5 Illustration: Sexual Assault

This illustration is based upon survey results that have yet to be officially released. Consequently, the data has been replaced with data simulated from a model that closely describes the original data. In addition, the survey questions have been modified to avoid confusion with the survey results.

The five questions in Table 11.6 were asked of a general population for the purposes of estimating the prevalence of rape in a population. An examination of the indicators suggested that indicators were bivocal with A, B, and C measuring X defined as a "victim of sexual violence in the past 12 months" and D and E measuring Y defined as "forced or pressured to have sex (e.g., raped) in the past 12 months." The path model of these relationships is shown in Figure 11.7. Note further that the cross-classification of X and Y defines four classes as shown in Table 11.7. However, the cell associated with $Y = 1$ and $X = 2$ is impossible because a person who has been raped has also been sexually assaulted by definition. Thus, there are only three LCs, and the constraint $\Pr(X = 2 | Y = 1) = 0$ must be added to the model as a *structural zero* cell in order for the model to be properly specified.

TABLE 11.6. Five Indicators for the Estimation of Rape Prevalence in a Population

Indicator	Definition
A	Have any of the following occurred in the past 12 months?
	Touched in a sexual manner with physical force?
	Touched in a sexual manner when uninvited?
	Forced you to perform sexual acts such as oral or anal sex?
	Forced or pressured you to have sexual intercourse (e.g., raped)?
	=1 if yes
	=2 if no
B	In the past 12 months, did someone use physical force, pressure you, or make you feel that you had to have any type of sex or sexual contact?
	=1 if yes
	=2 if no
C	How long has it been since someone used physical force, pressure you, or make you feel that you had to have any type of sex or sexual contact?
	=1 if past 12 months
	=2 if more than 12 months or never
D	In the past 12 months, did someone force or pressure you to have sexual intercourse or rape you?
	=1 if yes
	=2 if no
E	How long has it been since someone forced or pressured you to have sexual intercourse; that is, you were raped?
	=1 if past 12 months
	=2 if more than 12 months or never

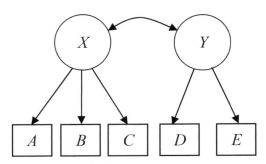

FIGURE 11.7. Relationship between the five indicators of rape.

TABLE 11.7. Four Classes Defined by Cross-Classifying X and Y

	$X = 1$ (Victim of Sexual Assault)	$X = 2$ (Not a Victim of Sexual Assault)
$Y = 1$ (raped)	Raped	Impossible
$Y = 2$ (not raped)	Victim of assault, but not raped	Never a victim of assault

TABLE 11.8. Results of the Model Building Process

		df	BIC	L^2	P	d
Model 0	$\{XY\}\,\{XA\ XB\ XC\ YD\ YE\}$	111	−53.5	1054.8	0.00	0.03
Model 1	$\{XYFG\}\{XFA\ XGA\}\{XFB$ $XGB\}$ $\{XFC\ XGC\}\{YFD\ YGD\}$ $\{YFE\ YGE\}$	82	−658.4	160.4	0.00	0.005
Model 2	Model 1+ $\{BD\}$	81	−654.5	154.3	0.00	0.004
Model 3	Model 2 + $\{CE\}$	80	−735.2	63.6	0.91	0.003

The model building process proceeded as follows. First, the model $\{XA\ XB$ $XC\ YD\ YE\}$ was fit denoted by Model 0 in Table 11.8. Next, a number of grouping variables were considered including the demographic characteristics, prior victimizations, and variables summarizing the respondent's attitude toward the interview, the interview setting (privacy, distractions, etc.). Two variables determined to be the most predictive of both the structural and measurement components of the model were added to the model to account for heterogeneity producing Model 1 in the table.

Then the LORC procedure was performed using the observed and Model 1 predicted frequencies for the $ABCD$ table. The first pass of the LORC found three statistically significant dependencies: AD, BD, and CE. Because the pair BD has the highest z-value, this direct effect was added to the submodel for

D to produce Model 2. The LORC procedure was repeated for Model 2, and the pair *AD* and *CE* were significantly dependent, with the latter having the highest *z*-value. Thus, *CE* was added to the submodel for *E* to produce Model 3. Applying the LORC test again for Model 3 turned up no further significant dependencies thus terminating further LORC testing. The LORC results are shown in Table 11.9.

Finally, the estimates of the sexual victimization and rape prevalence, classification error rates under Model 3, and the corresponding true, simulation parameters are compared in Table 11.10. The model building procedure was successful at identifying some local dependencies in the underlying model. Still the model parameter estimates are in fairly close agreement with the true parameters. The exceptions are estimates of the terms involving *Y* which exhibit fairly important biases. The likely reason for this is too few indicators of *Y*. Since three indicators are needed for the standard LC model to be

TABLE 11.9. *z*-Values for the LORC of Indicator Pairs for Three Models

Pair	Model 1	Model 2	Model 3
AB	0.63	0.8	−0.43
AC	−0.34	−0.35	−0.09
AD	**2.57**[1]	**2.92**[1]	1.80
AE	−0.25	−0.22	0.55
BC	−1.21	−1.15	−0.30
BD	**4.96**[1]	0.49	1.08
BE	0.77	0.86	1.94
CD	−1.35	−1.07	−0.68
CE	**4.05**[1]	**3.94**[1]	0.05
DE	0.04	0.37	0.45

[1]*z* is significant at $\alpha = 0.05$.

TABLE 11.10. Comparison of the Estimates for Model 3 and the True Model

	X	Y	A|X	B|X	C|X	D|X	E|X
Model 3—*{FGXY}{FAX GAX}{FBX GBX}{FCX GCX}{FDY GDY BD}* *{EY FE GE BE CE}*							
Prevalence	0.022	0.014					
False +			0.003	0.001	0.003	0.002	0.003
False −			0.166	0.360	0.181	0.559	0.376
True Model—*{FGXY}{FAX GAX}{FBX GBX}{FCX GCX}{FDY GDY BD AD}* *{FEY GEY CE DE}*							
Prevalence	0.022	0.022					
False +			0.003	0.001	0.003	0.002	0.002
False −			0.187	0.379	0.192	0.423	0.606

identifiable, estimates of Y parameters must draw information from A, B, and C which are only indirectly related to D and E through X and Y. Thus, the accuracy of the Y parameters depends largely on the strength of the correlation between X and Y. For example, if this correlation were 1, then all five indicators would be univocal and locally independent and, thus, the parameter estimates would be unbiased. As this correlation departs from 1, estimation accuracy deteriorates (Berzofsky, 2009). This suggests that if bivocality is unavoidable in the construction of multiple indicators, the underlying latent variables should be made as highly correlated as possible.

11.5   CONCLUSION

Although LCA is a theoretically sound and logically constructed modeling methodology, its subtleties and theoretical intricacies make valid use of this approach challenging, especially for the novice. Ideally, the LCA modeler should be well-versed in log-linear modeling and keenly aware of the estimation complexities and additional assumptions encountered with some of the variables are latent. This paper documents some common problems in LCA and provides guidance from the literature and our own experiences about how to deal with these problems. Space and the number of issues prevented more than a compressed review of LCA and potential problems and cures. Some problems were not presented at all. These include problems with the analysis of cluster correlated data and unequal probability sampling in complex surveys (Patterson et al., 2002; Vermunt and Magidson, 2007), issues associated with Markov LCA, potential problems when the dependent variable is ordinal, and other issues associated with the general field of log-linear modeling with latent variables.

Despite challenges, the prospects for applying LCA for questionnaire design and improvement are excellent. Applications are currently not prevalent in the literature, but the field is fairly new and interest is growing. Encouragingly, even pedestrian applications of LCA have yielded valuable information on item accuracy and data quality. We believe that, as a means for studying bias and validity in surveys, LCA is an essential tool for survey methodologists.

REFERENCES

Agresti A (1990). Categorical Data Analysis, 1st ed. New York: Wiley & Sons.

Agresti A (2002). Categorical Data Analysis, 2nd ed. New York: Wiley & Sons.

Agresti A, Yang M (1987). An empirical investigation of some effects of sparseness in contingency tables. Computational Statistics & Data Analysis; 5:9–21.

Aitkin M, Anderson D, Hinde J (1981). Statistical modeling of data on teaching styles. Journal of the Royal Statistical Society, Series A; 144:419–461.

Alwin DF (2007). Margins of Error: A Study of Reliability in Survey Measurement. Hoboken, NJ: John Wiley & Sons.

Bartholomew DJ, Knott M (1999). Latent Variable Models and Factor Analysis. London: Arnold.

Berzofsky M (2009). Survey classification error analysis: critical assumptions and model robustness. Presented at the 2009 Classification Error Society Conference, St. Louis, MO.

Biemer PP (2001). Nonresponse bias and measurement bias in a comparison of face to face and telephone interviewing. Journal of Official Statistics; 17(2):295–320.

Biemer PP (2009). Measurement errors in surveys. In: Pfeffermann D, Rao CR, editors. Handbook of Statistics, Vol. 29: Survey Sampling, Vol. 1. Elsevier BV.

Biemer P, Wiesen C (2002). Latent class analysis of embedded repeated measurements: an application to the National Household Survey on Drug Abuse. Journal of the Royal Statistical Society: Series A; 165(1):97–119.

Burnham KP, Anderson DR (2002). Model Selection and Multimodel Inference: A Practical-Theoretic Approach, 2nd ed. New York: Springer-Verlag.

Collins LM, Fidler PF, Wugalter SE, Long LD (1993). Goodness-of-fit testing for latent class models. Multivariate Behavioral Research; 28(3):375–389.

Cressie N, Read TRC (1984). Multinomial goodness-of-fit tests. Journal of the Royal Statistical Society, Series B; 46:440–464.

Dayton CM, Macready GB (1980). A scaling model with response errors and intrinsically unscalable respondents. Psychometrika; 45:343–356.

De Menezes LM (1999). On fitting latent class models for binary data: the estimation of standard errors. British Journal of Mathematical and Statistical Psychology; 52:149–158.

Dempster A, Laird N, Rubin D (1977). Maximum likelihood from incomplete data via the EM algorithm. Journal of the Royal Statistical Society: Series B; 39(1):1–38.

Efron B (1979). Bootstrap methods: another look at the jackknife. The Annals of Statistics; 7(1):1–26.

Efron B, Tibshirani RJ (1993). An Introduction to Bootstrap. London: Chapman and Hall.

Espeland MA, Handelman SL (1989). Using latent class models to characterize and assess relative error in discrete measurements. Biometrics; 45(2):585–599.

Feiser M, Lin Y (1999). A goodness of fit test for the latent class model when the expected frequencies are small. In: Sobel ME, Becker MP, editors. Sociological Methods. Oxford: Blackwell; pp. 81–111.

Formann AK (1985). Constrained latent class models: theory and applications. The British Journal of Mathematical and Statistical Psychology; 38:87–111.

Formann AK (1992). Linear logistic latent class analysis for polytomous data. Journal of the American Statistical Association; 87(418):476–486.

Galindo-Garre F, Vermunt JK (2006). Avoiding boundary estimates in latent class analysis by Bayesian posterior mode estimation. Behaviormetrika; 33:43–59.

Garrett ES, Zeger SL (2000). Latent class model diagnosis. Biometrics; 56(4):1055–1067.

Goodman LA (1973). The analysis of a multidimensional contingency table when some variables are posterior to the others. Biometrika; 60:179–192.

Goodman LA (1974a). Exploratory latent structures analysis using both identifiable and unidentifiable models. Biometrika; 61:215–231.

Goodman LA (1974b). The analysis of system of qualitative variables when some of the variables are unobservable. Part I—a modified latent structure approach. American Journal of Sociology; 7(5):1179–1259.

Haberman SJ (1977). Log linear models and frequency tables with small expected cell counts. The Annals of Statistics; 5:1148–1169.

Haberman SJ (1979). Analysis of Qualitative Data, Vol. 2: New Developments. New York: Academic Press.

Hagenaars JA (1988). Latent structure models with direct effects between indicators: local dependence models. Sociological Methods and Research; 16:379–405.

Hagenaars JA (1990). Categorical Longitudinal Data Log-Linear Panel, Trend, and Cohort. Newbury Park, CA: Sage Publications.

Hui SL, Walter SD (1980). Estimating the error rates of diagnostic tests. Biometrics; 36:167–171.

Kreuter F, Yan T, Tourangeau R (2008). Good item or bad—can latent class analysis tell? The utility of latent class analysis for the evaluation of survey questions. Journal of the Royal Statistical Society, Series A: Statistics in Society; 171:723–738.

Lazarsfeld PF, Henry NW (1968). Latent Structure Analysis. Boston: Houghton Mifflin.

Magidson J, Vermunt JK (2004). Latent class analysis. In: Kaplan D, editor. The Sage Handbook of Quantitative Methodology for Social Sciences. Thousand Oaks, CA: Sage Publications; pp. 175–198.

Mieczkowski T (1991). The accuracy of self reported drug use: an analysis and evaluation of new data. In: Weisheit R, editor. Drugs, Crime, and Criminal Justice System. Cincinnati, OH: Anderson; pp. 275–302.

Muthén LK, Muthén BO (1998). Mplus. Los Angeles, CA: Muthén & Muthén.

Noreen EW (1989). Computer Intensive Methods for Testing Hypotheses: An Introduction. New York: Wiley.

Patterson B, Dayton CM, Graubard B (2002). Latent class analysis of complex survey data: application to dietary data. Journal of the American Statistical Association; 97:721–729.

Qu Y, Tan M, Kuther MH (1996). Random effects models in latent class analysis for evaluating accuracy of diagnostics test. Biometrics; 53(3):797–810.

Sepulveda R, Vicente-Villardon JL, Galindo MP (2008). The Biplot as a diagnostic tool of local dependence in latent class models. A medical application. Statistics in Medicine; 27:1855–1869.

Sinclair M (1994). Evaluating reinterview survey methods for measuring response errors. Unpublished doctoral dissertation, George Washington University.

Turner C, Lessler J, Devore J (1992). Effects of mode of administration and wording of reporting of drug use. In: Turner C, Lessler J, Gfroerer J, editors. Survey Measurement of Drug Use: Methodological Studies. Washington DC: US Department of Health and Human Services.

Uebersax SL (2000). CONDEP Program. Available online at http://www.john-uebersax. com/stat/condep.htm.

van de Pol F, Langeheine R, de Jong W (1998). PANMARK User Manual, Version 3. Voorburg, The Netherlands: Netherlands Central Bureau of Statistics.

Vermunt JK (1997). LEM: A General Program for the Analysis of Categorical Data. Tilburg, Netherlands: Department of Methodology and Statistics, Tilburg University.

Vermunt JK, Magidson J (2005). Technical Guide to Latent Gold 4.0: Basic and Advanced. Belmont, MA: Statistical Innovations.

Vermunt JK, Magidson J (2007). Latent class analysis with sampling weights: a maximum-likelihood approach. Sociological Methods Research; 36(87):87–111.

Wang Y, Zhang NL (2006). Severity for local maxima for the EM algorithm: experience with hierarchical latent class models. Proceedings for the 3rd European Workshop on Probabilistic Graphic Models (PGM-06); pp. 301–308.

12 Response 1 to Biemer and Berzofsky's Chapter: Some Issues in the Application of Latent Class Models for Questionnaire Design

FRAUKE KREUTER
University of Maryland

12.1 INTRODUCTION

I am honored to discuss the chapter by Paul Biemer and Marcus Berzofsky. For a good decade, Biemer has been a leader in the application of latent class models (LCMs) to various aspects of survey error. His work with LCMs has improved our understanding of errors in several parts of the survey process, in particular coverage error (Biemer et al., 2001), nonresponse error (Biemer, 2001, 2009), and most prominently, measurement error (Biemer and Witt, 1996; Biemer and Bushery, 1999; Biemer and Wiesen, 2002).

As Biemer and Berzofskz make clear, LCMs are an attractive choice for studying measurement error. Instead of relying on error-free external data that is often not available, these models make use of multiple measures or multiple indicators (Alwin, 2007) of the same construct or characteristic to estimate error rates. Unlike other questionnaire development methods, such as expert reviews and cognitive interviews, results from LCMs provide item-specific quantitative measures.

Question Evaluation Methods: Contributing to the Science of Data Quality, First Edition.
Edited by Jennifer Madans, Kristen Miller, Aaron Maitland, Gordon Willis.
© 2011 John Wiley & Sons, Inc. Published 2011 by John Wiley & Sons, Inc.

However, these models can be difficult to use, and Biemer and Berzofsky address in their chapter challenges researchers face in applying LCMs. The challenges can be grouped into estimation issues and issues with the practical implementation of suitable designs in the questionnaire development context. Among the estimation issues, there is a subset which can be largely overcome with the proper use of latent variable modeling software. My discussion here is more focused on the practical implications. I will start with some examples from research I have done with colleagues (Kreuter et al., 2008; Yan et al., 2010) to highlight the main points. After that I will discuss the role of latent variable models in more general terms. Despite the challenges with respect to the use of LCMs in questionnaire development, for substantive analysis purposes, there is a strong appeal in designing questionnaires that allow the use of latent variable models. Thus, questionnaire developers should consider latent variable modeling requirements at the design stage. Finally, I will outline some recommendations about archiving latent class analysis results in the Q-Bank database of question evaluation reports.

12.2   PRACTICAL PROBLEMS WHEN APPLYING LCM

Violations of the local independence assumption are one issue practitioners must contend with. Biemer and Berzofsky point out that it can arise when "the questions for assessing X are asked in the same interview, respondents may provide erroneous responses to each question either deliberately or inadvertently causing errors to be correlated across indicators" or when "... comprehension errors caused by the wordings of questions may cause errors that are correlated across the indicators." Unfortunately from a questionnaire designer point of view, both situations seem quite likely. And so does bivocality, another potential violation of the local independence assumption. Here, items that measure different latent constructs are modeled as if they belong to the same latent construct.

Biemer and Wiesen (2002) give an example of correlated errors over time, with respondents being likely to underreport infrequent drug use due to social desirability concerns. Correlated errors due to social desirability concerns were also found in data from a University of Maryland (UMD) Alumni Survey, conducted in 2005 with a sample of alumni who have graduated since 1989. This survey included three items that were designed to tap essentially the same information about unsatisfactory or failing grades (Table 12.1).[1]

TABLE 12.1. The Three Items Included in the Survey

q12	Did you ever receive a grade of "D" or "F" for a class?
q18a	Did you ever receive an unsatisfactory or failing grade?
q18b	What was the worst grade you ever received in a course as an undergraduate at the University of Maryland?

TABLE 12.2. Bivariate Correlations

	q12	q18a	q18b
q12	1.0000		
q18a	0.5849	1.0000	
q18b	0.9093	0.5853	1.0000

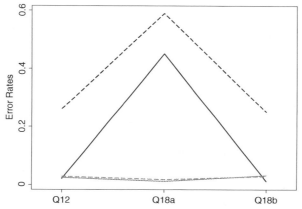

FIGURE 12.1. False positive and false negative response rates for the LCA and true score comparison: - - - -, true score, false negative; ———, LCA, false negative; - - - -, true score, false positive; ———, LCA, false positive (Kreuter et al., 2008).

The second question (18a) was deliberately more vague in its wording, in order to test the performance of LCMs in identifying the poor item. This study used records from the UMD registrar's office to calculate true measurement error.

As one would have expected, responses to the first and third indicator listed in Table 12.1 correlate highly. Responses of either the first or the third indicator are not as strongly correlated with the second indicator (Table 12.2).

Comparing all three binary indicators with the presumably true value from the UMD registrar data, we get quantitative error rates for each item. The false negative rates (i.e., reporting that the students did not get a D or F although the registrar data showed they did) for item 12 and 18b are of similar size about 25%. The false negative rate of item 18a on the other hand is 60%—more than twice as large (Fig. 12.1). Fewer differences are visible for the false positive rates, though item 18a shows fewer of those.

The latent class analysis shows the same pattern (solid lines in Fig. 12.1). Here too, a questionnaire designer would have identified question 18a as the flawed item, and picked up on a much higher false negative rate. Compared with the analysis done with registrar data, the overall size of the estimated measurement error is biased downward. Thus in this example, questionnaire designers would have made the same choice regardless of whether they use latent class analysis or record data to evaluate the items. However, the conclusions about the quantitative nature of the measurement error attached to each of the items would be different.

The Alumni Survey data could be seen as an example of bivocality: One could argue that the second item is tapping into a different construct. This second construct could be the respondents' own satisfaction with a certain grade, rather than an outside perspective of what is a satisfactory grade. One solution for bivocality mentioned by Biemer and Berzofsky is adding another latent variable to capture the second construct. Unfortunately in our data set, no further indicator variables are available to make such a model identified. We would need to add grouping variables, that is, covariates with certain restrictions on their relationship to class membership, to estimate the second construct. However, adding these grouping variables does require knowledge about subgroups and assumptions about the error properties.[2] The same concern with model identification arises in situations in which only two latent class indicator variables are available, and at least one grouping variable is needed for identification of this two-indicator LCM. In Kreuter et al. (2008) we examined several such grouping variables.

Two additional issues should be highlighted here. First, the examples discussed so far require the strong assumption that there are only two latent classes in the latent variable. This seems to be a fair assumption for certain items (i.e., being pregnant), but not necessarily for others. Biemer and Berzofsky have not touched upon the difficulties of class enumeration (deciding on the number of classes) in their presentation, but the more complicated the LCMs which are tested against each other, the less straightforward is a decision about the "right" number of classes (Nylund et al., 2007).

Second, among the modeling issues Biemer and Berzofsky mentioned are problems of data sparseness (i.e., empty cells) that can occur with small sample sizes or structural zeros. Additional sample size-related problems are boundary estimates and improper solutions. In the examples Biemer and Berzofsky give, sample size and empty cells are less of a problem. Their examples employed LCMs in the context of large surveys, such as the National Survey of Drug Abuse. In the questionnaire development context, sample size could be a crucial obstacle. In the first development phase of a questionnaire, samples are usually quite small. (However, other authors in this volume have made arguments for larger sample sizes at this stage.) Larger samples are often only available in the pilot stage in which the final flow of a questionnaire is tested, together with its length, and final wording problems. It will be interesting to discuss if this phase of the questionnaire development cycle could host multiple competing items that could be evaluated using LCMs. I could see a potential problem there with multiple measures distorting the overall flow of the questionnaire, one of the elements under examination in a pilot test.

12.3 THE ROLE OF LCM IN THE QUESTIONNAIRE DEVELOPMENT CONTEXT

Thinking about the early stages of questionnaire development, the question arises—how can LCM results compare to more traditional questionnaire development techniques? Given the large number of different evaluation methods and the differences between them, an important question is whether the different methods yield consistent conclusions about a set of items. Of course, there are different levels of agreement across methods and different reasons why the methods might disagree. An item deemed to have problems based on cognitive interviewing results may nonetheless show low error rates in an LCM (i.e., items can be reliable but can all show bias). In addition, the different methods may yield different conclusions about the specific nature of the problems with an item.

The organizers of this volume asked those of us who comment on the larger chapters to examine different questionnaire evaluation methods comparatively. Thus, it is a pleasure for me to use this opportunity to report results from a National Science Foundation research grant, in which Roger Tourangeau, Ting Yan, and I took on the task of comparing different questionnaire development techniques and the role of LCMs within them. In our attempt to compare LCM to other questionnaire testing techniques, we used expert reviews, cognitive interviewing, and two large web surveys that allow quantitative analyses of the items.

We had four experts rate five sets of questions that each consisted of three items tapping a single underlying construct. We told the experts that we were doing a methodological study that involved different methods of evaluating survey questions but did not give more specific information about the aims of the study. We asked them to say whether each item had serious problems or not (and, if it did, to describe the problems) and also to rate the item on a 5-point scale, ranging from "This is a very good item" to "This is a very bad item."

All three items within each triplet were intended to measure the same underlying construct. To give two examples, one triple asked about the respondent's views on skim milk:

- Please indicate how you feel about the following foods: apples; whole milk; skim milk; oranges (these items appeared in a grid, with a 10-point response scale; the end points of the scale were labeled "like very much" and "dislike very much")
- How much would you say you like or dislike skim milk? (like very much; like somewhat; neither like nor dislike; dislike somewhat; dislike very much)
- How much would you say you agree or disagree with the statement "I like skim milk." (agree strongly; agree somewhat; neither agree nor disagree; disagree somewhat; disagree strongly)

Another triplet asked about the respondent's concern about diet:

- On a scale of 0 to 9, where 0 is not concerned at all and 9 is strongly concerned, how concerned are you about your diet? (10-point scale, with verbally labeled end points)
- Would you say that you care strongly about your diet, you care somewhat about your diet, you care a little about your diet, or you don't care at all about your diet? (strongly; somewhat; a little; not at all)
- Do you worry about what you eat or do you not worry about it? (worry about what I eat; do not worry about what I eat)

In addition, cognitive interviews were conducted in the University of Michigan Survey Research Center's Instrument Development Lab in Ann Arbor Michigan. All fifteen of the items were tested in interviews carried out by five experienced cognitive interviewers. Three versions of the questionnaire were tested, each containing one item from each of the five triplets plus some additional filler items. Respondents were randomly assigned to get one version of the questionnaire. A total of 15 cognitive interviews were done on each version. An observer also watched each interview through a one-way mirror. The cognitive interviewers asked the respondents to think aloud as they formulated their answers, administered pre-scripted "generic" probes (such as, "How did you arrive at your answer?" or "How easy or difficult was it for you to come up with your answers?") and followed up with additional probing ("What are you thinking?" or "Can you say a little more?") to clarify what the respondents said or how they had arrived at their answers. We counted a respondent as having had a problem with an item if both the interviewer and the observer indicated the presence of a problem.

Finally, the triplets were issued in two web surveys. The first study used an opt-in panel, where the six questions of the first two triplets were spread throughout the questionnaire, which was administered to 3000 respondents. The second study using the two triplets shown above plus one additional one was fielded in a one-time web survey completed by 2410 respondents. Two of the nine target questions were spread throughout the questionnaire, with one item from each triplet coming at the beginning of the survey, one in the middle, and one at the end. Both web survey data sets provide us the opportunity to perform latent class analyses.

Further details on the data collection and estimation can be found in Yan et al. (2010), but I do want to share a small subset of the results for the two triplets mentioned above. We computed mean ratings of the experts for each item (with higher ratings indicating a worse item), the proportion of cognitive interviews in which the item was found to have a problem, and the error rates from the latent class modeling. For the statistics derived from the web survey data (i.e., the validity coefficients and the error rates from the LCMs), we used the random groups to calculate the standard errors for the statistics themselves as well as for the differences between pairs of statistics. Based on estimated

TABLE 12.3. Ranking of Items across Four Methods

	Expert Review	Cognitive Interview	LCA
Skim Milk	Average Rating	% Problems	Error Rates
Item 1	(2)	(1)	(3)
Item 2	(1)	(2)	(1)
Item 3	(2)	(2)	(2)
Diet	Average Rating	% Problems	Error Rates
Item 1	(1)	(3)	(1)
Item 2	(1)	(1)	(3)
Item 3	(1)	(2)	(2)

statistics and their corresponding standard errors we rank-ordered the items, with lower ranks indicating better items.

Table 12.3 has the overall ranking of each item that we derived from the results of each evaluation. (Equal ranks were assigned if error rates did not significantly differ from each other.) On the skim milk item we see all expert reviews and LCMs agree on their favorite item. On the diet items, there is no difference in the ratings of expert reviewers; the results of cognitive interviews and LCMs go in opposite directions.

Looking back at the statistics and across all 15 items, the correlation between the misclassification rates from the LCM and the expert ratings are of moderate size $r = 0.53$ and significant ($P < 0.05$). The correlation between the misclassification rates from the LCM and the proportion of items in which the cognitive interviewers found problems is of similar size $r = -0.57$; however, the correlation is in the opposite direction. That is to say, the higher the proportion of cognitive interviews revealing problems with the item, the lower is the misclassification rate according to the LCMs. These two measures disagree about which is the better item.

The different methods also vary with respect to the recommendations that questionnaire developer can draw from them. Unlike other methods, results from LCMs most often do not provide much insight into the nature of the problem. Using the survey response process model (Tourangeau et al., 2000) as a classification scheme, cognitive interviewing might, in addition to flagging an item as problematic, be able to distinguish comprehension problems from problems in making judgments or estimating quantities. The LCMs usually do not make such distinctions, though they can distinguish items with high rates of false positive responses from those with high rates of false negative answers. Thus, even if common LCMs flag a question as problematic, questionnaire designers are left to other techniques to determine how to fix the question. Exceptions are studies in which the multiple measures of the same latent construct are carefully crafted to rule out certain effects. For example, if the only difference between the three diet items were the answer scale, it is unlikely that the problem with one of the three items is due to a comprehension problem. Or likewise in the alumni survey example questions could be kept similar in structure and with only the reference period varying across them. In such situations recall problems could be identified. However, in most real-life situation, LCM results leave questionnaire designers with the sense that the item has a problem but not what that problem is.

12.4   DIFFERENT PERSPECTIVES ON THE USE OF LCMS

The various examples that Biemer and others have given us over the years suggest to me that questionnaire designers might benefit from a broader perspective on latent variable models in general. For one, questionnaire designers might be interested in broadening their general approach to measurement toward approaches that use multiple items instead of single indicators. Second, they might be interested in broadening the use of these models in the questionnaire development phase, by acknowledging that the local independence assumption will likely be violated and thus preemptively plan to expand the set of items used in the evaluation stage.

Latent class analysis are usually used to allow the measurement of an underlying latent variable through imperfect categorical indicators (Lazarsfeld and Henry, 1968). In this spirit when we bring LCMs into survey methodology, questionnaire developers could build multiple measures into the final instrument so that LCMs are possible during analysis of the response data. This would clearly be a shift in perception, as factor and latent variable models have usually been perceived as something that is reserved for things that cannot be measured with a single item (e.g., attitudes). Acknowledging that all of our attempts to measure can be flawed, a move toward multiple measures of many constructs seems reasonable.

We could, for example, have two independent listings of housing units and acknowledge that both can be erroneous. Usually the second listing is considered to be a "check" and is therefore seen as being better. However, if weather conditions, prior knowledge of the area, match between lister and area demographics, and so on (Eckman and Kreuter, 2010) play a role in the quality of the listing, then it would be a shortcoming to call one or the other listing better. It would be interesting to understand if perhaps estimated listing probability coming out of LCMs can provide useful information for field operations.

Multiple listings are expensive, but there are other examples in which multiple measures would not necessarily increase costs. Take, for example, questions with complex and difficult definitions (i.e., residence status). Instead of trying to explain to the respondent a complex definition, a series of questions could be asked that are easily answered with "yes" or "no." Combining those questions might provide a better measure of the underlying construct, household membership, than a single item would.

A similar logic is proposed by Ganzeboom (2009) to measure socioeconomic status variables, such as occupation or educational attainment.

Ganzeboom makes the point that indicators of social background variables are unreliable measures with a large random variance component. He showed for the German part of the European Social Survey (ESS) that models using a single measure of education (as independent or dependent variable) can be biased, but this bias is reduced if a second indicator of education, even a crude one, is added to the models.

There are already examples of an expanded use of latent variable models in questionnaire development, focusing so far on measurement properties of answer scales across a large set of surveys and topical areas. Such work has been done for several years now by Willem Saris and others using Multi-Trait-Multi-Measure (MTMM) designs. Work from Oberski et al. (2011), who is part of Saris' research group, shows application of LCMs to a series of experimental studies in the context of the ESS. The analyses also showed surprisingly few differences between the conclusions of the three models used (CFA, categorical CFA, LCM). In this context, it would be worthwhile exploring if the field can benefit from recent advances in latent variable modeling software that allow for hybrid models, blending not only categorical and continuous indicators but also categorical and continuous latent variables (Muthén, 2008). And likewise, it might be useful to examine extensions to multilevel LCMs (Vermunt, 2008; Muthén and Asparouhov, 2009). Those will be particularly useful in the presence of interviewer-induced correlated response variance (see, e.g., Biemer and Stokes, 1985).

12.5   GUIDELINES OR STANDARDS WHEN APPLIED AND USED FOR Q-BANK

In addition to discussing specific methods, the organizers of this volume asked us to provide guidelines or standards for Q-Bank when the method in question is used. On this note, Biemer and Berzofsky's chapter is an excellent source for researchers to turn to, when performing LCA. Multiple reads of their chapter, as well as Biemer's upcoming book on LCMs, should guide researchers in their modeling attempts. Both list in detail modeling assumptions that should be checked.

For Q-Bank to include latent class modeling results in its documentation material, I would recommend asking researchers for detailed information on which software was used, what features such software provides, and how these features were used in the modeling attempts (including settings and the random seeds). The question of random perturbed starting values is one example. Without a proper search across the parameter space, the risk of local maxima would be too high. Ideally, I would even say that information on convergence of the LCM should be provided.

A detailed description should also be given regarding the model specification, not just the models used, but also those rejected. Unfortunately, the set of model diagnostic tools is not as well developed for LCMs. However, there are several out there, and their use should be required before results are made part of the Q-Bank documentation. One should, for example, make it a habit to report residual diagnostics and discuss modification indices (for local dependence parameters).

Q-Bank should also consider employing a reproducible research policy (Gentleman and Lang, 2004; Anderson et al., 2005; Peng, 2009). Ideally, the analytic data from which the principal results were derived, should be made available on the Q-Bank Web site. Any computer code used to compute published results should also be provided on the Q-Bank Web site. If widely available software is used, references to that software will be sufficient. If authors write their own software to perform the analysis, such code should be provided as well. Further suggestions can be found at www.*reproducibleresearch.net*.

There could be separate storage for full code, and ideally the data used to run such models. If individual data cannot be provided for confidentiality reasons, one could at least take advantage of the categorical nature of these models and provide cross-tabulated frequencies. Those are sufficient to rerun many of the models. Providing code that matches the study design and published model graphs will also help to clean up confusion that might arise through the fact that researchers from different disciplines use different terms to describe the same models.

In closing, I want to compliment the Q-Bank team on their initiative to collect more information on all these methods and to learn how these techniques can contribute to the Q-Bank mission.

NOTES

1   "D" and "F" are the two worst grades given out at the University of Maryland.
2   Of course, technically, these subgroups could themselves be latent classes, but in regular government surveys without a large multiplicity of items for the same construct, it is unlikely that such models with two latent variables (each having several classes) could be estimated easily.

REFERENCES

Alwin DF (2007). Margins of Error. A Study of Reliability in Survey Measurement. Hoboken, NJ: John Wiley & Sons.

Anderson RG, Greene WH, McCullough BD, Vinod HD (2005). The role of data and program code archives in the future of economic research. FRB of St. Louis Working Paper No. 2005-014B. Available online at SSRN http://ssrn.com/abstract=763704.

Biemer P (2001). Nonresponse bias and measurement bias in a comparison of face to face and telephone interviewing. Journal of Official Statistics; 17(2):295–320.

Biemer P (2009). Incorporating level of effort paradata in nonresponse adjustments. JPSM Distinguished Lecture. May 8, 2009, College Park.

Biemer P, Bushery J (1999). Estimation of labor force classification error rates using Markov latent class analysis. Final analysis report for Census Bureau Project 6972-02.

Biemer P, Stokes S (1985). Optimal design of interviewer variance experiments in complex surveys. Journal of the American Statistical Association; 80(369):158–166.

Biemer P, Wiesen C (2002). Measurement error evaluation of self-reported drug use: a latent class analysis of the US National Household Survey on Drug Abuse. Journal of the Royal Statistical Society, Series A; 165(1):97–119.

Biemer P, Witt M (1996). Estimation of measurement bias in self-reports of drug use with applications to the National Household Survey on Drug Abuse. Journal of Official Statistics; 12(3):275–300.

Biemer P, Woltmann H, Raglin D, Hill J (2001). Enumeration accuracy in a population census: an evaluation using latent class analysis. Journal of Official Statistics; 17(1):129–148.

Eckman S, Kreuter F (2010). Confirmation bias in housing unit listing. Public Opinion Quarterly, Spring 2011, 75(1):1–12.

Ganzeboom H (2009). Multiple indicator measurement of social background. Keynote presentation at the European Survey Research Association (ESRA). Warsaw, July 1, 2009.

Gentleman R, Lang DT (2004). Statistical analyses and reproducible research. Bioconductor Project Working Papers. Working Paper 2.

Kreuter F, Yan T, Tourangeau R (2008). Good item or bad—can latent class analysis tell? The utility of latent class analysis for the evaluation of survey questions. Journal of the Royal Statistical Society, Series A; 171(3):723–738.

Lazarsfeld PF, Henry NW (1968). Latent Structure Analysis. Boston: Houghton Mifflin.

Muthén B (2008). Latent variable hybrids: overview of old and new models. In: Hancock GR, Samuelsen KM, editors. Advances in latent variable mixture models. Charlotte, NC: Information Age; pp. 1–24.

Muthén B, Asparouhov T (2009). Multilevel regression mixture analysis. Journal of the Royal Statistical Society, Series A; 172(3):639–657.

Nylund K, Asparouhov T, Muthén B (2007). Deciding on the number of classes in latent class analysis and growth mixture modeling: a Monte Carlo simulation study. Structural Equation Modeling; 14(4):535–569.

Oberski D, Hagenaars J, Saris WE (2011). The latent class MTMM model. In: Measurement error in comparative surveys. Tilburg: Tilburg University.

Peng R (2009). Reproducible research and biostatistics. Biostatistics; 10(3):405–408.

Tourangeau R, Rips L, Rasinski K (2000). The psychology of survey response. Cambridge: Cambridge University Press.

Vermunt J (2008). Latent class and finite mixture models for multilevel data sets. Statistical Methods in Medical Research; 17(1):33–51.

Yan T, Kreuter F, Tourangeau R (2010). Evaluating survey questions: a comparison of methods. Submitted.

13 Response 2 to Biemer and Berzofsky's Chapter: Some Issues in the Application of Latent Class Models for Questionnaire Design

JANET A. HARKNESS
University of Nebraska–Lincoln

TIMOTHY P. JOHNSON
University of Illinois at Chicago

13.1 INTRODUCTION

Latent class models (LCMs) are one of the most recently proposed analytic strategies for the evaluation of survey questionnaires. They are also possibly one of the least accessible to most questionnaire design experts. Consequently, the chapter in this volume by Paul Biemer and Marcus Berzofsky is an important contribution, since it provides for the first time a comprehensive, if somewhat technical, overview of this important methodology, its underlying assumptions, and strategies for using LCM to address survey question measurement problems.

Our chapter provides a twofold response to Biemer and Berzofsky. The discussion of latent and manifest entities in their chapter led us to focus in Sections 13.2–13.4 on latent and manifest aspects of question design that we feel are somewhat neglected both in design and in pretesting. Our discussion of these does not relate directly to LCM, but it does certainly consider aspects of questionnaires below the level of the actual wording, and how they may affect respondent understanding and response. In doing so, we draw attention to aspects of question design and pretesting that to date are not systematically addressed in either statistical testing procedures or more

Question Evaluation Methods: Contributing to the Science of Data Quality, First Edition.
Edited by Jennifer Madans, Kristen Miller, Aaron Maitland, Gordon Willis.
© 2011 John Wiley & Sons, Inc. Published 2011 by John Wiley & Sons, Inc.

qualitative questionnaire assessment procedures. In Sections 13.5–13.8, we contribute additional thoughts regarding the application of LCM to questionnaire design and testing.

13.2 LATENT AND MANIFEST ASPECTS OF QUESTIONNAIRE DESIGN, IMPLEMENTATION, AND TESTING

Our starting point in thinking about question design and its enactment in specific measurement occasions considers certain aspects of design as manifest or overt and other features as covert or, if you will, latent. It may be helpful to think of an iceberg in this respect; the part of the iceberg above the water is overt and manifest, the mass under the water is, by and large, covert. In like fashion, a given question formulation in a given mode can be viewed as the overt manifestation of its design, while the measurement goals the question encapsulates are one component of its numerous more covert properties.

Cross-cultural, cross-lingual research often leads one to attend to aspects of question design and implementation not usually of central concern in studies that are not specifically intended for implementation across populations. Deciding and testing question meaning is a good example. Certainly, respondents in any study may not understand a question in the way intended by researchers; checking comprehension is thus a standard part of cognitive question testing (cf Willis, 2005). However, in cross-lingual and cross-cultural studies, differences between how researchers intend questions to be understood and how they are actually understood by different populations are very much the order of the day. The extent to which a design team can build on expectations of "common ground," that is, assumed and actual shared understanding of the context and the world at large (Grice, 1989; Schwarz, 1996; Stalnaker, 2002),[1] is very much more limited in cross-cultural research. Correspondingly, the degree to which perceptions of researchers and respondents differ may be unexpectedly quite high. Numerous other linguistic and cultural factors, such as culturally prompted perception (cf Schwarz et al., 2010) and culturally preferred response styles (cf Yang et al., 2010), may further complicate design or administration. As a result, developmental procedures and the testing strategies that test their effectiveness should, in comparative research at least, include steps that accommodate different perceptual tendencies in different target populations. To date, this is not a systematic part of testing for comparative contexts.

In any survey research, the basic approach researchers adopt in designing questionnaires is to move first from theories about a research topic to the identification of indirectly measurable constructs associated with this. Then indicators are selected which are considered pertinent for collecting data about these constructs. Finally, questions are formulated which enable the data desired about the chosen indicators to be collected. This process is sometimes discussed in terms of concepts, constructs, and indicators (Bollen and

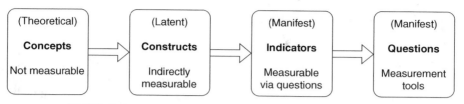

FIGURE 13.1. Concepts, constructs, indicators, and questions.

Long, 1987; Bohrnstedt and Knoke, 1988; Marcoulides and Schumacker, 1996; Harkness et al., 2003). Working in the comparative context, Harkness et al. (2010) found it useful to distinguish explicitly between indicators and the questions formulated about these, as presented in Figure 13.1 (Harkness et al., 2010, p. 42).[2]

In very general terms, survey instruments are designed to implement the questions resulting from such a model. Pretesting in its different forms is expected to test whether the various components making up the questionnaire and instrument design[3] are in fact performing as intended, in an efficient and reliable manner. In Section 13.4, we identify factors contributing to question-naire meaning which, although sometimes considered on an ad hoc basis, are not adequately accommodated in question design or testing as these are cur-rently practiced.

13.3   MEASUREMENT EVENTS AND MEASUREMENT OCCASIONS

Yang et al. (2010) describe the enactment of a survey question (e.g., when it is asked in an interview) as a measurement event. The measurement event is located within a measurement occasion, that is, the survey interview, or the occasion on which a respondent completes a self-administered questionnaire. A measurement event involves (1) a minimum set of instrument components (such as the question text and any answer options and instructions) and (2) the respondent's engagement with the questions either directly (as in self-completion modes) or indirectly, in working with an interviewer. Pretesting of instruments needs to consider all of the features of a (standardly envisaged) measurement event, since any of the factors involved, including characteristics of the instrument, the interviewer, the respondent, and the interview context, can affect how questions perform and how respondents react and respond (cf Willis, 2005).

The next section discusses question meaning in terms of what might seem manifest and unambiguous against what might actually be involved. It includes brief consideration of the contributions a measurement event may make to perception, meaning, and response. Examples are occasionally drawn for com-parative research for reasons explained earlier, but the points made are con-sidered to be relevant for question design and testing in general.

13.4 MEANINGS AND CONTEXTS

In considering various aspects of "question meaning," we make use of the terms *intended meaning* and *perceived meaning*, *linguistic* or *semantic meaning*, and *pragmatic meaning*. We also refer to *common ground* and to different kinds or levels of *context*. Although in language philosophy and in linguistic pragmatics, distinctions between some of these are much debated, for our present purposes, some broad distinctions and clarifications can be made that are of relevance in helping to think about designing questions successfully and testing them appropriately.

People use language as a major tool to produce communications they wish to make. The meaning that the producer of an utterance intends it to convey is its intended meaning. Some would prefer to speak here of communicative intention (cf Bach, 1987). This allows notions that, in using language, people are doing things (promising, threatening, requesting, etc.) to be included in discussion of "meaning" (cf Austin, 1962). In this sense, a sentence such as "Where have you been?" uttered in a given context to given recipients, could, for example, be intended as a warning, a comment on the person's appearance, or a joke. The meaning understood by receivers of the utterance in a given instance is its perceived meaning. Intended and perceived meaning always require processing of the linguistic content of the utterance in question. Thus, the formulation of survey questions is important. At the same time, neither intended nor perceived need be delimited by the linguistic content of sentences or expressions.

Sometimes, intended and perceived meaning match, in other instances they do not. The success or lack of success of a communication can be affected by multiple factors, including what is actually said and how it is said, if orally delivered, the participants in the communication and what they bring to it, and the context of utterance.

It is also commonly recognized that people often do not say, nor need to say, exactly what they mean. That is, the conventional linguistic meaning associated with an expression or a sentence may not be the same as the meaning its producer expects it to convey successfully. For example, although we could explicitly request to be informed about the time of day, we might also simply ask "Do you have a watch?" or "Do you have the time?" Neither of these are explicit requests to be informed about the time of day. The linguistic or semantic meaning of the first question, for example, is about the possession (in some undefined form) of some form of watch. In many English speech communities, however, it is *conventionally* used as a request to be informed about the time of day. Its pragmatic meaning is therefore not about watches but about the time of day. In addition, the interrogative is understood as a request, not simply as a question. The "yes" or "no" response possible to a question about possession of a watch would, for example, be an inadequate response to a request for time of day information.

The contribution an utterance makes to a given discourse beyond the linguistic meaning associated with it is not always established conventionally in this way. Nonetheless, it may be considered an appropriate contribution by its producer and be understood by recipients as intended. Notions of context and of the relevance of contributions to the given discourse are often evoked to explain how participants in a discourse bridge the gaps between what is said and what is meant. For example, formally the question "How many people live here" anticipates a numerical response. However, many other ways of answering this question could be appropriate in normal discourse, including the response "It's just me." This response, we note, does not explicitly provide a number of people, although many researchers would be satisfied that the question was adequately answered. That is, applying the generally accepted notion that conversation is based on cooperative interaction and follows conversational norms and (by and large) Gricean maxims (Grice, 1975, 1989), the response can be understood as an implicit way of indicating a count of one person. This is not a pragmatic meaning conventionally associated with the sentence; hence, it differs from "Do you have a watch." In a different situation, the same response might be used to answer the question "Who is there?", a question conventionally understandable as a request for whomever is addressed to identify himself/herself. "It's just me" as a response would here seem to flout the implicit request for identification. If we assume the participants are cooperating and posit that the person replying is confident that the person asking will recognize their voice, "It's just me" can be seen as a relevant and appropriate response to the request.

Research investigating standardized survey interview interaction abounds in examples of respondents providing what they consider to be pertinent responses to what they have understood the question to involve (e.g., Fowler and Mangione, 1990). These responses often do not fit the special requirements of survey responses which must be coded in terms of the scheme of response answers available (cf Dijkstra, 2002). The same research documents that respondents also offer answers which cannot in fact be negotiated toward a required survey response category (e.g., Suchman and Jordan, 1990; Schober and Conrad, 1997; van der Zouwen et al., 2010).

In trying to understand why respondents do not always perceive what researchers intend them to, mention is often made of a lack of common ground and the effects that context may have on respondent perception and processing (e.g., Cannell et al., 1981; Tourangeau and Rasinski, 1988; Tourangeau et al., 2000). In considering question design and its testing, it is useful to distinguish between different kinds and functions of context and the effects that may be associated with them. First, we can distinguish between immediate levels of context and more general levels. We can do so both with respect to a survey instrument and with respect to the measurement event and measurement occasion.

Research on question order effects, for example, has encouraged researchers to think about how the context of questions within a questionnaire (what material and questions go before and what comes after) may affect how respondents perceive question and answer options (Schuman and Presser, 1981; Tourangeau et al., 2000). This is one fairly immediate level of context within the framework of the instrument. A different immediate level of context, neglected in discussions of design, relates to the actual measurement event and measurement occasion. This context, for example, helps to determine the specific reference realized in a given utterance with deictic or context-sensitive linguistic expressions such as *this* week, *this* room, *you*, *this* neighborhood, and *now*.

At a somewhat more general level, the established norms of the discourse exchange in a survey (Suchman and Jordan, 1990) potentially empower interviewers to ask questions of respondents which in other contexts they would not normally ask of strangers. In terms of the specific measurement occasion, interactions may take place that build or weaken rapport or increase or lower engagement in the survey (Hill and Hall, 1963; Dijkstra, 1987). Questions may be asked that trigger social desirability or prompt response style or other skewed response behaviors, and respondents and interviewers can become tired, enervated, or confused. Each factor can affect perception and response. Any characteristic of an instrument or the measurement occasion as planned that is likely to result in systematic bias must, at the very latest, be discovered in testing and addressed as best possible. We suggest that a systematic consideration of different levels and forms of context could be useful in doing so.

Beyond these levels, other more general levels of context or situation contribute to how and what questions and answers are intended to mean and are understood. One factor is common ground, referred to earlier as the background information shared or assumed to be shared by participants in a conversation. This shared understanding may range from information immediately pertinent to the discourse to a general assumed understanding of the world and the language usage of a given speech community. Research has demonstrated how misleading it can be to assume shared understanding of even much used everyday expressions and questions in survey research (Belson, 1981; Suchman and Jordan, 1990; Schober and Conrad, 1997).

General context—for example, the current economic situation, recent social events, new government policies—can contribute to how respondents perceive surveys and their questions. A good example of unexpected perception is presented by Kortmann (1987). The question "Is your appetite good?" is often used as one of a series designed to measure depression. However, *appetite*, in contrast to *hunger*, implies a context where food is available. Kortmann (1987) reports that when it was asked in Ethiopia during a famine, respondents understood the question to be one about the availability of food. In this context, the question was certainly not useful as an indicator of depression.

One of the aims of this section has been to indicate that in developing questions and testing their meaning, we need to move beyond ascertaining and possibly testing the linguistic or semantic meaning associated with the chosen question wording. As other chapters in this volume reflect, there is a large body of survey research that points to and explores diverse aspects of design and implementation which may affect perception and response. Question format, question order, complicated or ambiguous formulations, visual aspects, answer response categories, and sensitive questions are mentioned here only by way of example. However, an integrated approach has still to be developed that explicitly addresses such manifest and latent aspects in instrument design and implementation design as mentioned here.

One concern in developing such an approach is the extent to which question design is able to anticipate and accommodate considerations that can contribute to the meaning respondents perceive during real measurement occasions. Obviously, the goal cannot be to anticipate the particular characteristics and experiences that individual respondents bring to measurement events. The focus must be more on features of question design and the implementation design that prompt groups or classes of respondents to process and react in desired or undesired ways. Testing can play an important role in establishing what these might be and what is to be learned for design. It does not seem to be too great a step to move from there to considering an approach to design that incorporates these.

A third important question is therefore how best to test for systematic bias stemming from such characteristics of questionnaire designs and/or measurement enactments. In the following sections, we consider the contributions which LCMs can make.

13.5   THE RELEVANCE OF LCMs FOR QUESTIONNAIRE DESIGN AND TESTING

LCM is essentially a highly advanced form of categorical data analysis (McCutcheon, 1987). It is unique among categorical analysis procedures, though, due to the integration of latent variables into its modeling strategy. Although seldom employed for purposes of developing or pretesting survey questionnaires, several strategies for the analysis of survey data via latent variable modeling have been developed over the past 60 years. One of the earliest discussions of latent variables was by Paul Lazarsfeld (1950), who first recognized their potential usefulness while confronting measurement problems when working on the post-WWII *The American Soldier* project (Henry, 1999). Early empirical work focused on development of latent measures via maximum likelihood-based confirmatory factor analytic techniques that used continuously measured observed indicators (Jöreskog, 1967). This strategy subsequently led to the development of structural equation models and the LISREL software (Jöreskog, 1970). Almost concurrently, Goodman (1974) introduced a maximum likelihood strategy for estimating LCMs for discrete variables, an accomplishment that quickly led to the first software program, MLSSA (Maximum Likelihood Latent Structure Analysis), for LCM analysis (Clogg, 1977). Today, a variety of well-documented LCM software programs are available, including Latent GOLD (Vermunt and Magidson, 2005), LEM (Vermunt, 1997), Mplus (Muthén and Muthén, 1998–2010), and SAS Proc LCA (Lanza et al., 2007).

Despite these developments, latent variable analyses today are only infrequently reported in the survey methodology literature, despite broad recognition that measurement error is one of its most serious problems (Biemer et al., 1991). This is in part because advanced statistical technologies have not historically been viewed as necessary for purposes of questionnaire development in some disciplines. For example, qualitative methods, reviewed elsewhere in this volume, are widely preferred within numerous disciplines and have been productively employed for survey question design and testing for many years. Marshalling advanced statistical models such as LCM to systematically investigate survey question measurement errors will require collaborations between qualitatively oriented questionnaire design experts and more quantitative statistical modeling experts. In the following section, we briefly identify several advantages of doing so.

13.6   LCM ADVANTAGES

The key reason for using LCM is its potential utility for identifying poor survey questions. One of Biemer and Berzofsky's examples, adopted from Biemer and Wiesen (2002), demonstrates how LCA's comparison of classification errors across three indicators could be used to identify a poor measure of past year marijuana use. The small body of literature on this topic that is now available, much of it also contributed by Biemer and colleagues, provides other useful examples of the application of LCM models to survey question design problems (Johnson, 1990; Biemer and Witt, 1996; Biemer, 2001, 2004; Biemer and Wiesen, 2002; Flaherty, 2002; Kreuter et al., 2008). They demonstrate how LCM can be applied to identify underlying cognitive processes that may be responsible for observed measurement errors. Biemer and Wiesen (2002), for example, discuss how their finding of independent response errors across three marijuana items suggested that the likely error source in their model was interpretation of meaning, rather than social desirability problems. Such examples also show how LCM is able to focus on the errors associated with individual survey items, rather than groups of items such as is the case with item response theory (IRT) models. As a result, LCM can potentially be used to examine any survey item.

Another important advantage that LCM can offer to questionnaire design experts is that it can be used to examine measurement quality when "gold" standard comparison measures are not available or do not exist. This is not an uncommon situation. In many instances, the auxiliary measures employed in lieu of "gold" standards to evaluate survey question measurement error have themselves been found to be imperfect indicators. Medical records (Sudman et al., 1997) and drug use tests (Fendrich et al., 2004) are examples that come quickly to mind. In the few instances where acceptable external measures may be available, accessing them is likely to require considerable time, effort, and cost. In contrast, the data required to conduct LCM analyses of survey question error can be built into survey instruments a priori, circumventing the need for external evaluation criteria and the costs associated with acquiring these. We believe that LCM will prove most fruitful in those instances where its use can be planned in advance, during the time when survey instruments are still being constructed. Doing so will enable investigators to employ an optimal set of indicators and avoid many of the limitations of secondary data analyses in which ideal measures are often unavailable. We note, however, that most of the LCM examples currently available in the published literature represent ex post facto applications; the Kreuter et al. (2008) paper is one notable exception.

13.7   CONSTRAINTS ON LCM AS AN ASSESSMENT TOOL

●There are also some important limitations to the use of LCM for the evaluation of survey items. Large sample sizes are recommended for LCM analyses, considerably larger than are commonly accrued during most survey pretests (Formann, 1984). Thus, realistically, LCM is most likely to be useful for the evaluation of survey questions *after* surveys have been completed and final survey data sets are available for analysis. Hence, the value of this technique for the actual pretesting of many survey instruments is very limited. This is indeed an important disadvantage. Nonetheless, LCM can still be of considerable value for evaluating the quality of questions that will be used on a recurring basis. Once survey data sets become available, assessments of survey item quality can be accomplished easily and inexpensively using this methodology, both in cases where specific plans for question evaluation were previously built into the design of survey instruments, as well as opportunistically where clever analysts identify the unintended availability of the multiple variables necessary to do so. Q-Bank (Miller, 2005) would seem to offer numerous opportunities for the latter.

LCM requires multiple indicators of the construct of interest. Thus, in cases where multiple indicators are not available, LCM cannot be utilized. In practice, it is rare to see multiple repetitions of the same or highly similar questions in a survey instrument. Doing so is typically not recommended because of concerns regarding respondent burden and the likelihood that respondents may become aggravated when they believe they are being asked the same question more than once, as well as the costs associated with the questionnaire space required to include these additional items. Consequently, there will be many surveys and even more individual survey constructs that will not have the data necessary to take advantage of LCM techniques.

13.8   FUTURE OPPORTUNITIES

There are some areas that we see as opportunities for expanding the utility of LCM for addressing survey measurement problems. First, LCM could be applied to comparative measurement problems via multigroup models. Determining the degree to which latent class measurement structures and findings can be replicated across multiple race and ethnic groups, for example, would be a valuable additional tool for use in cross-cultural studies of measurement comparability.

We believe that it is also important that more survey professionals be trained in the application of LCM models. Some of the resistance we see to the use of this and other advanced statistical techniques, we believe, comes from a lack of familiarity with them. Formal training in the application of these models to address survey measurement problems should be encouraged as part of academic programs, as well as via continuing education opportunities at professional conferences.

13.9 CONCLUSION

As with each of the questionnaire evaluation techniques reviewed in this volume, we believe LCA to be one more valuable tool that can be constructively used to evaluate the quality of survey questions when certain conditions are met. While there are many situations in which LCM will not be appropriate or useful for addressing applied survey measurement problems, particularly during the pretest phase, there will be other circumstances in which it has the potential to be one of the most valuable strategies available. To take advantage of this methodology, however, will require for many of us open-mindedness and a willingness to expand one's range of expertise. We believe it would be a mistake for questionnaire design professionals to avoid applying LCM in situations where it offers valuable information that might not be as easily or as quickly obtained using other strategies, solely because it requires developing new technical skills or forging collaborations with other experts who do.

Survey researchers know a great deal about designing and testing questions, but if we take stock of current theory and practice, there seem to be practical and theoretical gaps in frameworks currently used. Researchers such as Verba (1969, pp. 80–99) and Scheuch (1968, pp. 119), for example, could hardly have been clearer about the importance of (their forms of) context, but we have still not clearly articulated forms and levels of context for questionnaire design, nor addressed them in practice. Thus, we have also taken Biemer and Berzofsky's chapter as a welcome opportunity to talk briefly about some of the less obvious aspects of question meaning we feel need to be addressed more directly in future research.

NOTES

1 Stalnaker (2002) attributes the earliest use of the term "common ground" to Grice (1989). However, Stalnaker's own earlier discussions of meaning and possible worlds (e.g., 1973) are frequently referenced in language philosophy discussions of common ground.

2 One of the motivations in doing so is to be able to discuss design models for multilingual studies which involve numerous questionnaire versions. Indicators might, for example, be (much) the same across cultures, but the questions asked about the indicators could differ to small or larger degrees. These distinctions also facilitate discussion of design models in which both the indicators and the questions differ across cultures while the research topic and the latent constructs are assumed to be common to all.

3 In speaking here of both questionnaire and instrument design, we wish merely to reflect that a finished instrument design (ready for implementation) involves technical design features (including programming features) not usually incorporated at the stage at which questions are formulated and a blueprint questionnaire is developed (cf instrument technical design at /http://ccsg.isr.umich.edu/instrdev. cfm/).

REFERENCES

Austin J (1962). How to Do Things with Words. Oxford: Clarendon Press.

Bach K (1987). On communicative intentions: a reply to Recanati. Mind and Language; 2:141–154.

Belson W (1981). The Design and Understanding of Survey Questions. London: Gower.

Biemer PP (2001). Nonresponse bias and measurement bias in a comparison of face to face and telephone interviewing. Journal of Official Statistics; 17:295–320.

Biemer PP (2004). Modeling measurement error to identify flawed questions. In: Presser S, Rothgeb JM, Couper MP, Lessler JT, Martin E, Martin J, Singer E, editors. Methods for Testing and Evaluating Survey Questionnaires. New York: Wiley; pp. 225–246.

Biemer PP, Wiesen C (2002). Measurement error evaluation of self-reported drug use: a latent class analysis of the US National Household Survey on Drug Abuse. Journal of the Royal Statistical Society, Series A: Statistics in Society; 165:97–119.

Biemer PP, Witt M (1996). Estimation of measurement bias in self-reports of drug use with applications to the national household survey on drug abuse. Journal of Official Statistics; 12:275–300.

Biemer PP, Groves RM, Lyberg L, Mathiowetz N, Sudman S (1991). Measurement Errors in Surveys. New York: John Wiley.

Bohrnstedt GW, Knoke D (1988). Statistics for Social Data Analyses, 2nd ed. Itasca, IL: F. E. Peacock.

Bollen KA, Long JS (1987). Test for structural equation models. Sociological Methods and Research; 2:123–131.

Cannell C, Miller P, Oksenberg L (1981). Research on interviewing techniques. In: Leinhardt S, editor. Sociological Methodology. San Francisco: Jossey-Bass; pp. 389–437.

Clogg CC (1977). Unrestricted and restricted maximum likelihood latent structure analysis: a manual for users. Working Paper 1997–09, Population Issues Research Office, The Pennsylvania State University. University Park, PA.

Dijkstra W (1987). Interviewing style and respondent behavior: an experimental study of the survey interview. Sociological Methods and Research; 16:309–334.

Dijkstra W (2002). Transcribing, coding and analyzing verbal interactions in survey interviews. In: Maynard DW, Houtkoop H, Schaeffer NC, van der Zouwen J, editors. Standardization and Tacit Knowledge: Interaction and Practice in the Survey Interview. New York: Wiley; pp. 401–425.

Fendrich M, Johnson TP, Hubbell A, Spiehler V (2004). The utility of drug testing in epidemiological research: results from an ACASI general population survey. Addiction; 99:197–208.

Flaherty BP (2002). Assessing reliability of categorical substance use measures with latent class analysis. Drug and Alcohol Dependence; 68:S7–S20.

Formann AK (1984). Die Latent-Class-Analyse [Latent Class Analysis]. Weinheim: Beltz.

Fowler FJ, Mangione TW (1990). Standardized Interviewing: Minimizing Interviewer-Related Error. Newbury Park: Sage.

Goodman LA (1974). The analysis of systems of qualitative variables when some of the variables are unobservable. Part I: a modified latent structure approach. American Journal of Sociology; 79:1179–1259.

Grice HP (1975). Logic and conversation. In: Cole P, Morgan JL, editors. Syntax and Semantics. New York: Academic Press; pp. 41–58.

Grice HP (1989). Studies in the Way of Words. Cambridge, MA: Harvard University Press.

Harkness JA, Edwards B, Hansen SE, Miller D, Villar A (2010). Designing questions for multipopulation research. In: Harkness JA, Braun M, Edwards B, Johnson TP, Lyberg L, Mohler P, Pennell B-E, Smith TW, editors. Multinational, Multicultural and Multiregional Survey Methods. Hoboken, NJ: John Wiley; pp. 33–57.

Harkness J, Mohler P, van de Vijver FJR (2003). Comparative research. In: Harkness JA, van de Vijver FJR, Mohler P, editors. Cross-Cultural Survey Methods. Hoboken, NJ: John Wiley; pp. 3–16.

Henry NW (1999). Latent structure analysis at fifty. Paper presented at the annual Meeting of the American Statistical Association, Baltimore, MD. Available online at http://www.people.vcu.edu/~nhenry/LSA50.htm.

Hill RJ, Hall NE (1963). A note on rapport and the quality of interview data. Southwestern Social Science Quarterly; 44:247–255.

Johnson RA (1990). Measurement of Hispanic ethnicity in the U.S. Census: an evaluation based on latent-class analysis. Journal of the American Statistical Association; 85:58–65.

Jöreskog KG (1967). Some contributions to maximum likelihood factor analysis. Psychometrika; 32:443–482.

Jöreskog KG (1970). A general method for analysis of covariance structures. Biometrika; 57:239–251.

Kortmann F (1987). Problems in communication in transcultural psychiatry: the self-reporting questionnaire in Ethiopia. Acta Psychiatrica Scandinavia; 75:563–570.

Kreuter F, Yan T, Tourangeau R (2008). Good item or bad—can latent class analysis tell? The utility of latent class analysis for the evaluation of survey questions. Journal of the Royal Statistical Society, Series A: Statistics in Society; 171:723–738.

Lanza ST, Collins LM, Lemmon DR, Schafer JL (2007). PROC LCA: a SAS procedure for latent class analysis. Structural Equation Modeling; 14:671–694.

Lazarsfeld PF (1950). The logical and mathematical foundation of latent structure analysis. In: Stouffer SA, Guttman L, Suchman EA, Lazarsfeld PF, Star SA, Clausen JA, editors. Measurement and Prediction. Princeton, NJ: Princeton University Press; pp. 362–412.

Marcoulides GA, Schumacker RE (1996). Advanced Structural Equation Modeling: Issues and Techniques. Mahwah, NJ: Erlbaum.

McCutcheon AL (1987). Latent Class Analysis. Newbury Park, CA: Sage.

Miller K (2005). Q-Bank: development of a tested-question database. American Statistical Association 1996 Proceedings of the Section on Government Statistics. Washington, DC: American Statistical Association. Available online at http://wwwn.cdc.gov/qbank/Report/M 2005.pdf.

Muthén LK, Muthén BO (1998–2010). Mplus User's Guide, 6th ed. Los Angeles: Muthén & Muthén.

Scheuch EK (1968). The cross-cultural use of sample surveys: Problems of comparability. In: Rokkan S, editor. Comparative Research Across Cultures and Nations. Paris: Mouton; pp. 176–209.

Schober MF, Conrad FG (1997). Does conversational interviewing reduce survey measurement error? Public Opinion Quarterly; 61:576–602.

Schuman H, Presser S (1981). Questions and Answers in Attitude Surveys: Experiments in Question Form, Wording, and Context. New York: Academic Press.

Schwarz N (1996). Cognition and Communication: Judgmental Biases, Research Methods, and the Logic of Conversation. Mahwah, NJ: Lawrence Erlbaum.

Schwarz N, Oyerman D, Petcheva E (2010). Cognition, communication, and culture: implications for the survey response process. In: Harkness JA, Braun M, Edwards B, Johnson TP, Lyberg L, Mohler P, Pennell B-E, Smith TW, editors. Multinational, Multicultural and Multiregional Survey Methods. Hoboken, NJ: John Wiley; pp. 177–190.

Stalnaker R (1973). Presuppositions. Journal of Philosophical Logic; 2:447–457.

Stalnaker R (2002). Common ground. Linguistics and Philosophy; 25:701–721.

Suchman L, Jordan B (1990). Interactional troubles in face-to-face survey interviews. Journal of the American Statistical Association; 85:232–241.

Sudman S, Warnecke RB, Johnson TP, O'Rourke D, Jobe J (1997). Understanding the cognitive processes used by women reporting cancer prevention examinations and tests. Journal of Official Statistics; 13:305–315.

Tourangeau R, Rasinski K (1988). Cognitive processes underlying context effects in attitude measurement. Psychological Bulletin; 13:299–314.

Tourangeau R, Rips L, Rasinski K (2000). The Psychology of Survey Response. Cambridge: Cambridge University Press.

van der Zouwen J, Smit JH, Draisma S (2010). The effect of the question topic on interviewer behavior; an interaction analysis of control activities of interviewers. Quality and Quantity; 44:71–85.

Verba S (1969). The uses of survey research in the study of comparative politics: issues and strategies. In: Rokkan S, Verba S, Viet J, Almasy E, editors. Comparative Survey Analysis. The Hague: Mouton; pp. 56–106.

Vermunt JK (1997). LEM: A General Program for the Analysis of Categorical Data. Tilburg, The Netherlands: Department of Methodology and Statistics, Tilburg University.

Vermunt JK, Magidson J (2005). Technical Guide to Latent Gold 4.0: Basic and Advanced. Belmont, MA: Statistical Innovations.

Willis GB (2005). Cognitive Interviewing: A Tool for Improving Questionnaire Design. Thousand Oaks, CA: Sage Publications.

Yang Y, Harkness JA, Chin TY, Villar A (2010). Response styles and comparative research. In: Harkness JA, Braun M, Edwards B, Johnson TP, Lyberg L, Mohler P, Pennell B-E, Smith TW, editors. Multinational, Multicultural and Multiregional Survey Methods. Hoboken, NJ: John Wiley; pp. 199–221.

PART V
Split-Sample Experiments

14 Experiments for Evaluating Survey Questions

JON A. KROSNICK
Stanford University

14.1 INTRODUCTION

For 100 years, experiments have been conducted that allow researchers to compare different ways of asking the same question in measuring a single construct. These experiments have not always been conducted with the purpose of identifying flaws in question design or practices for obtaining the most accurate measurements. But regardless of their original purpose, these experiments have accumulated into a gigantic literature with implications for best practices in questionnaire design.

In this chapter, I review the experimental method for evaluating questions and identifying their flaws. I begin by reviewing the logic of experimentation and how experiments can be valuable for evaluating question functioning. I then talk about how experimental data are produced and analyzed. A discussion of the assumptions underlying the method is followed by a listing of the types of insights that can be gained from experiments, how problems with questions can be characterized, how the method can be conducted misleadingly, whether experiments are suitable to cross-cultural investigations, how other question evaluation methods can be coordinated with experimentation, and what methodological criteria should be used for including the method in Q-Bank.

Question Evaluation Methods: *Contributing to the Science of Data Quality,* First Edition.
Edited by Jennifer Madans, Kristen Miller, Aaron Maitland, Gordon Willis.
© 2011 John Wiley & Sons, Inc. Published 2011 by John Wiley & Sons, Inc.

14.2   THE LOGIC OF EXPERIMENTATION

14.2.1   Treatment versus Control Groups

Experimentation is one of the oldest methods of scientific investigation. In lay language, to experiment with something is to test it out and to explore how it works. But in science, experimentation is a formal method for investigating causal relations. The vast majority of experiments compare what occurs under one set of circumstances with what occurs under a different set of circumstances, to assess whether the change in circumstances is responsible for any observed change in events. For example, if one wanted to assess whether taking a particular drug causes a reduction in blood cholesterol levels, one group of research participants would take the drug regularly for an extended time period (called the "treatment group"), and another group of participants would not take the drug (called the "control group"). At the end of the time period, blood cholesterol levels of both groups could be compared, to see whether the former group's levels are lower than the latter's.

14.2.2   Confounds

A number of important principles guide optimal experimental design to maximize a scientist's ability to reach a valid inference from the study's result. For example, it is essential to minimize confounds in the design. A confound is a difference between the two groups of participants that could be responsible for observed differences between the treatment and control groups. One potential source of confounds is participant self-selection into the treatment and control groups. Imagine that a researcher allowed each participant to decide whether to be in the treatment group or the control group of the study. And imagine that people who have higher cholesterol levels before the study tend to choose to be in the treatment group (in the hope that the drug will lower their cholesterol), whereas people with lower preexisting cholesterol levels tend to choose to be in the control group more often (because they have no special motivation to lower their cholesterol and want to generously allow others who need such help to be in the treatment group). If the treatment group ended the study with a higher cholesterol level than the control group, this difference might have been present even before the study began. Therefore, experiments routinely involve random assignment of participants to the treatment and control groups to eliminate the potential for systematic self-selection-driven differences between the groups.

Other potential confounds in an experimental design can come from experiences that vary between the treatment and control groups that are not the intended treatment itself. Imagine, for example, that the treatment group in a drug experiment is asked to come to a doctor's office to get an injection each week, whereas the control group does not visit a doctor's office weekly, because they have no need to do so. If, at the end of the study, the treatment group differs from the control group in terms of any outcome measures, the difference could have been caused by the doctor visits experienced uniquely by the treatment group (e.g., because each visit brought the participants into contact with very sick people, and seeing them inspired them to live healthier lives in general), not to the drug itself. Therefore, experimental researchers work hard to minimize the extent of all possible differences between the treatment and control groups, other than the essence of the treatment itself.

Another potential source of confounds are participants' beliefs about the hypothesis being tested and their role in the experimental investigation. Imagine that members of the treatment group are told that the study is investigating whether a new drug improves cholesterol levels and that they will be receiving the drug, whereas the control group is told about the purpose of the study and is told that they will not be receiving the drug. Simply believing that one is receiving a new and potentially helpful drug might make people happier and more optimistic about their lives, and this general optimism might cause lifestyle changes that themselves improve cholesterol levels. Therefore, experimental researchers believe that it is important for research participants to be as uninformed as possible about which experimental group they have been assigned to, so as to minimize the risk that expectations will cause differences between the groups in health status.

It is interesting to note, however, that the scientists' desire to conduct an experiment with strong internal validity (meaning that he or she can reach a strong conclusion about whether the treatment of interest caused changes in the outcome variable of interest) can limit the external validity of the findings (meaning the degree to which the experiment's results describe what would happen to people experiencing the treatment outside the experiment in the course of everyday life). For example, when conducting an experiment, the researcher might prefer that participants not know whether they are in the treatment or control group, whereas in real life, people taking a particular drug weekly would be well-aware that they are doing so. As a result, the researcher must decide whether he/she is interested in learning about the chemical and biological impact of the drug per se, or whether he/she is interested in learning about that impact coupled with the impact of the knowledge that one is taking the drug.

14.2.3 Applying Experimental Methods to Studying Questionnaires

The use of experiments to study survey questions is based on this same basic philosophy. Such experiments can be done in one of two ways, either by manipulating a question or by manipulating the conditions under which a question is asked. In the first, answers obtained by one version of a question are compared to the answers obtained by a different version of the question. If answers vary, then that variation can be attributed to changes in the question. The second approach involves asking a question of all people but varying the context in which it is asked. Both approaches can yield results that will help to optimize principles of optimal question design.

Experiments with questions can be done employing either a between-subjects design or a within-subjects design. In a between-subjects design, each respondent is randomly assigned to be asked one version of a question or another version of it. Thus, differences between the questions are ascertained by comparing the answers provided by groups of people. This approach can detect aggregate patterns of change, collapsing across people. For example, if some respondents are asked whether they approve or disapprove of a policy, while others are asked whether they favor or oppose it, the proportions of people who offer a favorable response can be compared to ascertain whether one question yields more such responses. If different people's answers are changed in opposite directions (some shifting from favorable to unfavorable, and others shifting from unfavorable to favorable), net change may understate the number of respondents' answers that would have been different had they been asked the other question. With this design, it is also not possible to determine whether any one respondent's answer would have been different had he or she been asked the different question. However, if a researcher has hypotheses about particular subgroups of respondents being affected differently by a question manipulation, tests can be conducted by comparing responses to the different questions within subgroups of respondents.

In a within-subjects design, the same respondents are asked more than one version of the same question. In theory, this allows a researcher to identify people whose answers would or would not be different depending on which question they are asked. However, in order for this approach to be informative about how a respondent would answer a single question in isolation (since researchers usually wish to identify the single optimal version of a question to ask), carryover effects must be eliminated. One type of carryover effect would involve respondent fatigue: If respondents are asked one version of a question and are then asked a different but very similar version of it, they may doubt the competence or efficiency of the researcher, who might appear to be wasting their time by making the same inquiry twice. Such doubt could undermine respondent motivation to provide accurate answers to the survey questions, thereby compromising data quality in the remainder of the questionnaire.

A second type of carryover effect would occur if respondents not only remember that they were asked a previous, highly similar question, but also remember the answer they gave to that question. One of the core principles validated by decades of research in psychology is the notion of "commitment and consistency" (Festinger, 1957; Cialdini, 2008). Once people express a particular opinion or describe themselves in a particular way, they have a tendency to want to maintain an image consistent with that initial statement. Therefore, respondents' answers to a later, similar question may be derived from their recollection of their answer to the earlier question.

A third type of potential carryover effect derives from the rules of conversation (Grice, 1975). Everyday conversation is governed by a set of principles that people learn implicitly, that speakers follow, and that listeners assume that speakers follow. One such rule is that speakers ask or say only what is necessary. Therefore, a listener can normally presume that anything a speaker says is said because the speaker thinks it is necessary for the conversation to be effective. Asking the same question of a person twice would most likely violate this rule. For example, if I were to ask a person, "How are you?" and she answered "Fine," it would be very odd indeed for my next question to her to be, "How are you?", because she can presume that I already know the answer. If I were to ask the same question again, she might wonder whether I did not hear her initial answer, or she might think that I am asking again because her initial answer was an inadequate answer in my opinion, so I seek more elaboration. But if an interviewer were to ask a simple, dichotomous yes/no question in a face-to-face interview, obviously hear and record the respondent's affirmative answer, and then ask the same question again, this would clearly be an unexpected and puzzling violation of the rules of conversation. As a result, if a questionnaire were to ask a question once and then ask a subtly different version of it later, even if the researcher's intent was to measure the same judgment in a slightly different way, the respondent might be motivated to find a way to interpret the second question so that its meaning is as different as possible from the initial question's. That way, it would make sense for the questioner to ask the second question.

Consider, for example, a researcher who wants to compare answers to two questions measuring life satisfaction: "How satisfied are you with your life?" and "How happy are you with your life?" If each respondent is asked only one of these questions, all respondents might interpret both questions as having essentially the same meaning. But if a respondent were initially asked the first of these questions and was then asked the second, he or she might try to find a way to interpret the second question so that it has a different meaning than the first. Of course, the difference between the two questions is "satisfied" versus "happy." So a respondent might choose to answer the second question not by reporting overall satisfaction with his or her life but instead to report how often he or she feels happy or how happy he or she typically feels, which would be describing the frequency or intensity of a mood state, rather than the degree to which life meets some standard of acceptability, regardless of emotional state. Therefore, the two questions might acquire very different answers from the respondent, even though they were intended to measure the same construct.

If a respondent were to be asked both questions in the reverse order, then the interpretation process would most likely unfold in a very different way (because the respondent would most likely interpret the first, happiness question as tapping overall life satisfaction, then leaving the respondent puzzled about how to reinterpret the second, satisfaction question). None of this is desirable when conducting a within-subjects experiment to compare two questions. Therefore, if a researcher were to implement such an experimental design, he or she should assure that there is sufficient time passage (and perhaps distraction) between the administrations of the two similar questions so as to minimize the likelihood of contamination of answers to the second by answers to the first.

Unfortunately, it is difficult to know on theoretical grounds or based on practical evidence how long the necessary minimum time interval must be. Van Meurs and Saris (1995) reported evidence suggesting that 20 minutes is sufficiently long. If this is correct for most questions, and if a researcher's questionnaire is not naturally at least 20 minutes long, then achieving the passage of time of 20 minutes would require lengthening an interview or would require making a second contact with the respondent on another, later occasion. Either way, this would substantially increase respondent burden and researcher challenge and thereby increase the cost of implementing a within-subjects experiment. Consequently, the seeming efficiency that this design brings is likely to come at significant practical costs.

However, it seems unlikely that the needed time interval to assure forgetting of a prior question is uniformly 20 minutes. Most likely, this time interval varies as a function of the particular topics and forms of the questions and intervening events. For example, if during a long time interval filled with questions about a person's personal finances, I were to insert two questions, one early and one late, about whether he or she had ever run over a cat or a dog with a car by accident, the respondent seems likely to remember this unusual and emotionally provocative question long after 20 minutes have passed. Researchers could address the risk of this latter type of carryover by, at the time that respondents are asked the second question, also asking whether they remember being asked a similar question earlier, and if so, how that question was worded and how they answered it. But doing so would then require asking more questions of respondents and would therefore lengthen the interview further.

If researchers were to discover that a substantial number of respondents did remember the prior question and their answers to it, then the value of the experiment would be significantly compromised. The people who remember would need to be dropped from analysis, thereby reducing the representativeness of the sample in a biased way and reducing the effective sample size and statistical power. Or the experiment could be redesigned and rerun, thus throwing away all the data from the first execution, again costly in terms of researcher and respondent time, money, and perhaps other resources.

For all of these reasons, it is not obvious that a within-subjects design has practical or analytic advantages over between-subjects designs. And for me personally, the practical challenges of minimizing carryover effects are substantial enough so that I strongly prefer between-subjects designs. This preference brings with it not knowing which particular respondents would have answered the two forms of the same question differently. But most question evaluation and comparison can be done without knowing that information, because we usually seek to make generalizations about questions regardless of particular individuals.

14.2.4   Analyzing Experimental Data

When conducting question design experiments, researchers most often focus on two principal criteria: distributions of answers and correlates of answers. For example, one might wonder whether the order of presentation of the answer choices influences answers to a closed-ended question. To do so, half of the respondents in an experiment could be given the choices in one order, and the other half could be given the choices in the reverse order, and the distributions of answers to the two versions could be compared. No difference in distributions would suggest that there had been no impact of order. The less impact that order has on responses, the more valid the question might appear to be, because if a person's answer to a question is an accurate self-description, then that answer should be the same regardless of the order in which the choices are offered.

A researcher might also wish to assess the validity of measurements obtained by different versions of a question by observing their correlations with other variables. For example, imagine that a researcher were interested in identifying the most valid way to measure life satisfaction and that solid theory and evidence indicate that people who earn more money are happier, on average. A researcher could then measure earnings for a sample of people and randomly assign each of them to get one of two different versions of a question measuring life satisfaction. If answers to the two questions correlate equivalently with income, then they would appear to be equally valid according to this test. But if answers to one version correlate more strongly with income than do answers to the other version, that would suggest the former might be more effective at accurately assessing life satisfaction, because answers contain less random and/or systematic measurement error.

Yet another analytic option is to examine test–retest reliability of items through experiments. To do so, each respondent can be asked the same question two or more times, separated by a suitably long time interval of days, weeks, or months. Then, an analyst can estimate the consistency of answers to the question as an indicator of reliability of the measurements. If different respondents are randomly assigned to answer different versions of the question, then the reliabilities of the various items can be compared to one another to identify the superior item in terms of measurement quality. Refinement of the analytic approach can be enhanced if each respondent answers multiple questions tapping the same construct on each occasion, so that latent variable covariance structure modeling can be conducted to produce an estimate of the reliability of the items while controlling for any change that might have occurred in the construct between the measurement occasions. This can be accomplished even if only a single question is asked on at least three occasions (see, e.g., Krosnick, 1988).

14.2.5   Statistical Options and Issues

When analyzing data from question design experiments, a number of analytic methods can be used. First, when comparing the distributions of answers, it is common to create a contingency table cross-tabulating question version with answers. For example, in an experiment that varied the order in which response choices were offered, each row can correspond to an answer choice, the columns can distinguish the groups of respondents asked the question with the different orders, and the cell entries can be the percent of people in each experimental group (i.e., column) who gave each response, with each column totaling 100%. The statistical significance of the differences between distributions can be assessed with a χ^2 statistic. If a question offers interval-level response options (e.g., "On how many days during the last week did you eat bread?"), a researcher could compute the mean response to two versions of the question and compute a t-test to assess the statistical significance of the difference between the two means.

When computing such statistical tests, a researcher must grapple with some potential complexities due to the design of the study. One issue is whether to weight the data or not. Two types of weighting can be done. One is to reflect unequal probabilities of selection to participate in the survey. Imagine, for example, a study in which interviewers visited the homes of respondents and randomly selected one adult resident to interview in each household. Thus, a resident's chances of being selected into the survey are inversely proportional to the number of adults living in the household. In order to properly project a survey's results to the population, it is necessary to statistically adjust for such intended inequalities of probability of selection through weighting. This type of weighting follows unambiguously from the design of a sample and is therefore relatively straightforward to implement.

A second type of weighting is post-stratification. When a survey sample's demographics differ notably from the population of interest in terms of known distributions of benchmark variables, analyses can be done while increasing the weight assigned to respondents from underrepresented groups and reducing the weight assigned to respondents from overrepresented groups. This sort of weighting is much more an art than an exact, formulaic science. Sampling statisticians do not agree on a single, optimal approach to computing such weights.

Gelman (2007) argued that rather than doing post-stratification weighting using demographics, researchers should statistically control for all such demographics when estimating the parameters of regression equations. This advice might seem relevant to the analysis of question design experiments because this can be done with regression. For example, if a researcher wants to assess whether one version of a life satisfaction question yields more favorable answers than another, he or she could randomly assign respondents to be asked one of those two versions and regress answers (either coded continuously and analyzed with OLS regression or coded categorically and analyzed with multinomial regression) on a dummy variable differentiating people who were asked the two different versions of the question plus demographic control variables that could instead have been used to construct post-stratification weights.

In the context of an experiment, controlling for demographics is very unlikely to alter the parameter estimate for the effect of the question design manipulation, because random assignment will mean that the demographic controls will be essentially uncorrelated with the question design manipulation. But controlling for demographics is likely to improve the ability of the analysis to detect the statistical significance of the manipulation's effect if the demographics explain any of the variance in responses. This is likely because controlling for those demographics will reduce the error variance used in tests of statistical significance.

However, controlling for demographics in such an analysis to solve unrepresentativeness in the sample seems unlikely to yield the same outcome as post-stratification would unless the impact of the question manipulation is uniform across the sample. If instead, the size of the effect of the manipulation is moderated by a variable that would have been used in weighting, then additional computation must be done. By failing to take into account the deviations between the sample and the population in terms of the distribution of the moderating variable, people at various levels of that variable will not be represented in the analysis in their proper proportions.

Imagine a case in which a question manipulation has strong impact on answers provided by respondents who did not graduate from high school and has no impact at all on answers from people who did graduate from high school. And imagine that the participants in an experiment vastly underrepresent the proportion of people who did not graduate from high school. If the data were analyzed without post-stratification, the few people in the sample without a high school degree would be vastly outnumbered by the others, and the lack of responsiveness to the question manipulation among the latter individuals could prevent a researcher from seeing a statistically significant effect of the question manipulation in the sample as a whole.

Implementing post-stratification to correct the underrepresentation of people without high school degrees would greatly increase their presence in the sample and might therefore cause the full sample test of the question manipulation to yield a significant effect. Even if a researcher does post-stratification, controlling for all demographics and other variables that were used to generate the weights is likely to be a wise idea, again because if those benchmarks are related to answers to the manipulated question, then controlling for them will remove some systematic variance in answers and increase a researcher's ability to detect a real difference between the question forms as statistically significant. As long as assignment to question version is done truly randomly, controlling for weighting variables in this way is unlikely to cause an effect of question design to appear to be significant when in fact it is an illusory artifact of the statistical estimation procedure.

Using the approach of not post-stratifying and instead controlling for the variables in terms of which the sample is known to deviate from the population will produce a proper total experimental effect size only if the researcher makes post-estimation adjustments for variation in the experimental effect size across subgroups of the sample. The effect size must be computed within groups of people differing in their values on any moderating variable (e.g., at different levels of education), and the estimated effect sizes can be combined in a way that takes into account the proportions of the various groups of respondents in the population, so as to yield an overall effect size for the population. The only way to accomplish this is to check for variation in the effect size across all levels of all variables in terms of which the sample deviates from the population. In this light, post-stratification may be a simpler and effective way to accomplish the goal of producing an effect size estimate for the population, if that is of interest.

Such post-stratification does not come without a cost. Specifically, such weighting weakens statistical power because of uncertainty in the weights themselves. The more variable the weight values are from one another across respondents, the greater the "design effect" of the weighting. The larger the design effect, the more variance in answers is caused by the researcher's somewhat arbitrary decision about how to construct those weights. If data are analyzed properly with statistical software that recognizes the impact of the weights on statistical confidence, weighting reduces statistical power to detect a real difference between answers to two different versions of a question. The larger the design effect of a survey's weighting approach, the more risk the researcher takes of failing to detect a real influence of a question manipulation on answers.

Therefore, the decision about whether or not to weight data from a question design experiment must be made based upon a researcher's goal for the experiment, and there are at least two legitimate but importantly different possible such goals. The first goal is to ask whether the question manipulation had a reliable effect on the answers provided by the people who participated in the study. This is a legitimate question and in fact has probably been the default assumption made in the vast majority of experiments conducted across the social and physical sciences. The data from such experiments have routinely been analyzed asking whether the two or more experimental groups differed from one another due simply to chance alone due to the random assignment procedure, or whether the manipulations are likely to have caused observed differences between those particular groups of people.

Another equally legitimate goal would be to ask what the impact of this experiment's findings would have been if the study had been done with a fully representative sample of a particular population. To ask that question requires specifying that population, of course. And it is not always obvious what population should be used for this purpose. For example, if an experiment were to be conducted in 1970 in the United States and researchers wished to ascertain what its result would have been if all American adults had been interviewed at that time instead, it is reasonable to post-stratify the data to match the American adult population in 1970.

But if a researcher wants to make a broader statement about the difference between the question versions, he or she might be tempted to think that any conclusions reached from this experiment would apply in 1980 and 1990, not only in the United States but in other countries as well. Therefore, it might be tempting to analyze the data weighting them numerous different ways to reflect many different populations, including hypothetical future populations that do not yet exist. Clearly, there is an endless number of possible populations to which a researcher might weight if he or she wishes to make general statements about the impact of a question manipulation. So simply weighting to match one of these many populations is not likely to be especially intellectually satisfying. Nonetheless, doing so might be of some value to reassure scientists that the result of an experiment with a highly unrepresentative sample can be generalized to a very different population to be studied later.

Another analytic issue to consider with experimental data is clustering in the sample design. National, face-to-face surveys routinely keep costs under control by sampling cases in geographically defined clusters (e.g., primary sampling units). As a result, interview locations are not smoothly spread across the entire nation (which would require a lot of interviewer traveling or a huge number of interviewers) and instead are grouped in clusters. Because of the clustering, all people interviewed who live within a single cluster are likely to be more similar to one another than they are to people in different clusters. Routines for representing such nonindependence are offered by various statistical packages, and they should be used to properly adjust the error variance to reflect this known source of covariation.

14.2.6 Assessing Validity

Comparing the accuracy of measurements obtained by two versions of a question via predictive validity can be done by measuring the association of the two sets of answers with a criterion variable. Depending upon whether the variables involved have nominal response options or interval-level response options, a researcher can conduct OLS regression or logistic regression or multinomial logistic regression, in addition to other such techniques. Regression coefficients can be computed to gauge the associations between variables, and coefficients for different question versions can be compared.

If a regression is conducted predicting a criterion with answers to a target question, a dummy variable indicating which version of the target question each respondent was asked, and the product of those two predictors, the product tests the interaction and therefore effectively tests whether the relation of the target question with the criterion was significantly different depending on which version of the target question was asked. The version producing the stronger relation is presumed to have produced the more valid measurements.

The ideal criterion variable for comparing the validities of two versions of a question would be a perfect measure of the construct of interest. For example, if one wanted to measure the amount of exposure a person had to television news programs during the past week, it would be wonderful to correlate answers to two survey questions with a pure and completely accurate assessment of that television news program exposure. With such a measure, we could estimate the parameters of the following equations separately using two different questions measuring media exposure:

$$\Gamma_1 = b_1(T) + s_1 + e_1 \tag{14.1}$$

$$\Gamma_2 = b_2(T) + s_2 + e_2 \tag{14.2}$$

where Γ_1 is answers to one question asking about television news exposure, Γ_2 is answers to the second question assessing television news exposure, T is the true amount of television news exposure each respondent experienced, b_1 is the validity of Γ_1, b_2 is the validity of Γ_1, s_1 and s_2 represent systematic measurement error in answers to each question (such as a tendency for people to underreport exposure using a particular measure, either intentionally or accidentally because of misremembering), and e_1 and e_2 represent random measurement errors in answers to each question. If $b_1 > b_2$ and/or $e_1 < e_2$, that would suggest that the first question is a more valid and/or reliable measure of true exposure than the second question. And if $b_1 < b_2$ and/or $e_1 > e_2$, that would suggest that the first question is a less valid and/or reliable measure of true media exposure than the second question.

Unfortunately, no pure and completely accurate assessment of television news exposure or any other construct of interest in surveys yet exists. For example, to measure media exposure, many different approaches have been explored, including observation, diaries, experience sampling, and more, and a large literature has emerged highlighting advantages and drawbacks of them all (for reviews, see, e.g., Stipp, 1975; Engle and Butz, 1981; Webster and Wakshlag, 1985; Kubey and Csikszentmihalyi, 1990; Robinson and Godbey, 1997, pp. 61–62). Each of these methods is subject to unique sources of systematic measurement error (e.g., diaries are often not filled out daily but rather are filled out for the entire reporting period just before they must be turned in), so none is perfect. The validity of global self-reports can be assessed via correlations with such alternative measures, but recognizing that perfect measurement is impossible.

When no direct measure of the construct of interest is available, researchers can take an alternative approach by using a criterion variable that theory suggests should be correlated with the construct being measured by the target question. This approach was suggested by the American Psychological Association (1954) for gauging the validity of a measure: assessing construct validity, which focuses on the extent to which a target measure is related to measures of other constructs to which theory says it should be related (see also Messick, 1989).

The relation of two versions of a target question to such a criterion can be represented this way:

$$Y = b_3(\Gamma_t) + b_4(\Phi) + s + e \tag{14.3}$$

$$Y = b_5(\Gamma_p) + b_4(\Phi) + s + e \tag{14.4}$$

where Y is the criterion measure that is associated with television news exposure (e.g., a quiz assessing the amount of factual knowledge about politics that the respondent possesses), Γ_1 and Γ_2 are the two different measures of television news exposure, b_3 and b_5 are coefficients estimating the associations of Y with Γ_1 and Γ_2, Φ is a vector of other correlates of the criterion that have been measured in the survey, b_4 is a vector of coefficients reflecting the strength of impact of these other causes, s is systematic measurement error in assessments of the criterion, and e is random error in measurements of the criterion. b_3 and b_5 can be estimated in the two separate equations leaving Φ and s out of the equation, because the impact of other causes and systematic measurement error will be the same in both. Invalidity and random measurement error in the measures of television news exposure will attenuate b_3 and b_5. So if $b_3 > b_5$, that would suggest that the first question is a more valid and/or reliable measure of its construct than the second version. And if $b_3 < b_5$, that would suggest that the first question is a less valid and/or reliable measure of its construct than the second question.

Another analytic approach that can be employed to assess item validity in an experiment involves latent variable covariance structure modeling (e.g., Bollen, 1989). This approach can be implemented by collecting multiple identical measures of a construct and randomly assigning respondents to be asked one of various different versions of another measure of the construct. The various measures can then be treated as indicators of a single latent construct in a covariance structure analysis. Software such as LISREL, M-Plus, Amos, or EQS can be used to produce estimates of the validity and reliability of the different versions of the target measure, and tests can be computed to assess the significance of the difference between these parameters.

Such analysis can be done relatively simply by including just the latent construct of interest in the model and various measures of it. However, it is also possible to include measures of other constructs to which the construct of interest is likely to be related. So, for example, a study comparing the validity of various life satisfaction measures can administer all such measures to respondents and also measure their incomes. Income can be represented in the covariance structure model as a separate construct correlated to some unknown degree with life satisfaction. With this model structure (two correlated latent constructs: life satisfaction and income), the statistical estimation procedure has more information (i.e., the correlations of the various target measures of life satisfaction with income) to use in gauging the validity of the two alternative versions of the measure of life satisfaction.

When associations between questions are estimated using experimental data, researchers are often tempted to compute standardized measures of association between items, such as Pearson Product Moment Correlations. But comparing the magnitude of a correlation across different versions of a question can be misleading if the questions differ in the variability of responses to them. Consider, for example, two questions that are equally valid in tapping the construct of interest, but one question yields answers that are more variable than the other. This increase in variance will lead a predictive validity correlation to appear stronger for the latter question. It is therefore preferable to examine unstandardized regression coefficients to estimate the strength of associations between items, with all variables coded to range from 0 (meaning the lowest possible level of the construct) to 1 (meaning the highest possible level). Such coefficients are easy to interpret and are less impacted by differences between experimental conditions in the variance of the items involved and therefore provide a clearer comparison of the validities and/or reliabilities of the items.

14.2.7   Assessing Administration Ease

In addition to measuring the impact of question variations on response distributions, reliability, and validity, researchers can conduct experiments to compare the amount of time that it takes respondents to answer different versions of a question. Response latency, as psychologists call it, can be measured easily under conditions of computer administration by having the computer record the moment in time when a question appears on the screen and the moment in time when the respondent submits an answer to the question. In telephone or face-to-face interviews, the interviewer can push a key on a computer keyboard to mark the moment in time when he or she finishes reading a question aloud and can push the key again to mark the moment in time when the respondent begins to utter an answer to the question. For practical reasons, questions that people can answer more quickly are generally preferred to those that take longer to answer, because asking questions that can be administered more quickly allow for asking more total questions in a fixed interview time period. But in addition, answering a question more quickly may be an indication that the cognitive tasks of interpreting the question, retrieving information from memory to generate an answer, compiling a summary answer, and reporting it were easier for respondents to do, which would make the question desirable, because it would minimize respondent fatigue.

14.3   HOW EXPERIMENTS CAN BE VALUABLE FOR EVALUATING QUESTIONS

Experiments are valuable for evaluating questions because they offer a quantitative technique for assessing whether question variations cause changes in answers, either in terms of their distributions or their validities. If a researcher is uncertain about which question approach is preferable for measuring a construct, an experiment can be conducted to provide evidence with which to decide. And if a researcher has a hypothesis about a particular way in which a target question might bias or distort measurement of the construct of interest, it is possible to compare that target question to other versions of it in an experiment to test the hypothesis.

Next, I review illustrations of how experiments have been conducted to help optimize question design. Specifically, I describe experiments that tested for effects of response choice order, question balancing, acquiescence response bias, branching/labeling of bipolar rating scales, word choice, and social desirability response bias.

14.3.1 Response Order Effects

In a study of response order effects, Schuman and Presser (1981) asked a randomly selected half of a telephone survey sample this question:

> Some people say that we will still have plenty of oil 25 years from now. Others say that at the rate we are using up our oil, it will all be used up in about 15 years. Which of these ideas would you guess is most nearly right?

The other half of the respondents were asked instead:

> Some people say that at the rate we are using up our oil, it will all be used up in about 15 years. Others say that we will still have plenty of oil 25 years from now. Which of these ideas would you guess is most nearly right?

Thus, the order in which the two answer choices were read to respondents was varied between subjects. When this experiment was run in January, 1979, 64% of respondents chose the "plenty" option when it was presented first, whereas 77% of respondents chose it when it was presented last, a highly significant difference ($P < 0.001$). This is what is called a "recency" effect, because a response option is advantaged when it is presented last. This experimental evidence documented a source of systematic measurement error in responses to the question.

In another study, Holbrook et al. (2000) investigated how the order in which response choices are offered can either conform to or violate respondents' expectations based on conversational conventions and can thereby compromise the accuracy of the answers respondents provide. In their experiments, some respondents were randomly assigned to be asked this question:

> The federal government is considering raising the import tax on steel that comes into the United States from other countries. Raising the steel tax would protect the steel industry from foreign competition and create more jobs for American steel workers. However, it would also increase the prices Americans pay for

products made from steel. If you could vote on this, would you vote for raising the import tax on steel or would you vote against it?

Others were randomly assigned to answer this question:

The federal government is considering raising the import tax on steel that comes into the United States from other countries. Raising the steel tax would protect the steel industry from foreign competition and create more jobs for American steel workers. However, it would also increase the prices Americans pay for products made from steel. If you could vote on this, would you vote against raising the import tax on steel or would you vote for it?

Consistent with the notions that (1) asking whether one favors or opposes is the more natural way to phrase the question, and (2) asking whether one opposes or favors is counter-normative and distracting, Holbrook et al. (2000) found that people answered the first question more quickly than they answered the second (3.67 seconds vs. 4.04 seconds on average, $P < 0.05$) and that the predictive validity of questions using the conventional response option order was higher than that of questions employing the counter-normative response option order.

14.3.2   Question Balance

Shaeffer et al. (2005) explored whether minimal balancing of a closed-ended question is sufficient for producing unbiased responses, rather than having to implement full balancing. To do so, Shaeffer et al. (2005) randomly assigned respondents to be asked one of two versions of the same question:

Minimal Balance. "As it conducts the war on terrorism, do you think the United States government is or is not doing enough to protect the rights of American citizens?"

Full Balance. "As it conducts the war on terrorism, do you think the United States government is doing enough to protect the rights of American citizens, or do you think the government is not doing enough to protect the rights of American citizens?"

The distributions of answers to the two questions were the same, as were the strengths of the associations between these questions and criteria. Therefore, the authors concluded that the two question forms yielded equivalent data, so the minimal balancing approach (which entails fewer words) appears to be the preferable approach.

14.3.3   Acquiescence Response Bias

Acquiescence response bias is the tendency to provide affirmative answers to questions offering agree/disagree, true/false, or yes/no answer choices. One experiment illustrating acquiescence was conducted by Schuman and Presser (1981), who asked respondents whether they agreed or disagreed with one of the following two statements:

> Individuals are more to blame than social conditions for crime and lawlessness in this country.

> Social conditions are more to blame than individuals for crime and lawlessness in this country.

Of the people given the first statement, 60% agreed with it, which might lead one to expect that at least 60% of people given the second statement would disagree with it. But in fact, only 43% of people disagreed with the second statement, and 57% agreed with it. Thus, it appeared that a majority of respondents agreed with both a statement and its opposite. This identified a source of systematic measurement error present in answers to these questions.

14.3.4   Branching/Labeling of Bipolar Rating Scales

Krosnick and Berent (1993) explored the idea that when respondents are asked to make a rating on a bipolar dimension (with a zero point in the middle), reporting accuracy may be improved by decomposing the reporting task into two subtasks: reporting whether the respondent is at the midpoint, on one side of it, or on the other side of it, and then separately reporting how extreme the respondent is on his or her chosen side of the midpoint. To do so in one experiment, some respondents (chosen randomly) were asked:

> There has been a lot of debate recently about defense spending. Some people believe that the U.S. should spend much less money for defense. Suppose these people are at one end of a seven-point scale, at point number 1. Others feel that defense spending should be greatly increased. Suppose these people are at the other end of the scale—at point number 7. And, of course, other people have opinions somewhere in between, at points 2, 3, 4, 5, and 6. Where would you place yourself on this scale?

Others were instead asked this branched question sequence:

> There has been a lot of debate recently about defense spending. Do you think the U.S. should spend less money on defense, more money on defense, or continue spending about the same amount on defense? [If less:] "Would you say we should spend a lot less, somewhat less, or a little less?" [If more:] "Would you say we should spend a lot more, somewhat more, or a little more?

Each respondent answered the question twice, separated by about four weeks. Respondents also answered four other questions in the same format, either not branched or branched. Collapsing across the four questions, 39% of respondents gave the same answer to the same question when asked the nonbranching format, and 64% did so when asked the branching format, a highly significant difference ($P < 0.00001$). Thus, branching and labeling all response options with words improved test–retest reliability and therefore appears to improve measurement quality.

14.3.5   Question Wording

Chang and Krosnick (2003) investigated whether question wording affected the validity of measurements of exposure to political news through the media. Respondents in their study were randomly assigned to be asked one of two versions of an exposure question:

How many days in the past week did you watch the news on TV?

How many days in a typical week did you watch the news on TV?

Similar questions were asked about newspaper reading as well.

Chang and Krosnick (2003) gauged the validity of these items by estimating associations of answers to them with four measures assessing the amount of knowledge each respondent possessed about politics. This approach was based on the assumption that most people gain most of their political knowledge from exposure to the news media, so stronger relations between media exposure and knowledge volume would be an indication of greater validity of the measure of exposure. Political knowledge volume was measured by (1) asking respondents to provide a summary judgment of how informed they were about politics, and (2) giving respondents quizzes asking them to describe specific recent national and international events as best they could; coders graded the accuracy of the information each respondent had on each issue. Chang and Krosnick (2003) found greater predictive validity for the typical week questions than for the past week questions, suggesting that the former produced more accurate assessments of chronic levels of news media exposure.

14.3.6   Social Desirability Response Bias

Holbrook and Krosnick (2010) explored the impact of social desirability response bias on reports of past behavior. Their focus was on respondent reports of whether they voted in a recent national election. Many scholars have speculated that such reports are intentionally distorted by some respondents' desire to appear to have fulfilled their civic duty, since people might be embarrassed to admit that they did not vote. Thus, some people who did not vote might claim to have done so when asked to admit that aloud to an interviewer. Some studies have suggested that such social desirability pressures may be removed when respondents answer questions on a computer, without having to admit their failings aloud to an interviewer (e.g., Chang and Krosnick, 2010).

To test whether social desirability pressures distort reports of turnout, Holbrook and Krosnick (2010) randomly assigned some respondents in a telephone survey and in Internet surveys to answer a direct question asking them whether they voted in a recent election, such as:

> In talking to people about elections, we often find that a lot of people were not able to vote because they weren't registered, they were sick, or they just didn't have time. How about you—did you vote in the Presidential election held on November 7, 2000?

Other respondents were instead asked to report turnout using the Item Count Technique (ICT). Among these individuals, some (chosen randomly) answered this question:

> Here is a list of four things that some people have done and some people have not. Please listen to them and then tell me HOW MANY of them you have done. Do not tell me which you have and have not done. Just tell me how many. Here are the five things: Owned a gun; Given money to a charitable organization; Gone to see a movie in a theater; Written a letter to the editor of a newspaper. How many of these things have you done?

Other individuals were instead asked this version of the question:

> Here is a list of five things that some people have done and some people have not. Please listen to them and then tell me HOW MANY of them you have done. Do not tell me which you have and have not done. Just tell me how many. Here are the four things: Owned a gun; Given money to a charitable organization; Gone to see a movie in a theater; Written a letter to the editor of a newspaper; Voted in the Presidential election held on November 7, 2000. How many of these things have you done?

The average answer given to the first question can be subtracted from the average answer to the second question to yield the proportion of people who said they voted in the election. The purported advantage of the ICT is that it allows respondents to provide completely confidential reports. Consistent with this reasoning, Holbrook and Krosnick (2010) found in their telephone survey that 72% of respondents said they had voted when asked the direct question, but the estimated turnout rate according to the ICT measure was 52%, a statistically significant decrease ($P < 0.05$). In the Internet surveys, the direct self-report questions and ICT measurements yielded equivalent turnout rates, suggesting that social desirability pressures did not distort direct self-reports under these measurement conditions.

14.4   HOW CAN EXPERIMENTS BE MISUSED OR CONDUCTED INCORRECTLY

Two types of problems have sometimes occurred in experimental studies comparing different questioning approaches. One is the failure of random assignment in between-subjects studies. If respondents are given the opportunity to choose which version of a question they receive, then the selection may be based on a factor that influences answers. So comparisons of answers to the two question forms may not reveal the impact of the question form per se but may instead be attributable to preexisting differences between the groups of people who answered the two questions. It is therefore important that random assignment be done.

Doing so might seem easy to do in practice, but it has turned out not to be so easy sometimes. For example, I was involved in the design and conduct of a major national survey in which respondents were supposed to be randomly assigned to be asked one of various different versions of questions. After the data had been collected, we inspected the patterns of question assignments and found them to depart so significantly from what would be expected by chance alone that it led us to doubt the effectiveness of random assignment. The programmer who had been responsible for implementing the random assignment insisted repeatedly that the random assignment had been done properly until finally recognizing that in fact, the assignment had not been truly random. Thus, it is important to be vigilant about the details of the procedure for implementing random assignment.

Even if random assignment is properly implemented, there is a nonzero probability that people in different experimental conditions will differ from one another substantially in ways not due to the manipulations implemented. If those preexisting differences between the experimental groups are related to the outcome variable of interest, this can cause results to be misleading. Therefore, researchers should routinely check their experimental data to see whether people in different experimental groups are indeed identical in the aggregate in terms of variables that should not have been affected by a treatment. And if the groups are not identical, it is easy and sensible to statistically control for the impact of the unintentionally confounded variable when estimating the effect of the treatment.

Another problem that appears sometimes in write-ups of experiments is focusing on a single manipulation when it is perfectly confounded with another manipulation that could be responsible for apparent differences between experimental conditions. For example, consider the experiment described above by Krosnick and Berent (1993) in which respondents were randomly assigned to be asked a branching question or a nonbranching question measuring the same attitude. In that study, the two tested versions of the question differed not only in terms of branching but also in terms of the verbal and numeric labeling of the scale points. The nonbranching version presented most rating scale points with numeric labels only, whereas the branched version of the question did not label any response options with numbers and instead labeled them with words. There is reason to believe that verbal labeling of scale points might improve measurement quality, and other studies reported by Krosnick and Berent (1993) showed just that. They found that part of the data quality improvement attributed to branching in the above example was due to branching, and part was due to more verbal labeling of scale points. It would be inappropriate to ignore one of the question format variations and presume that differences between experimental conditions are attributable to the other format variation.

A final mistake that can be made in analyzing the effect of an experimental manipulation is controlling for a variable that was affected by the manipulation of interest. For example, imagine that a researcher is interested in whether adding an argument to a question changes the answers that people give to it. To do so, some respondents might be asked this question, which is taken from a *Time Magazine* poll done in June, 2008:

> There is a type of medical research that involves using special cells, called embryonic stem cells, that might be used in the future to treat or cure many diseases, such as Alzheimer's, Parkinson's, diabetes, and spinal cord injury. It involves using human embryos discarded from fertility clinics that no longer need them. Some people say that using human embryos for research is wrong. Do you favor or oppose using discarded embryos to conduct stem cell research to try to find cures for the diseases I mentioned?

In order to ascertain whether the penultimate sentence influenced answers, other respondents could be asked the question without that sentence:

> There is a type of medical research that involves using special cells, called embryonic stem cells, that might be used in the future to treat or cure many diseases, such as Alzheimer's, Parkinson's, diabetes, and spinal cord injury. It involves using human embryos discarded from fertility clinics that no longer need them. Do you favor or oppose using discarded embryos to conduct stem cell research to try to find cures for the diseases I mentioned?

Imagine that these questions were followed by another question that was asked identically of all respondents:

> In general, do you think medical research is ever immoral?

It is easy to imagine that answers to this latter question might be influenced by the presence or absence in the prior question of the sentence, "Some people say that using human embryos for research is wrong." Perhaps people who hear that sentence in the previous question are more likely to say that medical research is sometimes immoral. A researcher might be tempted to control for answers to the question about immorality when assessing the impact of the sentence about using embryos being wrong on answers to the question containing it. This temptation might occur because the researcher thinks it would be advantageous to control for the more general opinion when examining the measure of a more specific opinion. But doing so would be inappropriate, because the control variable is not purely exogenous and is instead influenced by the manipulation of interest.

14.5   COORDINATING EXPERIMENTATION WITH OTHER QUESTION EVALUATION METHODS

Although experimentation has yielded hundreds of valuable publications informing optimal questionnaire design on its own, it has the potential to be used in coordination with other question evaluation methods in constructive ways. Specifically, conventional pretesting, behavioral observation, and cognitive pretesting are all evaluation methods intended to identify suboptimalities in the design of questions. When a researcher collects data using such methods and identifies a potential deficiency in a question, he or she often then redesigns the question and pretests it again, to see if the change produced an improvement in the quality of the responses. And most often, this improvement is assessed in terms of whether it was easier to administer the question and/or whether respondents' stated interpretations of the question and/or their articulated thoughts conform to researchers' hopes or expectations.

Much rarer is to see the results of such pretesting subjected to experimental evaluation. Specifically, an experiment could be conducted in which some respondents are randomly assigned to answer the original, possibly flawed version of the question, and other respondents are instead asked the new, presumably improved version of the question. If the question alteration was indeed an improvement, then we should see differences between the experimental conditions indicating that, such as faster reaction time, greater test–retest reliability of answers, and/or greater predictive validity. This seems like a fruitful direction for future studies to coordinate the use of multiple question evaluation methods. Doing this sort of validation would certainly slow down the process of question design and refinement, but it would be valuable for validating the pretesting techniques themselves.

14.6   METHODOLOGICAL CRITERIA FOR INCLUDING RESULTS OF EXPERIMENTS IN Q-BANK

Experiments are likely to produce valuable insights into optimizing question design as long as (1) random assignment is done properly, (2) manipulations are done so that key elements of question design are unconfounded from irrelevant variables, and (3) data are analyzed properly to identify the impact of questions on administration difficulty and/or data quality. Some scholars have shown a preference for studies done of representative samples of populations over studies done with convenience samples, such as groups of college students. I believe that there is scientific value in experimental studies of many different types of respondent groups, even when they are not representative samples of populations. If data are collected from a convenience sample for one experiment, it is incumbent on the researcher to attempt to replicate the findings of that study with representative samples of populations. If replication occurs and findings are consistent across the various methods, this increases confidence in any conclusions drawn about best practices in question design. Q-Bank can be a place to gather reports of all such experimental studies and compare their results to draw general conclusions about best practices.

14.7   CONCLUSION

Experimentation has always played a central role in all types of scientific investigation, and it has played a central role in research optimizing question-naire measurement as well. I look forward to many more decades producing hundreds more experiments and just as many valuable insights that will help social science to fulfill its potential by testing interesting and important theories using maximally accurate measurements of the constructs of interest.

REFERENCES

American Psychological Association (1954). Technical recommendations for psychological tests and diagnostic techniques. Psychological Bulletin; 51(Pt 2):1–38.

Bollen KA (1989). Structural Equations with Latent Variables. New York: Wiley.

Chang L, Krosnick JA (2003). Measuring the frequency of regular behaviors: comparing the "typical week" to the "past week." Sociological Methodology; 33:55–80.

Chang L, Krosnick JA (2010). Comparing oral interviewing with self-administered computerized questionnaires: an experiment. Public Opinion Quarterly; 74:154–167.

Cialdini R (2008). Influence: Science and Practice. Needham Heights, MA: Prentice Hall.

Engle PL, Butz WP (1981). Methodological issues in collecting time use data in developing countries. Paper presented at the Biennial Meeting of the Society for Research in Child Development, Boston, MA, April 2–5.

Festinger L (1957). A Theory of Cognitive Dissonance. Stanford, CA: Stanford University Press.

Gelman A (2007). Struggles with survey weighting and regression modeling. Statistical Science; 22:153–164.

Grice HP (1975). Logic and conversation. In: Cole P, editor. Syntax and Semantics, 9: Pragmatics. New York: Academic Press; pp. 113–128.

Holbrook AL, Krosnick JA (2010). Social desirability bias in voter turnout reports: tests using the item count technique. Public Opinion Quarterly; 74:37–67.

Holbrook AL, Krosnick JA, Carson RT, Mitchell RC (2000). Violating conversational conventions disrupts cognitive processing of attitude questions. Journal of Experimental Social Psychology; 36:465–494.

Krosnick JA (1988). Attitude importance and attitude change. Journal of Experimental Social Psychology; 24:240–255.

Krosnick JA, Berent MK (1993). Comparisons of party identification and policy preferences: the impact of survey question format. American Journal of Political Science; 37:941–964.

Kubey R, Csikszentmihalyi M (1990). Television and the Quality of Life: How Viewing Shapes Everyday Experience. Hillsdale, NJ: Lawrence Erlbaum.

Messick S (1989). Validity. In: Linn R, editor. Educational Measurement. New York: Macmillan; pp. 13–103.

Robinson JP, Godbey G (1997). Time for Life: The Surprising Ways Americans Use Their Time. University Park, PA: Pennsylvania State University Press.

Shaeffer EM, Krosnick JA, Langer GE, Merkle DM (2005). Comparing the quality of data obtained by minimally balanced and fully balanced attitude questions. Public Opinion Quarterly; 69:417–428.

Schuman H, Presser S (1981). Questions and Answers in Attitude Surveys. New York: Academic Press.

Stipp HH (1975). Validity in social research: measuring children's television exposure. PhD dissertation, Columbia University.

van Meurs L, Saris WE (1995). Memory effects in MTMM studies. In: Saris WE, Münnich A, editors. Multitrait Multimethod Approach to Evaluate Measurement Instruments. Budapest: Eötvös University Press; pp. 89–103.

Webster JG, Wakshlag J (1985). Measuring exposure to television. In: Zillmann D, Bryant J, editors. Selective Exposure to Communication. Hillsdale, NJ: Erlbaum.

15 Response 1 to Krosnick's Chapter: Experiments for Evaluating Survey Questions

JOHNNY BLAIR

Abt Associates

15.1 INTRODUCTION

One illuminating effect of Krosnick's chapter (this volume) is to remind one how much more essential experiments are to survey research than survey application. Experimental design is indispensable to several research areas, in particular the study of response effects. However, when experiments are employed in survey development, it is seldom, aside from some large government surveys, in support of instrument development. This is especially interesting considering that response effects research addresses problems of measurement which are also a central concern of pretesting.

Our understanding of how question wording, features of response scales and question context, among other factors, affect survey response has been advanced largely by the use of experimental design. Krosnick provides an overview of the logic and principles of experimental design for research. He discusses the advantages and disadvantages of between-subjects and within-subjects designs, and concerns with carryover effects and confounds. Krosnick argues for the superiority of between-subjects designs, and goes on to describe how such designs use statistical tests on distributions of answers and correlates of answers to test response effect hypotheses. He notes that experiments to study survey questions can be done in two ways: manipulating the question or manipulating the conditions (e.g., the context) under which the question is asked. In either case, the analysis usually focuses on distributions of answers or correlates of answers.

Krosnick describes two types of problems that sometimes occur in experiments comparing different questioning approaches. One is the failure of random assignment—in between-subjects studies. The other is the error of

Question Evaluation Methods: Contributing to the Science of Data Quality, First Edition.
Edited by Jennifer Madans, Kristen Miller, Aaron Maitland, Gordon Willis.
© 2011 John Wiley & Sons, Inc. Published 2011 by John Wiley & Sons, Inc.

focusing on a single dimension when another factor could be responsible for observed differences between the experimental conditions. He provides several examples of studies where these types of problems seriously undercut the strength and clarity of the findings. The research on evaluating questions that he cites is concerned with the science of survey response, for which internal validity is primary, not improving survey questions for a particular survey, where external validity is paramount.

Experimental design is not explicitly addressed in most texts on survey design. There is no particular utilization of experiments that is considered an industry best practice. When experiments are employed in designing a survey, the focus is often on factors that influence costs or response rates (Biemer and Lyberg, 2003, pp. 283–284). Experiments to determine the impact of administration mode or respondent incentives or data collection strategies are more common than experiments to improve question design—again with the exception of the redesign of some major government surveys.[1]

In "Survey Research and Experimental Design," Lavrakas and Traugott (unpublished manuscript) call for:

> survey researchers. . . . [to] become more creative and careful about thinking how experimentation can benefit their research goals. And, . . . this consideration should be an explicit step in planning a survey, on par with traditional planning steps such as deciding upon one's sampling frame, sample size, or data collection mode.

Lavrakas and Traugott describe many ways that experiments have been used to improve survey design, and also relate some missed opportunities.

While the method of experiments is underutilized to develop and pretest survey questions, there are examples of effective use. For example, in addition to free-standing, embedded field experiments, the use of experiments in conjunction with other pretest methods has been recommended (see Fowler, 2004), and experiments employing vignettes have been effective (Martin, 2004).

One obstacle to utilizing experiments more often in survey development may be the difficulty of implementing an experimental design with the rigor Krosnick describes, along with concerns about the consequences of violating experimental design principles. However, practitioners may profit from some of the strengths of experimental design even though their applications do not meet the standards of basic research.

For example, in some circumstances, experiments may use small samples (e.g., of the sizes typical in conventional field pretests) that do not permit the types of quantitative analysis Krosnick notes as the main strength of experimental design, but still be of value in questionnaire development by employing qualitative outcome measures.

15.2 EXPERIMENTS FOR PRETESTING

The need for basic research findings to be generalizable and replicable requires strongly controlled conditions. Experiments for pretesting are mainly concerned with external validity—that the results will hold in survey implementation—and do not require understanding the causal mechanism of observed effects. In addition to the primary focus on external validity, experiments for research and for pretests differ in a number of other ways (Fig. 15.1), for example, an experiment that compares two versions of a question that differ in multiple ways. If the experiment finds that one of the versions performs better—for example, in comparing answers to a validation source to assess accuracy—it is not possible to know which aspect of the question accounts for the difference. Such a confound would be a serious problem in a basic research study, but for pretesting, the finding of improved performance would still be valuable. The conditions under which an experiment for question development is conducted may not be replicable, but its findings may be very informative for the survey designer's purposes.

These different requirements and goals permit more flexible uses of experiments for questionnaire design, including the use of qualitative data to assess outcomes—a possibility that does not preclude, and can sometimes complement, the use of statistical methods. These observations are not to argue for looser standards for survey pretesting experiments, but simply to note that design weaknesses that are unacceptable for research may not be so for questionnaire development.

Once these differences are explicitly acknowledged, more possibilities become available for employing experiments for survey development generally, and for pretesting in particular. The approach examined here is the use

Experiments for research	**Experiments for pretesting**
Discover knowledge	Improve questions
Strongly controlled conditions	Conditions controlled as much as feasible
Findings derived statistically	Findings derived by multiple means, both quantitative and qualitative
Must be replicable	Usually need not be replicable
Results reported to scientific community (in peer-reviewed publication)	Results reported to questionnaire developer (Documented in project methodology report)

FIGURE 15.1. Differences between experiments for pretesting and research. (Adapted from figure in Sharp et al., 2007.)

of cognitive interviews in experimental design for pretesting. More specifically, the following exploratory discussion considers how, for some purposes, verbal protocol analysis (VPA) may facilitate the use of cognitive interviews in experimental designs for pretesting.

Cognitive interviews are typically used for problem discovery. The modus operandi is the progression from an initial draft question to successive improved versions. But consider the situation when one has alternative versions of a question created because one does not know how best to address a particular measurement problem. In that case, neither question version is thought to be flawed. The versions differ mainly in that they apply different approaches to the problem, or focus on different components of the desired measurement. The instrument developer is faced with choosing between imperfect alternatives.

Such a situation can occur, for example, when the response task is especially difficult. One is concerned with examining methods to ease the response task, while maintaining and conveying the original item intent, and obtaining accurate measures. These sorts of complex cognitive response tasks may go beyond problems of comprehension. For example, in contingent valuation or willingness-to-pay studies, survey questions frequently require respondents to hold large amounts of information in memory, to compare hypothetical program outcomes, or to decide the monetary value of things not typically thought of as having a price. Such complex response tasks often require the respondent to keep relevant factors in mind, that is, in working memory, while constructing their answer. Experimental designs using cognitive interviews may be an effective method to compare alternative strategies for these types of survey questions.

Moreover, some procedures used in VPA, whose think-aloud technique was the original basis for cognitive interviewing, may be adaptable to investigating the performance of these and similar response tasks. After an overview of VPA, we consider how that method could be applied to cognitive interviews in an experimental design.

15.3   USING VPA IN A COGNITIVE INTERVIEW PRETEST EXPERIMENTAL DESIGN

Ericsson and Simon's (1993) VPA utilizing think-aloud reports was the main impetus to developing the survey cognitive interview. However, the labor-intensive coding and analysis of verbal reports that is central to VPA has generally, though not universally, been considered incompatible with the objectives and practical constraints of survey pretesting. Furthermore, cognitive interviewers use probes about question meaning, and other techniques to assess survey question performance, that are at odds with VPA methods for eliciting verbal reports. While this approach has resulted in cognitive interview practices freed of the costs of VPA, it has in the process also lost the advantages of the method's rigor.

15.4   AN OVERVIEW OF VPA

VPA relies on the think-aloud method, with instructions to subjects intended to elicit verbal reports that can be treated as data to test hypotheses about cognitive processes. Most VPA research is concerned with studying higher-level cognitive processes, such as solving a puzzle, playing a game, or applying expert knowledge to make a decision.

VPA begins with a model of how the process at issue may be carried out. That model may be based on psychological theory or based on a task analysis. The task analysis is most appropriate to the objectives of survey question development. Ericsson (2006) notes that:

> One of the principle (sic) methods [of the information processing approach is task analysis]. Task analysis specifies the range of alternative procedures that people could reasonably use, in light of their prior knowledge of facts and procedures, to generate correct answers to a task.

Task analysis specifies particular actions and their sequence. For example, if the process being studied is how a subject solves a puzzle, a task analysis-based model would specify the different steps (and the order of steps) that one can feasibly use to solve the puzzle. Ideally, this "problem space" will represent all the ways the puzzle can be solved. In an experimental design, the hypothesis may be that subjects with more experience solving this or similar puzzles will tend toward using particular paths to a solution that differ from the paths selected by novice puzzle solvers.

The model is used to specify a coding framework. The research then collects verbal protocols using the think-aloud method. Each protocol is examined for evidence of using a step specified in the model, and is coded as such. The codes for the experts and novices can then be compared to test the hypothesis. If certain paths were posited to be taken under some conditions than others (or certain paths more likely to be taken by experts vs. those taken by novices), the protocol analysis permits testing that hypothesis.

An experimental design employing VPA procedures with cognitive interviewing might follow the process outlined below.

15.5   THE STRUCTURE OF A COGNITIVE INTERVIEW PRETEST EXPERIMENT

(a) Do a task analysis to decompose the question response task into components one might expect to be separately identifiable in the cognitive interview verbal report.

(b) Develop a question-specific coding plan based on the task analysis.

(c) Develop a cognitive interview protocol that uses thinking aloud and administers the same set of probes (and/or specifies conditions under which to probe) for both question versions.

(d) The question versions should ideally differ in only a single aspect.

(e) Randomly assign respondents to question versions (and to interviewers).

(f) Code the verbal reports.

(g) Analysis.

The next sections describe this approach generally and illustrates how it might be applied.

15.6   SURVEY QUESTION TASK ANALYSIS

Several descriptive models of the survey question response process have been proposed. Most of the models seem best suited to autobiographical rather than attitude questions. Jobe and Herrmann (1996), in reviewing seven such models note that there was, at that time, little research on whether the stages specified by the models actually occur. Tourangeau et al. (2000) review six such models, comparing their basic categories. While the survey models describe the response process, the coding schemes developed from the models use the process mainly as a framework for defining question flaws or response problems. In contrast to most VPA, the survey pretest practitioner's primary interest is not in the response process but focuses on when and why the process breaks down, resulting in the respondent being unable to answer at all or producing an inadequate or inaccurate answer.[2]

One of the models (Conrad and Blair, 1996) highlights the cognitive interview pretest goals by indicating the usual response stages (comprehension, recall, judgment and reporting), but with three additional features to represent some types of failures.

1. Each stage provides information to the respondent to proceed to the next stage.
2. In the face of inadequate information, a respondent may choose to return to an earlier stage.
3. A respondent may "exit" at any stage, conceding inability to continue.

For example, a difficulty at a performance stage (e.g. uncertainty about recall accuracy) could result in the respondent reexamining her understanding of the question (i.e., returning to an earlier response stage, giving up trying to answer, or proceeding despite the flawed performance). The coding scheme included types of problems at each stage of the process, including problems severe enough to send the respondent back to an earlier stage or induce the respondent to give up trying to answer.

Any one of these general response models could provide a framework for a question-specific task analysis that will

(a) Define the response problem in terms of the specific survey question

(b) Develop a coding scheme for an experiment.

Some kinds of survey questions, for example, questions about experiences or behaviors, can be thought of as a kind of instruction to the respondent to perform a series of tasks that will result in an answer to the question. As such, the tasks can be deconstructed in such a way that evidence for performing each task can be coded, for example:

Think about the time period [ ]
count or estimate how many times you did X in this time period
where X is defined as [ ] to include [ ] and to exclude [ ]
Report the total.

15.7   EXAMPLE APPLICATION: PUBLIC HOUSING RENT SYSTEMS SURVEY

In a survey of public housing residents and people on waiting lists for public housing, one section of the questionnaire asked about different systems that could be used for setting the initial amount of rent, and changes in rent when the resident's income changed. This section required responses to questions comparing real and hypothetical situations, or two hypothetical questions. These questions posed a number of problems for respondents. The section followed a lengthy interview of more typical survey questions; so the section presented the respondent with a major change in type of response task. Examples of these questions are given in Figures 15.2 and 15.3.

The Rent Systems cognitive interviews used think-aloud and prescripted probes. Three interviewers conducted the pretest. The experiment came about inadvertently. The initial version was tested on a couple of respondents and, while not a disaster, respondents clearly had problems with the response task. Based on these two interviews, we identified some of the tasks the respondents had problems with.

In addition to comprehension issues, such as understanding the words used, such as "inflation" or "net income," we noted the following general components, which are very much like the structure of a task analysis.

- understanding that the basic task was to compare a hypothetical system to the current system; or to compare two hypothetical systems
- understanding each of the two systems options
- not confusing the two options; or, in some instances, comparing the options in the survey with the present option they lived under
- keeping relevant information in mind
- selecting one of the offered choices

Alternative versions were developed for each of the questions (and of the transition statements preceding the questions) in the Rent Systems section.

PREF2 For the next few questions, I am going to ask you about different systems that can be used to determine the level of rent families pay to live in public housing or to rent a housing unit with a voucher.

I will read two choices for each question. Please tell me which one of the two choices you would prefer for yourself. There are no right or wrong answers, we just want to know what you think.

Would you prefer a rent system with: (Option A) or a rent system with (Option B)

Option A	Option B
Lower rent, but a longer time waiting for assistance	Higher rent, but a shorter time waiting for assistance
Rent that does not change when your income goes up or down	Rent that goes up when your income goes up and down when your income goes down
Rent that increases each year you receive assistance	Rent that stays the same, but has a time limit on the number of years you can receive assistance
A lower rent, but the housing authority verifies your income every year.	A higher rent, but the housing agency does not verify your income after the first year.

FIGURE 15.2. Example A: rent system question.

15.7.1   Coding Plan

A coding scheme could have been developed, but was considered unnecessary given a small sample (n = 8). Such a coding frame might start with these categories, and would make comparison of larger samples in an experiment more systematic.

15.7.2   Rent System Question Response Codes Categories

- Understanding general task
- Understanding option A
- Understanding option B
- Identifying the key elements that differ between options
- Remembering information
- Selecting one of the offered options

An example from another field illustrates how this can be done with cognitive interviews in an experimental design, though for a different purpose from question development. Brown (1995) used think-aloud verbal reports in an experiment on estimation strategies of event frequency. In his study, 40

PREF3   I'm going to ask you some more detailed questions about your preferences for a rent system for the public housing and voucher program. Just like the previous question, I'll present two options and you tell me which one of the two choices you would prefer for yourself. Would you prefer.....

	Option A	Option B
PREF3a	The current income-based rent system—30 cents for each dollar of income—and the same amount of time you spend waiting for housing assistance	A rent system where you paid extra $100 in monthly rent, but one year less time waiting for housing assistance
PREF3b	The current assisted housing income-based rent—30 cents for each dollar of income—and the same amount of time you spend waiting for housing assistance	A rent system where you paid extra $100 in monthly rent, but two years less time waiting for housing assistance
PREF3c	A rent of $250 adjusted only for inflation that does not change when your income goes up or down.	Rent that is 30 cents for each dollar of net income and changes when your net income goes up or down.
PREF3d	A rent of $350 adjusted only for inflation that does not change when your income goes up or down.	Rent that is 30 cents for each dollar of income and changes when your net income goes up or down.
PREF3e	Rent that starts at $200 and increases by $50 each year. For example, it would be $250 in the second year and $300 in the third year.	Rent that is 30 cents for each dollar of income and changes when your income goes up or down. You would only be allowed to (stay in your public housing unit/use your voucher) for 6 years.

FIGURE 15.3. Example B: rent system question.

respondents—randomly assigned to one of two type-of-context conditions—were presented with a study list of 260 word pairs, and each pair included a category name. After studying the list, they were presented with 36 category names and their task was to estimate the number of times each category name had appeared on the study list. They were required to think aloud as they constructed their frequency estimates for each category name.

Two judges, working together, used a coding scheme—based on his theoretical model of alternative ways to produce the estimate—to code the verbal reports. The codes included classifications for such things as mention of a vague quantifier, such as "It showed up quite often," an assertion that the target word had not appeared in the study list, and a mention of frequency-relevant information. Using these codes, the verbal reports were assigned to "response types." A table of the distributions of response types by the two experimental conditions (same context; different context) was generated that permitted standard statistical analyses.

There are two points of interest for a survey application along these lines. First, the sample size, while exceeding present standards for cognitive interview pretests, is not very large. Second, and more important, a coding system created prior to the test was based on classifications for expected verbal report content. This permitted quantitative comparison of the experimental conditions. Third, samples of this size could be possible in some survey pretest circumstances, particularly if only a section of the questionnaire was the subject of the experiment.

15.7.3 Cognitive Interview

VPA relies mainly on subjects thinking aloud to elicit the verbal reports. Ericsson and Simon (1993) describe three levels of verbalization. Classic Think Alouds produce level 1 (L1) and level 2 (L2) verbal reports that are *incomplete*, though sequential, traces of working memory contents that can be reported while performing the specified task. VPA instructions for thinking aloud are intended to avoid level 3 (L3) verbalizations.

15.7.4 Levels of Verbalization

Level 1   Verbalizations that do not need to be transformed before being verbalized during task performance.

Level 2   Verbalizations that only need to be transformed before being verbalized during task performance.

Level 3   Verbalizations that require additional cognitive processing beyond that needed for task performance or verbalization.

Cognitive interviews seldom rely solely on think-aloud verbal reports but are supplemented with item-specific probes (among other methods). An item-specific probe is intended to determine how the respondent comprehends a question, or some part of it, or how the respondent performs a response task. The fact that L1 and L2 reporting is incomplete provides some theoretical justification for using probes to supplement classic verbal protocol elicitation.

For example, many survey questions include a reference period. In thinking aloud, the respondent may not mention the reference period. Absence of mention is not evidence that the reference period was not properly used (or not used at all). It may be that use of the reference period is an L1 or L2 verbalization but is simply not reported, or describing use of the reference period may require additional cognitive processing (L3) to report.

15.7.5   Experiment Design and Random Assignment

The alternative question versions *varied*:

(a) The amount of detail in the introductory description of the task
(b) Whether an explicit opportunity for having the question or pair of options repeated was included
(c) Several changes in the description of the options, for example, "30 cents for each dollar" versus "30%"

An additional eight respondents were randomly selected from a pool of residents and applicants for housing and randomly assigned to two question versions and the three interviewers. This is not an ideal design for several reasons. Given the small sample, this between-subjects design is not very strong. Observed differences could easily be attributable to differences between the two sets of respondents. There was no explicit analysis plan. We had an idea of what types of problems to look for, based on the list above. But we did not code the verbal reports.

Additional information could have been obtained with even a moderately larger sample. For example, with a larger sample it would be possible to compare versions that vary problematic components of each question systematically. Versions could be systematically varied on dimensions such as:

Longer, more detailed versus shorter introduction to question section
Explicit opportunity for the respondent to request that a question be repeated
Alternative descriptions of system options
Order within each set of preference questions
Location of preference questions in questionnaire

Despite the limitations, this experience suggested that an experiment using cognitive interviewing with qualitative analysis could be informative. This experience also provided an opportunity to consider if VPA might be combined with survey cognitive interviewing within an experimental design.

15.8   DISCUSSION

This sketch outlining one possibility for expanding the use of cognitive interviews is intended only to suggest a path for more extensive exploration. Experimental design using cognitive interviews can be done with relatively small samples when the analysis methods are qualitative. With small samples, between-subjects designs are feasible but, as in the Rent Systems example, may be weak. More information per case is possible if a within-subjects design is used. For example, alternative versions of the Rent Systems items could have been presented early and later in the cognitive interview. With somewhat larger samples, as illustrated in the Brown study, cognitive interview experiments can apply quantitative methods as well. A qualitative data collection procedure does not, in itself, preclude some types of quantitative analyses. Such approaches may bring the additional benefit of addressing criticisms of the sometimes impressionistic nature of cognitive interview analysis.

Certainly, there are obstacles to moving in this direction. Cognitive interview pretests often plan on small samples, but that is not invariably the case. And experimental design can be made more practical when experiments can be focused on sections of a questionnaire as a way to keep costs down. When resources are sufficient, as is often the case for federal survey development, these designs can be strengthened in a number of ways.

- Alternative question versions can be varied systematically, one element at a time to create discrete experimental conditions.
- Probes can be written explicitly to explore specific potential problems.
- Problem codes can be constructed using task analysis within a general response model, such as Conrad and Blair (1996) used, that explicitly allows for failure points in the response process.

More innovative applications of cognitive interviews in experimental design should also be considered. For example, when similar types of response problems, or even similar questions are encountered over time, some results (if planned for in advance) may be cumulated across a series of small samples at an organization. Without any special procedures, Q-Bank could be a platform for such experiments over time since its tools would support the necessary cross-pretest linking.

On a more general note, the cost of instrument development is clearly a factor in pretest design, but it has not been a subject of investigation in the same way as sampling, data collection, and other aspects of survey design. From this perspective, the consideration of alternative cognitive interview experimental designs for question evaluation is but a part of a larger goal to learn more about how to optimally combine different pretest/evaluation techniques. Work is needed to study uses of experimental design alone or in combination with other methods, as part of learning more generally how to effectively use these methods in concert.

NOTES

1   Another fairly recent impetus for project-specific experiments is the need for effective and consistent question performance in multi-language, multi-culture surveys.

2   An exception is the model used in research by Sudman and his colleagues (e.g., Blair et al., 1991), where the categories and coding frames are concerned with strategies respondents use to answer behavioral frequency questions.

REFERENCES

Biemer PP, Lyberg LE (2003). Introduction to Survey Quality. Hoboken, NJ: Wiley.

Blair J, Mennon G, Bickart B (1991). Measurement effects in self versus proxy responses to survey questions: an information processing perspective. In: Biemer PP, Groves RM, Lyberg LE, Mathiowetz NA, Sudman S, Forsman G, editors. Measurement Errors in Surveys. New York: John Wiley.

Brown NR (1995). Estimation strategies and the judgment of event frequency. Journal of Experimental Psychology: Learning, Memory and Cognition; 21(6):1539–1553.

Conrad F, Blair J (1996). From impressions to data: increasing the objectivity of cognitive interviews. American Statistical Association 1996 Proceedings of the Section on Survey Research Methods. Washington, DC: American Statistical Association; pp. 1–9.

Ericsson KA (2006). "Protocol Analysis and Expert Thought: Concurrent Verbalizations of Thinking during Experts' Performance on Representative Tasks," in The Cambridge Handbook of Expertise and Expert Performance Ericsson KA, Charness N, Feltovich PJ, Hoffman RR, editors. New York: Cambridge University Press; pp. 223–242.

Ericsson A, Simon H (1993). Protocol Analysis: Verbal Reports as Data, 2nd ed. Cambridge, MA: MIT Press.

Fowler FJ (2004). The case for more split-sample experiments in developing survey instruments. In: Presser S, Rothgeb JM, Couper MP, Lessler JT, Martin E, Martin J, Singer E, editors. Methods for Testing and Evaluating Survey Questionnaires. New York: Wiley; pp. 173–188.

Jobe JB, Herrmann DJ (1996). Implications of models of survey cognition for memory theory. In: Herrmann D, Johnson M, McEvoy C, Hertzog C, Hertel P, editors. Basic and Applied Memory Research: Vol. 2. Practical Applications. Hillsdale, NJ: Erlbaum, 1996; 193–205.

Lavrakas PJ, Traugott MW (Forthcoming). Survey research and experimental designs. Unpublished manuscript.

Martin E (2004). Vignettes and respondent debriefing for questionnaire design and evaluation. In: Presser S, Rothgeb JM, Couper MP, Lessler JT, Martin E, Martin J, Singer E, editors. Methods for Testing and Evaluating Survey Questionnaires. New York: Wiley; pp. 149–171.

Sharp H, Rogers Y, Preece J (2007). Interaction Design: Beyond Human-Computer Interaction. New York: Wiley.

Tourangeau R, Rips L, Rasinki K (2000). The Psychology of Survey Response. Cambridge: Cambridge University Press.

16 Response 2 to Krosnick's Chapter: Experiments for Evaluating Survey Questions

THERESA DeMAIO
U.S. Bureau of the Census

STEPHANIE WILLSON
National Center for Health Statistics

16.1 INTRODUCTION

Randomized experiments embedded within probability surveys provide an important tool that can profitably and best be used in conjunction with other techniques designed to evaluate and refine questionnaire content. In this chapter, we would like to focus on two aspects of the Krosnick chapter (this volume). First, we focus on the divergence in the measurement challenges encountered by the largely subjective measurements presented in the Krosnick chapter, and the generally nonsubjective measurements made in most federal demographic and health surveys. While one may quibble about subjective aspects of these measurements, there is quite a bit of difference between the measurement challenges posed by ascertaining whether a respondent has ever been diagnosed with Type 2 diabetes and ascertaining whether the respondent is "very happy, somewhat happy, or not happy at all."

Second, we focus on the integration of other evaluation methodologies with rigorous experimentation conducted within probability surveys, and we will argue that this is the key to future improvements in the quality of our survey measurements.

Question Evaluation Methods: Contributing to the Science of Data Quality, First Edition.
Edited by Jennifer Madans, Kristen Miller, Aaron Maitland, Gordon Willis.
© 2011 John Wiley & Sons, Inc. Published 2011 by John Wiley & Sons, Inc.

16.2   FACTUAL VERSUS SUBJECTIVE MEASUREMENTS

The most salient difference between factual and subjective questions involves conceptualizing what "validity" means for these two types of questions. In the first instance, the conventional notions of a "true value" might be determined without any additional theorizing and relatively simple strategies for validation abound (e.g., checks of physician records, health insurance claims). These arise because—although the respondent may not be a perfect informant—diagnosis is a social event for which other informants and written records can provide validation. That is not to say records exist for all factual measures, but in theory, there could be a way to validate this information by comparing the responses to some type of record. In contrast, happiness is a latent construct to which only respondents have direct access. The concept itself and its validation, if possible, requires a theoretical superstructure to relate responses to survey questions to other manifest behaviors (DeVellis, 2003). Validating such a measure is fraught with uncertainties that do not present themselves when one is trying to ascertain whether a respondent has been diagnosed with diabetes.

The difference between these types of measurements has serious implications for the types of analyses conducted on experimental data. Since Krosnick's chapter focuses mainly on attitudes and other subjective measures, he focuses on such analytic strategies as test–retest reliability in which the respondent is asked the same question later in the questionnaire to see if the answers are the same. He mentions that 20 minutes is enough time for the respondent to forget that he or she has already been asked the same question. He talks about using criterion variables based on theory when no direct measure of the construct is available. And he mentions using latent variable covariance structure modeling, which involves collecting multiple measures of a construct and using statistical estimation procedures to determine the validity of the measures. Response latency is another method he mentions, which is mostly relevant to psychological measures but which can also be used to provide information about the cognitive burden imposed in asking factual questions.

This response to the Krosnick chapter presents a different analytic approach that can be used to measure the reliability and validity of factual questions, and it is used at the U.S. Census Bureau. As a component of questionnaire design experiments, a second data collection—called a content reinterview survey—is frequently incorporated as an evaluation methodology. This content reinterview survey evaluates the quality of some, but not usually all, of the questions in the original experiment—and it can be used to evaluate a single questionnaire such as the decennial census form as well. The content reinterview questionnaire contains questions that are either identical to the original survey questions, or questions that collect more detailed information than the original questions. When the same questions are asked, a measure of reliability is obtained, and when other, more detailed questions are asked, information about validity is obtained. Data from the content reinterview are used to cal-

culate statistics such as the Index of Inconsistency and the net and gross difference rates, which provide information about the relative levels of bias and
response consistency in the alternative treatments (see U.S. Census Bureau,
1993, chapter 2). So for example, in an experiment conducted to evaluate
proposed revisions of the disability questions for the American Community
Survey (ACS), a variant of the Index of Inconsistency called the simple
response variance was calculated and was used to determine the reliability of
the experimental versus the control questions by measuring the consistency
of the responses in the two data collections (see Brault et al., 2007). In contrast,
evaluation of proposed new employment status questions used an expanded
series of questions in the reinterview—the Current Population Survey unemployment status questions that are considered the "gold standard"—and the
net difference rate was used to provide measures of the validity of the experimental questions (see Holder and Raglin, 2007). These were used in conjunction with the distribution of responses and item nonresponse rates, which the
paper notes are important analytic measures.

Another method of evaluating results of experimental manipulations is to
audio-record the interviews—or a sample of the interviews—and use behavior
coding to evaluate the problems the interviewer had in administering the
questions and the respondents had in answering them. Comparing the results
of behavior-coded interactions across multiple versions of survey questions
adds a different perspective to the analysis. This method provides information
about the reliability and possibly the validity of responses to the questionnaire.
Besides item nonresponse rates, distribution of responses, and reliability measures from a content reinterview survey, behavior coding data can provide an
indication of whether one version of a question is more difficult to administer
correctly or cause problems for respondents than another version. A research
survey developed by the survey methodology staff at the Census Bureau used
all four of these measures to evaluate two alternative ways of structuring a
demographic survey, depending on the sequence in which the questions are
asked about all the household members.

One treatment used the sequence used in most of the Census Bureau's
interviewer-administered surveys—all the questions about a household
member are asked at one time, followed by all the questions about the second
household member, and so on. This is known as the person-level approach.
This was tested against a household-level design, in which a screener question
first asked if anyone in the household had a specific characteristic of interest,
and then asked follow-up questions only for household members who possessed the characteristic. Behavior coding was used to identify desirable and
undesirable respondent and interviewer behaviors by question topic. Across
five general topic categories (demographic characteristics, functional limitations, health insurance, program income sources, and asset ownership), the
behavior coding component of this evaluation did not suggest that one treatment was better than the other. Respondents were equally able to produce
adequate answers in both versions. Each treatment showed strengths and

weaknesses regarding interviewer behavior, but no trend was evident (see Hess et al., 2001, 2002).

In addition to differences in analysis of the results of subjective and factual questions, there are differences in the types of issues that can be investigated. Attitudinal questions are frequently asked about only one person per household. In contrast, the person-based versus household-level design of structuring a questionnaire previously mentioned is relevant when asking for the same information about all members of a household, as many federal surveys do. Some of the things that Krosnick mentions—acquiescence response bias, question balance, and branching of bipolar rating scales—are more pronounced for attitude or opinion questions. In contrast, other phenomena, such as response order effects, social desirability response bias, and question wording effects, are quite relevant for all types of questions. The flexibility of the experimental method and the range of issues that can be investigated can be demonstrated by work that staff in the Census Bureau's Center for Survey Measurement has done using a continuing-research survey developed by Census Bureau survey methodologists.[1]

Joanne Pascale focused on the topic of health insurance coverage, and—over the course of several years—was able to conduct systematic research into such issues as the impact of the length of the reference period, the addition of a question to verify that no one in the household was uninsured, question sequencing to measure potential effects of asking about public health insurance coverage before private health insurance coverage and vice versa, order effects of asking about Medicaid or Medicare coverage first, the sequence of response options in asking about health status and level of physical activity, and alternative structures of asking questions about type of health plan. A survey is currently in the field to investigate differences between health insurance coverage measures from surveys that use different reference periods (current coverage in the American Community Survey vs. any time during the previous calendar year in the Current Population Survey) as well as a redesigned version that includes refinements based on previous experimentation. Content reinterview, respondent debriefing, behavior coding, and comparison with administrative records have all been useful in analyzing the results of these various experiments, in addition to the standard indicators such as item nonresponse and response distributions (Pascale et al., 2008/2009; Pascale, 2009; Pascale et al., 2009).

Although there are differences in the treatment of experiments that consist mainly of subjective versus factual questions such as those that comprise the majority of federal surveys, this does not mean that these experiments are mutually exclusive. Measurements of subjective phenomena can of course be made in conjunction with factual data. The health insurance coverage research program mentioned above includes a subjective measurement of health status—do you consider your health to be excellent, very good, good, fair, or poor—that is relevant to the subject of interest.

Experiments designed to evaluate questionnaire design and question wording, whether factual or subjective, can be conducted in different ways, depending on the amount of time and resources that can be devoted to data collection. For example, Fowler (2004) describes small-scale experiments conducted with random-digit-dialed (RDD) national samples and very little follow-up for refusal conversion to increase response rates. This is the approach taken by the Census Bureau in conducting the QDERS mentioned previously. This survey effort, designed strictly for research purposes, provides a vehicle for researchers to expand their methodological research interests, particularly those that emerge from results of their qualitative research, independent of the Census Bureau's ongoing production surveys. Although Fowler mentions experiments involving as few as 100 cases, the Census Bureau's experiments have involved sample sizes of about 1000–4000. Investigations of varied research topics have been accomplished through this vehicle, including person-level versus household-level measurement of demographic characteristics, topic-based versus person-based measurement of income, alternate methods of requesting respondent consent for linking survey data with administrative records, obtaining a direct measure of nonmarital cohabitation, and alternate methods of measuring where people should be counted according to census residence rules, have been accomplished through this vehicle.

The National Center for Health Statistics (NCHS) has also collaborated on small-scale experiments to investigate questionnaire design issues. Working with the Center for Survey Research at the University of Massachusetts-Boston, Paul Beatty has conducted several experiments using RDD telephone survey samples of less than 500 cases, incorporating behavior coding as an evaluative method. Rather than focusing on a specific survey or a specific subject matter, Beatty focused on a specific type of survey question, namely complex questions from federal health surveys that require specific pieces of data (see Beatty et al., 2006).

On the other hand, larger scale experimentation is also possible, and the Census Bureau has done this as well. In preparation for making changes to the questionnaire for the American Community Survey (ACS), the Census Bureau has conducted several tests using nationally representative samples of 30,000–60,000 cases, incorporating the mail, telephone, and personal visit modes that comprise data collection in the ACS. After cognitive testing was conducted to refine the treatments included in the experiment, new topics have been added (such as questions about marital history and the field of a person's bachelor's degree), revised questions have been tested in a number of areas (including disability, military service, employment status, educational attainment, number of rooms and bedrooms, and mortgage-related topics), and layout issues for the mailed questionnaire have been investigated (including a test of precoded response categories vs. write-in spaces for rent, property value, and number of vehicles, and comparison of a matrix or grid structure for the basic demographic characteristics against a sequential design that included questions and answers in a format that contained a separate column for each person in the household). These examples all represent the use of experimental designs beyond those described by Krosnick.

16.3   INTEGRATION OF EXPERIMENTS WITH OTHER METHODOLOGIES

The second point we would like to make in this discussion relates to the context in which the experiments are conducted. Krosnick notes that it is rare for the results of pretesting, through cognitive testing and behavioral observation, to be subjected to experimental evaluation. We agree that it is a good idea to think of experiments as best used in conjunction with other methods, and we provide examples of the experimental evaluation of pretesting results. Although rare, it has been done, and we feel that this combination of methods can improve our understanding of survey question performance.

Cognitive interviewing offers the qualitative advantage of being able to provide an understanding of the meaning behind the number produced by a survey question. In essence, it provides insight into construct validity. However, as it is typically conducted for questionnaire pretesting, it cannot provide a good measure of magnitude. For example, a cognitive interview study might show that a survey question is understood several different ways by respondents, but we have no way of knowing which interpretation occurs most often or, by extension, whether or not interpretations vary enough to impact survey estimates. On the other hand, experiments offer the quantitative advantage of understanding the degree to which a difference is significant. This method can test to see if two questions, worded differently but attempting to measure the same construct, produce similar estimates. However, if those two questions do not produce the same estimate, the experiment can do little to tell us why, or which question does a better job at measuring our concept. Indeed, even if there is no difference between the two questions, we still may be left wondering what underlying construct is being measured by each.

A fruitful path for improving data quality in survey estimates would be to begin with a cognitive test of survey questions followed by a split-ballot experiment to test the hypotheses generated from the findings. For example, Willis and Schechter (1997) use a controlled experiment to test whether alternatively worded survey questions would produce variations in estimates as predicted by results of cognitive testing. They found that results of the cognitive testing did, indeed, predict the direction of the estimates. This study lends support to the idea that a mixed-method approach could be of benefit in evaluating survey data quality.

The Questionnaire Design Research Laboratory at NCHS recently conducted a cognitive interview study on the self-administered questionnaire for the Pregnancy Risk Assessment Monitoring System (PRAMS). PRAMS is a state-specific, population-based self-administered survey of women with children between the ages of 2 months and 9 months. Topics center around health issues before, during, and after a woman's most recent pregnancy and include smoking, the use of assisted reproductive technologies, gestational diabetes, and Caesarean deliveries among others. A disproportionate number of apparent false positive response errors were noted in the "before pregnancy" portion of the questionnaire because some women appeared to be telescoping—that is, including behaviors that occurred outside the specified time frame of 3 months before pregnancy. Moreover, the women who tended to do this were those who had not planned their pregnancy. This led to the realization that the "before pregnancy" section of the instrument often had an embedded assumption that the respondent had in fact planned her pregnancy. As a result, the questions were modified in an attempt to eliminate this assumption.

To think about this project from a mixed-method perspective, it would be ideal to embed the original survey question and the modified question in a split-ballot experiment. The research hypothesis—generated from the cognitive test—would be that the original question will yield higher estimates than the modified question. In other words, we would expect that the number of false positive responses would decrease in the revised question as women with unintended pregnancies would no longer telescope and include behaviors they engaged in during (not before) their pregnancy. The Willis and Schechter (1997) experiment mentioned above follows this logic. In this case, the experiment, interesting as it might be, is unlikely to be conducted because no convenient vehicle exists to facilitate this experiment.

However, there are other instances in which the mixed-method approach we advocate has been employed. All of the questions that were tested in the ACS Content Test were subjected to cognitive testing prior to their quantitative testing. The question wording treatments evolved through recognition of specific types of errors that were committed by individual respondents, and subsequent revisions were made to the questions to eliminate the errors. For example, the 2007 ACS Content Test included a new question requested by the National Science Foundation to elicit the field of degree for each person who reported having a bachelor's degree. Cognitive testing was conducted using three versions of the question—two involved precoded response categories and included lists of major fields of study with a yes or no response required for each one. The third version was an open-ended write-in that allowed the respondent to enter the major field(s) in their own words. The cognitive testing showed that respondents had several difficulties with the precoded lists. First, they had trouble figuring out where the major fields of study fit in the lists they were presented with. Second, they did not know whether they were supposed to mark just one category or whether they could mark more than one, if they had a double major or their major was interdisciplinary. And third, some respondents marked only the yes responses, leaving the other boxes empty. The write-in version of the question was much preferred, especially in the Computer Assisted Telephone Interviewing (CATI) mode, because it was easier, more efficient, and respondents did not have to listen to long lists. The only problem observed was that the three lines listed after the question gave respondents the impression that additional detail was requested, and they added such things as whether the degree was a BS or a BA. Minor changes were made based on the cognitive testing, and one precoded question was tested against the write-in version in the Content Test. Analysis of the experiment validated the results of the cognitive testing. Regarding the second problem above, the precoded question resulted in significantly higher levels of multiple reporting, especially for respondents in the CATI mode. Item nonresponse to the individual fields of study was not analyzed, so it was impossible to determine whether the failure to mark the *no* boxes was replicated in the field. However, the respondent preference for the write-in question version translated into more reliable data in this version, based on the results of a content reinterview. People answering the categorical version overreported the number of categories, and in the reinterview they did not replicate the same categories they reported in the original interview. As a result, the write-in version was added to the ACS (see Rothgeb and Beck, 2007; Raglin et al., 2008).

Similarly, some of the experimental testing conducted within QDERS involved the combination of methods, as this emerged from cognitive research projects conducted by survey methodologists. The time constraints of questionnaire pretesting in federal surveys do not always permit the luxury of conducting experiments to validate the results of cognitive interviews or try out new ideas to solve problems. However, the off-line nature of the QDERS research survey facilitates follow-up of research results or experimentation with new hypotheses to see if the hypotheses formulated during the cognitive testing hold up under controlled experimental conditions. An example of the importance of this type of testing occurred with the 2010 Census Coverage Measurement program, which involves several surveys and is designed to determine where people were living on Census Day. The Person Interview, conducted 5–6 months after the census, is an independent measure to identify where the people at a sample of addresses were living on Census Day and obtains information about other places they might have been counted. These other places are an important factor in today's increasingly mobile society, where some people commute between cities to work, some people leave their "home" to spend winters in warmer climates, and children in joint custody travel back and forth between parents, among others. When a person has another place to live or stay, the residence-rule questions for the Person Interview are defined in terms of "cycles," that is, time periods where a person may stay at another place (e.g., 3 days a week, 5 months a year, every other week). This strategy was used in Census 2000 and previous censuses, but cognitive interviewing has shown that questions formulated on those rules do not work well for people who have three or more residences or for people who do not have a regular pattern of stay.

When there is some potential disagreement between the census and the Person Interview in terms of where a person should be counted, the Person Followup interview is conducted about 10 months after the census. For the

2006 Census Test, researchers suggested a different approach—the "dates" approach—for the Person Followup as a way to improve upon the data collected using the "cycles" approach. This approach obtained the dates during the year when the person was at a different address. Researchers speculated that this approach would work for people with two or more addresses and for people with and without a consistent pattern of stay. The results of cognitive testing of the "dates" approach suggested that this would in fact be the case. This approach was incorporated into the 2006 Person Followup interview for the 2006 Census Test, but before that, the "dates" and "cycle" approaches were tested against each other in the 2006 QDERS. The results (Childs et al., 2007; Nichols et al., 2008) revealed that the dates approach worked well and corroborated the decision to use it in the Census Test.

These examples of cognitive interviewing results being used to inform the design of experiments demonstrate how the mixed-method approach is superior to either method used alone, because it offers a multidimensional understanding of construct validity. The qualitative findings tell us why and how the questions are different, while the quantitative findings show to what extent they are different and whether this difference is statistically significant. Ultimately, then, this mixed-method approach allows us to understand what survey questions are actually measuring, and then make better decisions about which questions to field on our surveys. Along with Jon Krosnick, we believe this mixed-method approach should be used more frequently than it is.

NOTE

This survey, called the Questionnaire Design Experimental Research Survey or QDERS, was conducted every 18 months or so from 1999 to 2004.

REFERENCES

Beatty P, Cosenza C, Fowler FJ (2006). Experiments on the structure and specificity of complex survey questions. American Statistical Association 2006 Proceedings of the Section on Survey Research Methods. Washington, DC: American Statistical Association.

Brault S, Stern S, Raglin D (2007). Evaluation covering disability. 2006 American Community Survey Content Test Report P.4. January 3, 2007.

Childs JH, Nichols E, Dajani A, Rothgeb J (2007). A new approach to measuring residence status. American Statistical Association 2007 Proceedings of the Section on Survey Research Methods. Washington, DC: American Statistical Association.

DeVellis R (2003). Scale Development: Theory and Applications, Applied Social Research Methods Series, Vol. 26, 2nd ed. Thousand Oaks, CA: Sage.

Fowler FJ (2004). The case for more split-sample experiments in developing survey instruments. In: Presser S, Rothgeb JM, Couper MP, Lessler JT, Martin E, Martin J,

Singer E, editors. Methods for Testing and Evaluating Survey Questionnaires. New York: Wiley; pp. 173–188.

Hess J, Moore J, Pascale J, Rothgeb J, Keeley C (2001). Person- vs. household-level questionnaire design. Public Opinion Quarterly; 65:574–584.

Hess J, Moore J, Pascale J, Rothgeb J, Keeley C (2002). The effects of person-level vs. household-level questionnaire design on survey estimates and survey quality. Statistical Research Division Research Report #2002-05. March 11, 2002.

Holder K, Raglin D (2007). Evaluation report covering employment status. 2006 American Community Survey Content Test Report P.6.a. January 3, 2007.

Nichols E, Childs JH, Linse K (2008). RDD versus site test: mode effects on gathering a household roster and alternate addresses. American Statistical Association 2006 Proceedings of the Section on Survey Research Methods. Washington, DC: American Statistical Association.

Pascale J (2008/2009). Assessing measurement error in health insurance reporting: a qualitative study of the current population survey. Inquiry Journal; 45:422–437.

Pascale J (2009). Findings from a pretest of a new approach to measuring health insurance in the current population survey. Proceedings of the Federal Committee on Statistical Methodology Research Conference. 2–4 Nov 2009.

Pascale J, Roemer M, Resnick D (2009). Medicaid underreporting in the CPS: results from a record check study. Public Opinion Quarterly; 73:497–520.

Raglin D, Zelenak MF, Davis M, Tancreto J (2008). Testing a new field of degree question for the American Community Survey. 2007 American Community Survey Content Test Report. May 2008.

Rothgeb J, Beck J (2007). Final report: cognitive interview research results and recommendations for the National Science Foundation's proposed field of degree question for the American Community Survey. Statistical Research Division Research Report #2007-15. June 7, 2007.

U.S. Census Bureau (1993). Content reinterview survey: accuracy of data for selected population and housing characteristics as measured by reinterview, 1990 Census of Population and Housing research and evaluation reports, 1990-CPH-E-1. Available online at http://www.census.gov/prod/cen1990/cph-e/cph-e-1.pdf.

Willis G, Schechter S (1997). Evaluation of cognitive interviewing techniques: do the results generalize to the field? Bulletin de Sociologie Methodologique; 55:40–66.

PART VI
Multitrait-Multimethod Experiments

17 Evaluating the Reliability and Validity of Survey Interview Data Using the MTMM Approach[1]

DUANE F. ALWIN[2]

Pennsylvania State University

In order to properly assess reliability [validity] one needs, first, a model that specifies the linkage between true and observed variables, second, a research design that permits the estimation of parameters of such a model, and, third, an interpretation of these parameters that is consistent with the concept of reliability (validity). (Alwin and Krosnick 1991, p. 141)

17.1 INTRODUCTION

This chapter summarizes an approach that I and others have applied to using structural equation model (SEM) techniques for modeling measures from *multitrait-multimethod* (MTMM) designs (e.g., Alwin, 1974, 1989, 1997, 2007; Alwin and Jackson, 1979; Andrews, 1984; Saris and van Meurs, 1990; Saris and Andrews, 1991; Scherpenzeel, 1995; Scherpenzeel and Saris, 1997; Saris and Gallhofer, 2007). This chapter will not focus a great deal of attention on the details of the MTMM approach per se, as these are covered adequately in the literature on this subject. Instead I prefer to discuss the origins of the approach, its original purpose, and the path-breaking work of Frank Andrews, Willem Saris, and others, who have applied these techniques. Most of my attention in this chapter will be on clarifying the meaning of some important concepts—for example, distinguishing what I call "MTMM validity" from other concepts of validity, and discussing the relationship of both concepts of validity to the

Question Evaluation Methods: Contributing to the Science of Data Quality, First Edition.
Edited by Jennifer Madans, Kristen Miller, Aaron Maitland, Gordon Willis.
© 2011 John Wiley & Sons, Inc. Published 2011 by John Wiley & Sons, Inc.

concept of reliability. I also discuss the generic form of the MTMM as an extension of *classical true-score theory* (CTST) in order to clarify its assumptions, and I evaluate the merits of the typical survey designs applied to estimate components of variance. The latter topic leads me to evaluate the *active role of memory* in the production of survey responses, and I present some relevant data on that subject. Finally, I discuss some alternatives to current MTMM practices.

17.2   BACKGROUND

In a recent research monograph (Alwin, 2007) I investigated two alternative (but compatible) approaches that have been proposed to address some of the limitations of traditional methods of estimating the quality of survey data (e.g., Cronbach, 1951; Greene and Carmines, 1979; Marquis and Marquis, 1977): the MTMM/confirmatory factor analysis approach using cross-sectional survey designs and the *quasi-simplex* approach using longitudinal data. Both of these approaches involve a kind of "repeated measuring" that is increasingly viewed as a requirement for the estimation of the reliability of survey measures (Coleman, 1964, 1968; Goldstein, 1995, p. 142). What the methods share in common is their link to the traditional assumptions of CTST, in that both make the *assumption of the independence of errors* that is critical for the estimation of reliability and validity.

In that book I reviewed the advantages and disadvantages of these two approaches, particularly with respect to meeting the assumption of the independence of errors. I have employed both approaches in my own research and consider both to be valuable. I argued that each needs to be considered within the context of its limitations, and the estimates of parameters presumed to reflect reliability and validity need to be interpreted accordingly. The MTMM design relies on the use of *multiple indicators* measured within the same interview, using different methods or different types of questions for a given concept.[3] This design represents an adaptation of the "MTMM" design that was first proposed by Campbell and Fiske (1959) as an approach to establishing convergent and discriminant evidence for "construct validity" in psychology and educational research—an approach I review in the next section.

The MTMM approach acknowledges the fact that it is difficult, if not impossible, to obtain repeated or *replicate* measures of the same question—what I have elsewhere referred to as *multiple measures*—in the typical survey interview (Alwin, 2007). Rather, the basic MTMM design employs *multiple indicators* that systematically vary the method of measurement across different concepts.[4] As applied to multiple survey questions measuring the same concept, the approach starts with the use of confirmatory factor analysis, implemented using SEMs, to estimate three sources of variance for each survey question: that attributable to an underlying concept or "trait" factor, that attributable to a "method" factor, and the unsystematic or random residual. The residual is assumed to reflect random measurement error. The strategy is then to conduct meta-analyses of these estimates to look for characteristics of questions and of respondents that are related to the proportion of variance that is reliable and/or valid.

A second approach to estimating reliability uses *multiple measures* (as distinct from *multiple indicators*) in multiple-wave survey reinterview (i.e., test–retest) or panel designs. Such longitudinal designs also have problems for the purpose of estimating reliability. For example, the test–retest approach using a single reinterview must assume that there is no change in the underlying quantity being measured. With two waves of a panel study, the assumption of no change, or even perfect correlational stability, is unrealistic, and without this assumption little purchase can be made on the question of reliability in designs involving two waves. The analysis of panel data must be able to cope with the fact that people change over time, so that models for estimating reliability must take the potential for individual-level change into account. Given these requirements, techniques have been developed for estimating measurement reliability in panel designs where $P \geq 3$ in which change in the latent true score is incorporated into the model. With this *multiple measures* approach, there is no need to rely on multiple indicators within a particular wave or cross section in order to estimate the measurement reliability. This approach is possible using modern SEMs for longitudinal data, and is discussed at length in Alwin (2007).

17.3 THE MTMM MATRIX[5]

In 1959, psychologists Donald Campbell and Donald Fiske published what would become a classic paper on the topic of *construct validity* in *Psychological Bulletin* (Campbell and Fiske, 1959). Their purpose was to develop a strategy that would help establish the validity of measurement of psychological constructs. They reasoned that not only was it important to establish the *convergent validity* of measures (i.e., the agreement among diverse measures of the same thing), it was also essential to establish their *discriminant validity* as well. In other words, it was important to show that measures *do not* correlate with other measures they *should not* correlate with. The idea they developed was that measures assessed using the same methods may correlate precisely because they are measured using the same methods and that *method variance*, that is, method factors, may contribute to their correlation.

The general MTMM matrix involves the measurement of K traits by each of Q methods of measurement. The correlation matrix in Table 17.1 presents the general form of the MTMM matrix for K = 3 (traits X, Y, and Z) and Q = 3 (methods 1, 2, and 3). The correlations among traits all of which are measured by the same method are included in the *monomethod blocks*—in Table 17.1 there are three such monomethod blocks. There are two types of entries in the typical monomethod block: the monotrait-monomethod (MoTMM) values

TABLE 17.1. Multitrait-Multimethod Matrix for Three Traits and Three Methods

	Trait	Method 1			Method 2			Method 3		
		X	Y	Z	X	Y	Z	X	Y	Z
1	X	$r_{X_1 X_1}$								
	Y	$r_{X_1 Y_1}$	$r_{Y_1 Y_1}$							
	Z	$r_{X_1 Z_1}$	$r_{Y_1 Z_1}$	$r_{Z_1 Z_1}$						
2	X	$\mathbf{r_{X_1 X_2}}$	$r_{X_2 Y_1}$	$r_{X_2 Z_1}$	Monomethod Blocks					
	Y	$r_{X_1 Y_2}$	$\mathbf{r_{Y_1 Y_2}}$	$r_{Y_2 Z_1}$						
	Z	$r_{X_1 Z_2}$	$r_{Y_1 Z_2}$	$\mathbf{r_{Z_1 Z_2}}$						
3	X	Heteromethod Blocks			$\mathbf{r_{X_3 X_2}}$	$r_{X_3 Y_2}$	$r_{X_3 Z_2}$	$r_{X_3 X_3}$		
	Y				$r_{X_2 Y_3}$	$\mathbf{r_{Y_2 Y_3}}$	$r_{Y_3 Y_2}$	$r_{X_3 Y_3}$	$r_{Y_3 Y_3}$	
	Z				$r_{X_2 Z_3}$	$r_{Y_2 Z_2}$	$\mathbf{r_{Z_2 Z_3}}$	$r_{X_3 Z_3}$	$r_{Y_3 Z_3}$	$r_{Z_3 Z_3}$

Note: Values in validity diagonals (MTHM) are in boldface type.

Source: Alwin (1974)

and the heterotrait-monomethod (HTMM) values. The monotrait values refer to the reliabilities of the measured variables—a topic to which we return below—and the heterotrait values are the correlations among the different traits within a given method of measurement. There are .5K(K-1) HTMM values in the lower triangle of each monomethod block.

Correlations among trait measures assessed by different methods comprise the *heteromethod blocks*, which also contain two types of entries: the monotrait-heteromethod (MTHM) values and the heterotrait-heteromethod (HTHM) values. The MTHM values are also referred to as *validity values* because each is a correlation between two presumably different attempts to measure a given variable. The HTHM values are correlations between different methods of measuring different traits. In Table 17.1 there are .5Q(Q-1) heteromethod blocks, which are symmetric submatrices, each containing K MTHM values in the diagonal.

Given the MTMM matrix in Table 17.1 and the terminology outlined here, Campbell and Fiske (1959, pp. 82–83) advance the following criteria for *convergent* and *discriminant* validity. First, "the entries in the validity diagonal should be significantly different from zero and sufficiently large to encourage further examination of validity." That is, if two assessments of the same trait employing different methods of measurement do not converge, then there is probably little value in pursuing the issue of validity further. Second, "a validity diagonal value (MTHM values) should be higher than the values lying in its column and row in the heterotrait-heteromethod triangles." In other words, a "validity value" for a given trait should be higher than the correlations between that measure and any other measure having neither trait nor method in common. Third, a measure should "correlate higher with an independent effort to measure the same trait than with measures designed to get at different traits which happen to employ the same method." Thus, if different traits measured by the same method correlate more highly with one another than any one of them correlates with independent efforts to measure the same thing, then there is a clear domination of method variation in producing correlations among measures. Finally, "a fourth desideratum is that the same pattern of trait interrelationship be shown in all of the heterotrait triangles of both the monomethod and heteromethod blocks."

The Campbell and Fiske (1959) criteria for establishing the construct validity of measures have become standards in psychological and educational research (see AERA/APA/NCME, 1999). Indeed, it was the failure of conventional approaches to validation in the psychometric literature, which emphasized primarily convergence criteria (e.g., Cronbach and Meehl, 1955), that stimulated the efforts of Campbell and Fiske to propose a broader set of criteria. They viewed the traditional form of validity assessment—convergent validity—as only preliminary and insufficient evidence that two measures were indeed measuring the same trait. As operational techniques these criteria are, however, very difficult to implement and are not widely employed.

While most investigators agree that there is a commonsense logic to these criteria, some fundamental problems exist with their application. An alternative approach that developed in the early 1970s (see Werts and Linn, 1970; Jöreskog, 1971, 1974; Alwin, 1974; Browne, 1984)—formulated originally within a path-analytic framework—was to represent the MTMM measurement design as a common factor model that could be estimated using confirmatory factor analysis (or what we today refer to as SEM). Later on, the technique was found to be useful by survey methodologists interested in partitioning the variance in survey measures collected within an MTMM design into components of variance (e.g., Andrews, 1984; Scherpenzeel, 1995; Alwin, 1997, 2007; Scherpenzeel and Saris, 1997). I review the details of this approach, its advantages and limitations, later in the chapter, but first I discuss how the MTMM matrix approaches the question of validity of measurement. Before I turn to this approach, I want to make sure that certain concepts from the original MTMM matrix introduced by Campbell and Fiske (1959) are clarified. In the following two sections, I discuss (1) the meaning of the term "construct validity" and how the concept of "MTMM validity" or "trait validity" differs from more standard meanings of the term, and (2) a clarification of how the MTMM approaches to decomposing variance in measures relates to the approach of CTST.

17.4   THE MEANING OF CONSTRUCT VALIDITY

According to Campbell and Fiske (1959), both concepts of reliability and validity require that agreement between measurements be obtained—*validity* is supported when there is correspondence between two efforts to measure the same trait through *maximally different* methods; *reliability* is demonstrated when there is agreement between two efforts to assess the same thing using *maximally similar*, or replicate, measures (Campbell and Fiske, 1959, p. 83). In order to appreciate the importance of this distinction, it is necessary for our purposes to distinguish between *observed* (or manifest) variables and *unobserved* (or latent) variables and to conceptualize the linkage between the two levels of discourse. This distinction between latent and manifest variables allows us to conceptualize the concepts of reliability and validity in terms that are familiar to most researchers. For my purposes, it is important to think of a response to a survey question to be an "observed" variable that reflects some "latent" variable or construct of interest, and our interest is in conceptualizing the relationship between the two. If we define the response variable as a random variable (Y), observed in some population of interest, that has a latent counterpart (T), which is also a random variable in that population, and which in part is responsible for the production of the response, that is $T \rightarrow Y$, we can begin to conceptualize the concepts of reliability and validity.

As defined, the observed variable in this case is literally what we can think of as a "survey response," and the "latent" or "unobserved" variable is what the survey question is intended to measure. We may think of T as a part of (or a component of) Y. There is no necessary identity between the two, however; as we know, there are a number of types of "survey measurement errors," which we can define globally using the notation E, that are also a part of Y. In this sense, we can conceptualize Y as having two major parts that contribute to the survey response—what we are attempting to measure and our inability to do so, or errors of measurement, that is $T \rightarrow Y \leftarrow E$. To this point, other than the population reference, that is, that these are random variables in some population of interest, we have no constraints on our theorizing about the properties and relationships of these three processes, Y, T, and E.

Note that while Y is observed, both T and E are unobserved, or latent, variables. This type of formulation allows us to think in terms of two general considerations when examining the variation observed in survey responses; one is the variation in the underlying phenomenon one is trying to measure, for example, income or employment status, and the other is the variation contributed by "errors of measurement." The consideration of each of these parts—the part concerning the construct one is trying to measure, as well as the part concerning survey measurement errors—is a worthy effort. This is an

FIGURE 17.1. Path diagram of the classical true-score model for a single measure. Source: Alwin (2007).

effort that many people have joined, and sorting out the nature of these contributions of error has been defined as an important area of research (e.g., Groves, 1989; Biemer et al., 1991; Lyberg et al., 1997; Alwin, 2007; Saris and Gallhofer, 2007).

It is important to note here that our formulation is intended to be very general, allowing for a range of different types of Y and T random variables: (1) Y and T variables that represent discrete categories, as well as (2) Y and T variables that are continuous variables. Specifically, if Y (observed) and T (latent) are continuous random variables in a specified population of interest, for example, and if it is assumed that E is random with respect to the latent variable, that is, T and E are uncorrelated determinants of Y, we have the case that is represented by CTST, where the T latent variable is referred to as a true score, and measurement error is simply defined as the difference between the true score and the observed score, that is $E = Y - T$ (see Alwin, 2007). In the CTST tradition, a model for the gth measure of Y is defined as $Y_g = T_g + E_g$, and the variance of Y_g is written as $VAR[Y_g] = VAR[T_g] + VAR[E_g]$. Note that here we assume that $COV[T_g, E_g] = 0$, or as formulated here, in the population model the true- and error-scores are independent.

In general, there is no constraint on the covariance properties of E and T, although as noted in the CTST case, E is considered to be random, that is, uncorrelated with T, as illustrated in the path diagram in Figure 17.1. This assumption underlies the development of all psychometric approaches to estimating reliability.[6] Although the CTST tradition is quite valuable in many situations, it has certain limitations, especially when it comes to assessing the reliability and validity of categorical data. For this reason, we insist on a more general conceptualization, which permits T to be either a set of latent nominal classes, a set of latent ordered classes, or a latent continuous variable, *and* a conceptualization that permits Y to also be a nominal, ordinal, or interval response variable (see Alwin, 2007, pp. 128–130).

17.4.1   Reliability and Validity

On the simplest level the concept of reliability is defined in terms of the consistency of measurement. Consider a hypothetical *thought experiment* in which a measure of some quantity of interest (Y) is observed—it could be a child's height, it could be the pressure in a bicycle tire, it could be a question inquiring about family income in a household survey, or it could be a question assessing an attitude or belief about "banks" in a similar survey. Then imagine repeating the experiment, taking a second measure of Y, under the assumption that nothing has changed, that is, neither the measurement device nor the quantity being measured has changed, and under the additional assumption that there is *no memory* of the prior measurement. If across these two replications one obtains consistent results, we say the measure of Y is reliable, and if the results are inconsistent, we say the measure is unreliable. Of course reliability is not an "either-or" variable, and ultimately we seek to quantify the degree of consistency or reliability in social measurement (see Alwin, 2007, pp. 36–41).

I should note at this point that, although some will mistake it for validity, when traditional psychometric approaches refer to the reliability of measurement *on a conceptual level*, they refer not to the consistency across measures, but to the closeness of the correlation between Y and T. Thus, when one is dealing with measurements that are reliable, it is assumed that the measurement error part of Y is minimal, and in a "variance accounting" framework, mostly due to latent "true" variation (as we shall see below, these two ideas about reliability converge when we try to estimate this correlation between Y and T). By contrast, when measures are unreliable, it is because there is a great deal more "error of measurement" than "true variation," that is, much more "noise" than "signal," to use a metaphor from radio telegraphy.

It turns out that the best estimate of this correlation between Y and T is the correlation between two identical measures of T, say Y_1 and Y_2, and hence we tend to think of reliability in terms of the way we assess it, namely as the consistency in two identical efforts to measure the same thing—recall the observation made by Campbell and Fiske (1959, p. 83) that evidence for reliability comes from agreement between *maximally similar* measures. The path diagram in Figure 17.2 illustrates the relationship between two measures, both measuring the same true score. In this case, the way we estimate the reliability of either measure is derived from the consistency (i.e., the covariance) between them (see Alwin, 2007, pp.40–41). Before developing these ideas further, however, it is important to add more complexity, in order to eventually incorporate the concept of reliability into the discussion of validity. Indeed, in addition to reliability, and perhaps of greater interest, is the concept of validity—the extent to which the investigator is measuring the theoretical construct of interest. In a more general sense, measurement validity refers to the extent to which measurement accomplishes the purpose for which it is intended, that is, are you measuring what you think you are measuring. Validity is a somewhat more elusive concept than reliability, given the absence of adequate criteria for its

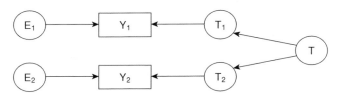

FIGURE 17.2.   Path diagram of the classical true-score model for two tau-equivalent measures. Source: Alwin (2007).

evaluation (see AERA/APA/NCME, 1999). Although matters of validity are often thought to be capable of solutions on an abstract level, as with the notion of "content" or "face" validity, on an empirical level, the issue of validity can only be defined with respect to some criterion that is related to the purpose of measurement. When the validity of measurement is at issue, this is often assessed with respect to its correlation with criterion variables, often outside the particular survey, which are linked to the purpose of measurement. In addition, validity is often defined at the level of the construct, that is, the relationship between T (or the latent variable being measured) and the construct of interest, C, that is, C → T. Of course, without a direct measure of this "construct of interest," one often must fall back on notions of criterion validity.

Normally, we think of measurement error as being more complex than the random error model developed on the basis of CTST. Some people will argue, justifiably, that not all measurement error in surveys is random. This, of course, does not invalidate the idea that there is random or unsystematic error; but it does push those developing models of error to include systematic components as well. The relationship between random and systematic errors can be clarified if we consider an extension of the classical true-score model, where the random variable M_g is a source of systematic error in the observed score, as shown in Figure 17.3. This model directly relates to the one given above, where $Y_g = T_g + E_g$ for the gth measure of Y, and where $T_g = T_g^* + M_g$. The idea, then, is that the variable portion of measurement error contains two types of components, a random component and a nonrandom, or systematic component. Within the framework of this model, the goal would be to partition the variance in Y_g into those portions due to T_g^*, M_g, and E_g, that is $VAR[Y_g] = VAR[T_g^*] + VAR[M_g] + VAR[E_g]$. Note that here we assume that $COV[T_g^*, Mg] = 0$, or as formulated here, in the population model the "trait" and "method" components of the true score are uncorrelated. It is frequently the case that systematic sources of error increase estimates of reliability. This is, of course, a major threat to the usefulness of CTST in assessing the quality of measurement (see Alwin, 2007, pp. 41–42, 80–82). It is important to address the question of systematic measurement errors, but that often requires a more complicated measurement design. As we shall see, this can be implemented using an MTMM measurement design along with confirmatory factor analysis, the method proposed by Andrews (1984) and his colleagues, a topic to which I return in a subsequent section.

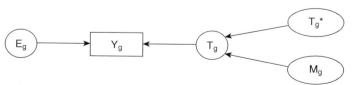

FIGURE 17.3.   Path diagram of the classical true-score model for a single measure composed of one trait and one method. Source: Alwin (2007).

Finally, it is important to understand the relationship between reliability and validity. Among social scientists, almost everyone is aware that without valid measurement, the contribution of the research may be more readily called into question. At the same time, most everyone is also aware that the *reliability of measurement* (as distinct from validity) is a necessary, although not sufficient, condition for valid measurement. In the CTST tradition, the *validity* of a measurement Y is expressed as its degree of relationship to a second measurement, some *criterion* of interest (Lord and Novick, 1968, pp. 61–62). It can be shown that the *criterion validity* of a measure—its correlation with another variable—cannot exceed the reliability of either variable (Lord and Novick, 1968; Alwin, 2007). There is a mathematical proof for this assertion (Alwin, 2007, pp. 291–292), but the logic that underlies the idea is normally accepted without formal proof: If our measures are unreliable, they cannot be trusted to detect patterns and relationships among variables of interest.

To summarize, for our present purposes we take reliability to refer to the relationship between Y and T (as stated above), whereas validity has to do with the relationship between T and C, that is, between the latent variable being measured and the theoretical construct of interest. When one has a "record of the variable" or a "gold standard" for this theoretical construct, then one can examine the relationship between T and C. Such designs are relatively rare, but when available, they can be very useful (see, e.g., Alwin, 2009). Both measures of "gold standards" and of "alternative methods" contain measurement error, so such correlations or measures of consistency must always be interpreted in light of reliability of measurement in both variables.

Establishing the validity of survey measurement is difficult because within a given survey instrument, there is typically little available information that would establish a criterion for validation. However, there is a well-known genre of research situated under the heading "record check studies," which involves the comparison of survey data with information obtained from other record sources. Although rare, such studies can shed light on survey measurement errors; however, as noted earlier, correlations among multiple sources of information are limited by the level of reliability involved in reports from either source (Alwin, 2007, pp. 48–49). As noted earlier, in the psychometric tradition, the concept of validity has mainly to do with do with the utility of measures with respect to getting at particular theoretical concepts. This concept of validity is difficult to adapt to the case of survey measures, given the absence of well-defined criteria representing the theoretical construct, but several efforts have been made. One important design that has been used for studying validity involves record check studies, where the investigator compares survey reports with a set of record data that may exist for the variables in question in a given population (e.g., Marquis, 1978; Traugott and Katosh, 1979, 1981; Bound et al., 1990). As we note below, the MTMM design represents another important design for getting at validity within a single cross-sectional survey, but the MTMM approach substitutes the notion of "trait validity" for "construct validity," and typically does not follow the dictum of Campbell and Fiske

(1959, p. 83) that validity requires the use of "maximally different" methods of measurement.

17.5 COMPONENTS OF VARIANCE

The purpose of designs aimed at the estimation of reliability of measurement is to decompose the variance of a set of observed variables into components that reflect *true-score* variance and *error-score* variance. The classical approach to estimating the components of response variance requires *multiple* or *replicate measures* within respondents. Two general strategies exist for estimating the reliability of single survey questions: (1) using the same or similar measures in the same interview, or (2) using replicate measures in reinterview designs. The application of either design strategy poses problems, and in some cases the estimation procedures used require assumptions that are inappropriate (see Alwin, 1989; Saris and Andrews, 1991). Estimating reliability from information collected within the same interview is especially difficult, owing to the virtual impossibility of replicating questions.

17.5.1 Multiple Measures versus Multiple Indicators

While the above model is basic to most methods of reliability estimation, including *internal consistency* estimates of reliability (e.g., Cronbach, 1951; Heise and Bohrnstedt, 1970; Greene and Carmines, 1979; Marquis and Marquis, 1977), the properties of the model very often do not hold. This is especially the case in *multiple indicators* (i.e., observed variables that are similar measures within the same domain, but not so similar as to be considered replicate measures) as opposed to *multiple measures*. Before discussing the MTMM approaches in greater depth, however, I first review the distinction between models involving *multiple measures* and those involving *multiple indicators* in order to fully appreciate their utility. An appreciation of the distinction between *multiple measures* and *multiple indicators* is critical to an understanding of the difficulties of designing survey measures that satisfy the rigorous requirements of the MTMM extension of the CTST model for estimating reliability. Following this discussion, I then talk about the value of the common factor model for understanding the nature of the MTMM approach and its relation to the CTST model.

Let us first consider the hypothetical, but unlikely, case of using replicate measures within the same survey interview (but see the discussion of the MTMM approach below). Imagine the situation, for example, where three measures of the variable of interest are obtained, for which the random measurement error model holds in each case, as follows:

$$Y_1 = T_1 + E_1$$

$$Y_2 = T_2 + E_2$$

$$Y_3 = T_3 + E_3$$

In this hypothetical case, if it is possible to assume one of the measures could serve as a record or *criterion of validity* for the other, it would be possible to assess the validity of measurement, but we do not place this constraint on the present example. All we assume is that the three measures are replicates in the sense that they are virtually identical measures of the same thing. This model assumes that each measure has its own true score, and it is assumed that the errors of measurement are not only uncorrelated with their respective true scores, they are also independent of one another. This is called the assumption of *measurement independence* referred to earlier (Lord and Novick, 1968, p. 44).

It is important to realize in contemplating this model that it does not apply to the case of multiple indicators. The *congeneric* measurement model (or the special cases of the *tau-equivalent* and *parallel* measures models) is consistent with the causal diagram in Figure 17.4a. This picture embodies the most basic assumption of CTST that measures are *univocal*; that is, that they each measure one and only one thing which completely accounts for their covariation and which, along with measurement error, completely accounts for the response variation. This approach requires essentially asking the same question multiple times, which is why we refer to it as involving *multiple* or *replicate measures* (we consider the multiple indicators approach below). We should point out, however, that it is extremely rare to have multiple measures of the same variable in a given survey. It is much more common to have measures for *multiple indicators*, that is, questions that ask about somewhat different aspects of the same things.

The *multiple indicator* model derives from the traditional *common factor model*, a psychometric tradition that predates CTST. In such models, the K latent variables are called common factors because they represent common sources of variation in the observed variables. The common factors of this model are responsible for covariation among the variables. The unique parts of the variables, by contrast, contribute to the lack of covariation among the variables. Covariation among the variables is greater when they measure the same factors, whereas covariation is less when the unique parts of the variables dominate. Indeed, this is the essence of the common factor model—variables correlate because they measure the same thing(s).

The common factor model, however, draws attention not only to the common sources of variance, but to the unique parts as well. A variable's *uniqueness* is the complement to the common parts of the data (the *communality*) and is thought of as being composed of two independent parts, one representing *specific variation* and one representing *random measurement error* variation. Specific variance is *reliable* variance, and thus the reliability of the variable is not only due to the common variance, but to specific variance as well. Unfortunately, because specific variance is thought to be independent (uncorrelated with) sources of common variance, it becomes confounded with measurement error variance. Because of the presence of specific variance in most measures, it is virtually impossible to use the traditional form of the common factor model as a basis for reliability estimation (see Alwin and

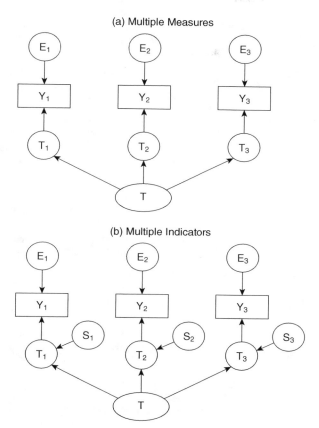

FIGURE 17.4. Path diagrams for the relationship between random measurement errors, observed scores, and true scores for the multiple measures and multiple indicator models. Source: Alwin (2007).

Jackson, 1979; Alwin, 1989), although this is precisely the approach advocated in the work of Heise and Bohrnstedt (1970), and followed by the MTMM/confirmatory factor analytic approach to reliability estimation we discuss below (see Andrews, 1984). This model is illustrated in Figure 17.3b.

The problem here—which applies to the case of multiple indicators—is that the common factor model typically does not permit the partitioning of the variance of u_j into its components, variance due to the specific component, s_j, and the measurement error component, e_j. In the absence of specific variance (what we here refer to as the "multiple measures" model), classical reliability models may be viewed as a special case of the common factor model, but in general it is risky to assume that the disturbance variance is equivalent to the measurement error variance (Alwin and Jackson, 1979). The model assumes that while the measures may tap (at least) one common factor, it is also the case that there is a specific source of variation in each of the measures.

A second issue that arises in the application of multiple indicators is that there may be correlated errors, or common sources of variance, masquerading as true scores. Imagine that respondents' use of a common agree-disagree scale is influenced by a "method" factor, for example the tendency to agree versus disagree (called "yea-saying" or "acquiescence" in the survey methods literature). In this case there are two common factors at work in producing T_g. In such a case of *nonrandom measurement error*, we can formulate the model as follows: $Y_g = T_g + E_g$, where $T_g = T_g^* + M_g$. This indicates that the covariance among measures, and therefore the estimates of reliability are inflated due to the operation of common method factors. This was illustrated in the path diagram in Figure 17.3 presented earlier.

The *multiple measures* model can, thus, be thought of as a special case of the *multiple indicators* model in which the latent true scores are linear combinations of one another, that is perfectly correlated. Unfortunately, unless one knows that the multiple measures model is the correct model, interpretations of the "error" variances as solely due to measurement error is inappropriate. In the more general case (the multiple indicators case) one needs to posit a residual, referred to as specific variance in the factor analytic tradition, as a component of a given true score to account for its failure to correlate perfectly with other true scores aimed at measuring the same construct. Within a single cross-sectional survey, there is no way to distinguish between the two versions of the model; that is, there is no available test to detect whether the congeneric model fits a set of G variables, or whether the common factor model is the more appropriate model.

To summarize our discussion up to this point, it is clear that two problems arise in the application of the CTST model to cross-sectional survey data. The first is that specific variance, while reliable variance, is allocated to the random error term in the model, and consequently, to the extent specific variance exists in the measures, the reliability of the measure is underestimated. This problem could be avoided if there were a way to determine whether the congeneric (multiple measures) model is the appropriate model, but as we have noted, within a single cross-sectional survey, there is no way to do this. The second problem involves the assumption of a single common factor, which is a problem with either version of the model shown in Figure 17.3. In this case the problem involves the presence of common method variance, which tends to inflate estimates of reliability. The latter problem is one that is also true of the multiple measures model in that it is just as susceptible to the multiple factor problem as the multiple indicator model. The problem is actually more general than this, as it involves any source of multiple common factors, not simply common method factors.

17.5.2   The MTMM Model as an Extension of CTST

There is an increasing amount of support for the view that shared method variance of the type discussed by Campbell and Fiske (1959) (and above) inflates estimates of reliability, representing one of the limitations of traditional *internal consistency* approaches to estimating reliability (see Alwin, 2007, chapter 3). One approach to dealing with this is to reformulate the CTST along the lines of a multiple common-factor approach and to include sources of systematic variation from *both* trait variables and method factors. With *multiple indicators* of the same concept, as well as different concepts measured by the same method, it is possible to formulate an MTMM common-factor model. The idea is basically to specify a critical component of the unique variance—namely variance associated with the method of measurement—and create a design that allows one to isolate and estimate its effects. In general, the measurement of K traits measured by each of Q methods (generating G = KQ observed variables) allows the specification of such a model.

17.6   RECENT RESEARCH USING THE MTMM APPROACH

In the 1970s the MTMM approach was reformulated as a confirmatory factor model in which multiple common factors representing trait and method components of variation were employed (see Werts and Linn, 1970; Alwin, 1974; Jöreskog, 1974, 1978; Browne, 1984). Over the past few decades, researchers have done extensive analyses of the quality of survey measures using this approach, one that has relied primarily, although not exclusively, on the analysis of cross-sectional data involving models that incorporate method variation in repeated measurement within the same survey (Alwin and Jackson, 1979; Andrews, 1984; Andrews and Herzog, 1986; Rodgers et al., 1988; Alwin, 1989, 1997; Saris and van Meurs, 1990; Rodgers et al., 1992; Saris and Andrews, 1991; Scherpenzeel, 1995; Scherpenzeel and Saris, 1997). Andrews (1984) pioneered the application of the MTMM measurement design to large-scale survey data in order to tease out the effects of method factors on survey responses. His research suggested that substantial amounts of variation in measurement quality could be explained by seven survey design characteristics, here listed in order of their importance in his meta-analysis of effects on data quality:

- The number of response categories.
- Explicit offering of a "Don't Know" option.
- Battery length.
- Absolute versus comparative perspective of the question.
- Length of the introduction and of the question.
- Position of the question in the questionnaire.
- The full labeling of response categories.

Andrews' (1984, p. 432) work suggested that several other design features were insignificant in their contributions to measurement error. He indicated that the mode of administering the questionnaire, for example telephone, face-to-face, or group-administered, was unimportant, stressing the encouraging nature of this result. He noted also that the use of explicit midpoints for rating scales, whether the respondent was asked about things they had already experienced or about predictive judgments with respect to the future, or the substantive topic being asked about in the survey question had only slight effects on data quality (Andrews, 1984, pp. 432–433).

Andrews' (1984) innovative application of the MTMM approach—based upon Heise and Bohrnstedt's (1970) factor analytic assumptions—to the evaluation of data quality in surveys has been extended by a number of researchers. Saris and van Meurs (1990) presented a series of papers from a conference on the evaluation of measurement instruments by meta-analysis of MTMM studies. These papers (including Andrews' original paper) addressed several topics of interest to researchers seeking to implement this research strategy. Saris and Andrews (1991) review many of these same issues, and Groves (1991) summarizes the applicability of the model to assessing nonrandom errors in surveys. In addition, several other researchers have contributed to the legacy of Andrews' work and have made important contributions to understanding the impact of method factors to response variance. There are several studies that summarize the results of the application of the MTMM to the detection of method variance in large-scale surveys (e.g., Andrews and Herzog, 1986; Alwin, 1997, 2007; Rodgers, 1989; Rodgers et al., 1988; Rodgers et al., 1992). Recently, Scherpenzeel (1995; see also Scherpenzeel and Saris 1997) extended Andrews' (1984) work to several survey studies developed in the Netherlands. In my own work, I have used the MTMM method to investigate a number of issues in the measurement of life satisfaction (see Alwin, 1989, 1997, 2007).

17.7   A CRITIQUE OF THE MTMM DESIGN

The MTMM model has clear advantages over the multiple indicators approach inasmuch as it attempts to model one source of specific variance. However, a skeptic could easily argue that it is virtually impossible to estimate reliability from cross-sectional surveys because of the failure to meet the assumptions of the model and the potentially biasing effects of other forms of specific variance. While I am a strong supporter of the MTMM tradition of analysis and have written extensively about its advantages (see Alwin, 1974, 1989, 1997; Alwin and Jackson, 1979), these designs are very difficult to implement given the rarity of having multiple traits each measured simultaneously by multiple methods in most large-scale surveys. Moreover, due to the assumptions involved, the interpretation of the results in terms of reliability estimation is problematic. To be specific, the reliability models for cross-sectional data place several constraints on the data that may not be realistic. We must face up to the potential disadvantages.

There are two critical assumptions that I focus on here. First is the assumption that the errors of the measures are independent of one another, and second that the measures are *bivocal*, that is, there are only two sources of reliable variation among the measures, namely the trait variable and the method effect being measured. These assumptions rule out, for example, the operation of other systematic factors that may produce correlations among the measures. For example, in cross-sectional survey applications of the MTMM model, it may be impossible to assume that measurement errors are independent, since even though similar questions are not given in sequence or included in the same battery, they do appear in relative proximity. The assumptions for reliability estimation in cross-sectional studies rule out the operation of other types of correlated errors. And while the MTMM model allows for correlated errors that stem from commonality of method, this does not completely exhaust the possibilities. Obviously, it does not rule out memory in the organization of responses to multiple efforts to measure the same thing. Respondents are fully cognizant of the answers they have given to previous questions, so there is the possibility that memory operates to distort the degree of consistency in responses, consistency that is reflected in the *validity* component of the MTMM model.

To make the argument somewhat more formally, the reader should recall that the MTMM approach relies on the assumptions of the common factor model in equating the communality with reliability (see also Heise and Bohrnstedt, 1970). In the common factor model, the variance of any measure is the sum of two variance components—communality and uniqueness. A variable's *uniqueness* is the complement to the common parts of the data (the *communality*) and is thought of as being composed of two independent parts, one representing *specific variation* and one representing *random measurement error* variation. Using the traditional common factor notation for this, this can be restated as follows. The variable's *communality*, denoted h_j^2, is the proportion of its total variation that is due to common sources of variation. Its *uniqueness*, denoted u_j^2, is the complement of the communality, that is, $u_j^2 = 1.0 - h_j^2$. The uniqueness is composed of specific variance, s_j^2, and random error variance, e_j^2. Specific variance is *reliable* variance, and thus the reliability of the variable is not only due to the common variance, but to specific variance as well. In common factor analytic notation, the reliability of variable j can be expressed as $r_j^2 = h_j^2 + s_j^2$. Unfortunately, because specific variance is thought to be independent (uncorrelated with) sources of common variance, it becomes confounded with measurement error variance. Because of the presence of specific variance in most measures, it is virtually impossible to use the traditional form of the common factor model as a basis for reliability estimation (see Alwin and Jackson, 1979; Alwin, 1989). The problem is that the common factor model typically does not permit the partitioning of u_j^2 into its components, s_j^2 and e_j^2. In the absence of specific variance (what we here refer to as the "multiple measures" model), classical reliability models may be viewed as a special case of the common factor model, but in general it is risky to assume that $u_j^2 = e_j^2$ (Alwin and Jackson, 1979).

There have been other criticisms of the MTMM approach. One of these, advanced by Schaeffer and Presser (2003), is that the order of presentation of measurement forms is typically the same across respondents and that the method of analysis does not take this into account. This may appear on the face of it to be a trivial objection, given that "the order of presentation" issue affects all studies of method effects in survey (e.g., the question form studies by Schuman and Presser [1981] and virtually all survey results generally). However, it may be viewed as particularly problematic in "within-subjects" designs, for the reasons given above, although hardly any implementation of the survey method of which I am aware randomizes the order of presentation of survey questions to the respondent. However, as noted above, the nonbalanced nature of the design does not permit distinguishing between form of measurement and measurement context, in that there may be some form of conditioning effect. For example, respondents may have the answers to the earlier questions still relatively available in memory, and this may affect their responses on the later tasks. This argument is compatible with the objections given above regarding the potential for memory to heighten the overall assessed consistency of measurement, that is, reliability, in the MTMM design.

17.8   THE ROLE OF MEMORY

As scientists, we are taught to find potential alternative explanations for our results and attempt to rule them out. A presumptive alternative interpretation in the case of the MTMM design is that memory plays a role in the organization of responses to multiple efforts to measure the same thing. As respondents search for internal cues in the production of a response, one of the critical features of their self-awareness is their cognizance of the answers they have given to previous questions, so there is the possibility that memory operates to distort the degree of consistency in responses. In this case the operation of memory would serve to enhance the estimation of trait validity. Hence, such "memory-enhanced consistency" is reflected in the *trait validity* component of the MTMM model. As a result, one may be overestimating the relative contribution of trait variation and underestimating the presence of method variance.

I examined the potential biases in estimates of the validity component (and consequently estimates of reliability) in MTMM studies by examining the measurement of peoples' memory in a nationally representative sample of men and women in the Health and Retirement Study (HRS) conducted in 1996 (Juster and Suzman, 1995). The HRS is a biennial survey of the American public that began in 1992. The original HRS was a national panel study of men and women aged 51 to 61 assessed in 1992 (N = 9824) and reinterviewed every 2 years thereafter through 2008 and beyond. For present purposes, I employ data from the 1996 wave of the study, which means that the respondents were 55–65 years of age at the time of the study.[7]

I rely on two measures of word recall as a way of testing the operation of memory in the survey interview: (1) *Immediate Word Recall*—The HRS immediate word recall test involved a set of 10 words read to the respondent early in the interview and asked to recall as many as possible.[8] Respondents received one of four different word lists, and were given up to 2 minutes to recall the words on that list. Over four waves (8 years) of the HRS, respondents never received word lists identical to that used in any other wave, so there is no cross-wave contamination of word recall; (2) *Delayed Word Recall*—The delayed word recall task built upon the immediate word recall task. Later in the interview (about 10 minutes) the respondent was asked to recall the list of words that had been read to them earlier and which they had been asked to repeat.[9] Again, the respondent was allowed up to 2 minutes to recall as many of the 10 words as possible. The scores used for the immediate and delayed word recall tests are the number (from 0 to 10) of words the respondent correctly identified.

The results of the immediate and delayed word recall tasks are presented in Table 17.2. The typical word on these four lists (40 words in all) is recalled by 60% of the HRS respondents. Once a word is recalled in the immediate word recall task, the typical word is recalled by 75% of the respondents. There are variations across words, of course; some words are more easily recalled in both tasks. These results suggest that, at least within the context of this survey, persons' memories for the words in the word lists are pretty good. The typical respondent recalls about 6 out of 10 words on the immediate recall task and about 5 words on the delayed recall task. Thus, there is some memory decay across the 10–15 minutes elapsed between the two tasks. However, if one looks at the delayed task and focuses solely on those words produced in response to the immediate recall task, the impression one gets is that within the context of the survey, people remember what they said earlier. This goes against the claims of Saris and van den Putte (1988) who argued that repeated measurements at the beginning and end of a 20-minute-long questionnaire can be considered free of memory effects and therefore a basis for estimating reliability (see also van Meurs and Saris, 1990). Those claims were, however, based on conjecture rather than on empirical evidence, and it is my point here that rather than simply assuming it away, the operation of memory in these studies needs to be ruled out with empirical evidence.

Based on the HRS results, it looks to me like persons' memories are active during the course of a survey interview and that it is difficult, if not impossible, to argue that in response to a question form assessing the same content in a previous form, that people completely forget what their earlier response was. Whether or not respondents who do remember such things strive for consistency in their responses is another matter; but at least I would suggest that it is fairly bold (or possibly naïve) to think that the MTMM approach is free of the operation of memory and its possible role in the production of responses. What is needed here is a systematic approach to measuring the role of memory in MTMM studies.

TABLE 17.2. The Measurement of Episodic Memory in the Health & Retirement Study (HRS)–1996 (N = 7944)

List	N of Cases	Word	Immediate Recall	Percent[1]	Delayed Recall	Percent[2]
1	2057	Book	1406	0.684	779	0.554
1	2057	Child	1461	0.710	1236	0.846
1	2057	Gold	1202	0.584	987	0.822
1	2057	Hotel	1784	0.867	1444	0.809
1	2057	King	929	0.452	608	0.654
1	2057	Market	776	0.377	535	0.689
1	2057	River	1645	0.800	1337	0.813
1	2057	Skin	1006	0.489	753	0.749
1	2057	Tree	1309	0.636	1083	0.827
2	2024	Butter	1353	0.668	841	0.622
2	2024	College	953	0.471	689	0.723
2	2024	Dollar	849	0.419	615	0.724
2	2024	Earth	987	0.488	715	0.724
2	2024	Flag	1107	0.547	835	0.754
2	2024	Home	1167	0.577	825	0.707
2	2024	Machine	762	0.376	568	0.745
2	2024	Ocean	1560	0.771	1233	0.790
2	2024	Sky	1801	0.890	1534	0.852
2	2024	Wife	1316	0.650	1125	0.855
3	1993	Blood	1180	0.592	964	0.817
3	1993	Corner	919	0.461	740	0.805
3	1993	Engine	1212	0.608	736	0.607
3	1993	Girl	1470	0.738	1286	0.875
3	1993	House	1259	0.632	926	0.736
3	1993	Letter	647	0.325	469	0.725
3	1993	Rock	1446	0.726	1053	0.728
3	1993	Shoes	1087	0.545	779	0.717
3	1993	Valley	1110	0.557	794	0.715
3	1993	Woman	1830	0.918	1643	0.898
4	1870	Baby	1470	0.786	1147	0.780
4	1870	Church	1502	0.803	1189	0.792
4	1870	Doctor	1173	0.627	892	0.760
4	1870	Fire	1004	0.537	792	0.789
4	1870	Garden	842	0.450	618	0.734
4	1870	Palace	657	0.351	484	0.737
4	1870	Sea	1072	0.573	857	0.799
4	1870	Table	1101	0.589	656	0.596
4	1870	Village	840	0.449	599	0.713
4	1870	Water	1640	0.877	1393	0.849

[1]Percent of respondents recalling a given word on the immediate recall task.
[2]Percent of respondents recalling a given word on the delayed recall task (among those recalling the word on the immediate recall task).

17.9   ARE THERE ALTERNATIVE APPROACHES?

If there are problems with the MTMM approach, what can be done to remedy them? Is there an alternative to the MTMM approach? Although I do not discuss it at length here, there is an approach to estimating reliability of single items that uses the reinterview approach within longitudinal, or panel designs, rather than cross-sectional data. I do not discuss this approach at length in this chapter, because it is not the focus of this workshop, but it deserves consideration. This approach moves the discussion in the direction of the classical "test–retest" approach, but it recognizes a fundamental limitation of the test–retest approach. In order to estimate reliability using a single reinterview, one must assume there is no change in the underlying quantity being measured, and in designing such studies, one must be sensitive to the potential role of memory in producing consistency of responses over time (see Alwin, 2007, pp. 96–101).

To address the issue of taking individual-level change into account, Coleman (1968) and Heise (1969) developed a technique based on three-wave *quasi-simplex* models within the framework of a model that permits change in the underlying variable being measured (see Jöreskog, 1970; Wiley and Wiley, 1970). This same approach can be generalized to multi-wave panels, as illustrated in Figure 17.5.

One of the main advantages of the reinterview design is that under appropriate circumstances it is possible to eliminate the confounding of the systematic error component discussed earlier, if systematic components of error are not stable over time. In order to address the question of stable components of error, the panel survey must deal with the problem of memory, because in the panel design, by definition, measurement is repeated. So, while this overcomes one limitation of cross-sectional surveys, it presents problems if respondents can remember what they say and are motivated to provide consistent responses.

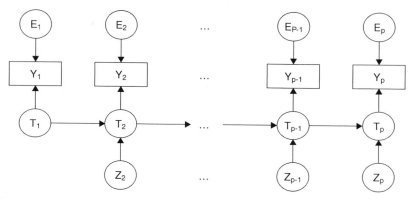

FIGURE 17.5. Path diagram of the quasi-Markov simplex model—general case $(P > 4)$. Source: Alwin (2007).

If reinterviews are spread over months or years, this can help rule out sources of bias that occur in cross-sectional studies. Given the difficulty of estimating memory functions, the conceptualization of reliability from reinterview designs makes sense only if one can rule out memory as a factor in the covariance of measures over time, and thus, the occasions of measurement must be separated by sufficient periods of time to rule out the operation of memory.[10]

In my recent work mentioned earlier, I employed the three-wave version of this model, as shown in Figure 17.6, in order to address several issues regarding the reliability of survey data (see Alwin, 2007). The design requirements set for this study included the following: (1) the use of longitudinal studies with at least three waves of measurement, (2) that are representative of known populations, and (3) which have reinterview intervals of *at least 2 years*. The last requirement is implemented in order to rule out the effects of memory in longitudinal studies.[11] I selected data from six panel studies that met these requirements, selecting nearly 500 survey questions that spanned the range of content, including measures of factual and nonfactual content. Four of the six were based on probability samples of the United States. Two were large-scale surveys of a sample originating in the Detroit metropolitan area. All the data collection in these surveys was conducted by the Survey Research Center of the University of Michigan. The studies are: (1) the 1956-1958-1960 National Election Study (NES) panel, (2) the 1972-1974-1976 NES panel, (3) the 1992-1994-1996 NES panel, (4) the 1986-1989-1994 American's Changing Lives (ACL) panel study, (5) the 1980-1985-1993 Study of American Families (SAF) sample of mothers, and (6) the 1980-1985-1993 Study of American Families (SAF) sample of children. This study demonstrated the feasibility of employing such alternative designs involving large-scale surveys to estimate the reliability of survey measurement in order to examine the contributions of various design features to the quality of data. Also, in terms of design, it should be pointed out that the MTMM approach has been used exclusively in the study of nonfactual content, for example, attitudes, beliefs, and self-

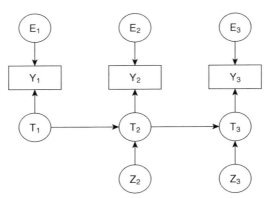

FIGURE 17.6.   Path diagram for a three-wave quasi-Markov simplex model. Source: Alwin (2007).

perceptions, and not applied to factual content measured in surveys, whereas the longitudinal approach can be applied to all types of survey data regardless of content.

17.10 CONCLUSION

Among social scientists, almost everyone agrees that without valid measurement, there may be little for social science to contribute in the way of scientific knowledge. A corollary to this principle is that *reliability of measurement* (as distinct from validity) is a necessary, although not sufficient, condition for valid measurement. The logical conclusion of this line of thinking is that whatever else our measures may aspire to tell us, it is essential that they are reliable; otherwise they will be of little value to us. There is a mathematical proof of this assertion (see Lord and Novick, 1968), but the logic that underlies these ideas is normally accepted without formal proof. An implication of this, of course, is that if our measures are *unreliable*, they are of little use to us in detecting patterns and relationships among variables of interest, and reliability of measurement is therefore a *sine qua non* of any empirical science.

Assuming that reliability of measurement is a legitimate approach to evaluating one aspect of survey data quality, the question then comes to how to proceed. I began this chapter with a quote from a paper Jon Krosnick and I wrote nearly 20 years ago (Alwin and Krosnick, 1991, p. 141). We suggested three desiderata for assessing the reliability of survey data:

- First, a model that specifies the linkage between true and observed variables.
- Second, a research design that permits the estimation of parameters of such a model.
- Third, an interpretation of these parameters that is consistent with the concept of reliability.

In my closing comments, I wish to use these three criteria for evaluating where we are with respect to the MTMM approach.

On the first point, there is little question that the extension of the CTST model of random error to include systematic sources of error attached to the method of measurement represents a vast improvement over some of our traditional ideas about reliability. These models have been around since the early 1970s (see Werts and Linn, 1970; Jöreskog, 1971, 1974; Alwin, 1974; Browne, 1984). Their contribution—formulated originally within confirmatory factor analysis framework—was to represent the MTMM measurement design as a common factor model. The application of these models to survey data was pioneered by Andrews (1984), and later on the technique was found to be useful by survey methodologists interested in partitioning the variance in survey measures collected within an MTMM design into components of variance (e.g., Alwin, 1989, 1997, 2007; Scherpenzeel, 1995; Scherpenzeel and Saris, 1997).[12] There is little disagreement that these are potentially very useful models for the reasons spelled out in the above presentation.

The areas where there is disagreement involve the second and third issues mentioned above—namely those having to do with design and interpretation. In this chapter I have reviewed the details of Andrews' approach, noting first that there is a difference between conventional psychometric ideas about "construct validity" and the notion of "trait validity" exemplified in the MTMM approach. These traditional notions about validity emphasize the need to employ external criterion variables, which is almost impossible to implement in the typical cross-sectional survey, and this is one of the reasons why the MTMM approach has a certain degree of appeal. In developing the MTMM matrix, however, Campbell and Fiske (1959) stressed the need to employ *maximally different* measures. In this sense, the idea that the "trait validity" embodied in the MTMM approach of Andrews and others may seem to fall far short of the conventional notions regarding "construct validity." I leave it to the readers to draw their own conclusions regarding the extent to which the MTMM approach meets these standards, but to my way of thinking, it remains a design issue that most people would agree needs to be addressed if the MTMM approach is to be a useful addition to the arsenal for evaluating the validity of survey data.

Using the MTMM designs, methods have been developed for separating validity and invalidity in the reliability estimates for survey reports, but the interpretation of components of variance in measures has not taken seriously the possibility that the consistency between multiple measurement formats as reflected in estimates of "trait validity" are *not* due to correlated errors that stem from the operation of memory in the organization of responses to multiple efforts to measure the same thing. The data I have presented here suggest that respondents are fully cognizant of the answers they have given to previous questions, so there is the possibility that memory operates to distort the degree of consistency in responses, consistency that is reflected in the *validity* component of the MTMM model.

In conclusion, it is clear that the quality of survey measures is thought to be tied to the quality of the question construction, questionnaire development, and interviewing techniques (Krosnick, 1999). Many guidelines for developing good survey questions and questionnaires represent the "tried and true" and aim to codify a workable set of rules for question construction. It is hard to argue against these guidelines. But, there is little hard evidence to confirm that following these rules improves survey data quality. The argument set forth by Saris and Gallhofer (2007), which I heartily endorse, is that we can assess these issues and evaluate the link between question/questionnaire characteristics and the quality of data by using levels of measurement error, or reliability and validity, as a major criterion of evaluation for examining differences in measurement quality associated with a particular question, specific interviewing practices, modes of administration, or types of questionnaires. This is clearly not the only criterion for survey quality—there are also questions of validity that ultimately must be addressed, but without a high degree of reliability, other matters cannot be addressed. In addition, there are other indicators of the quality of survey data that have to do with the quality of the procedures for obtaining coverage of the population, methods of sampling, and levels of nonresponse—but minimizing measurement error is an essential one (Alwin, 1991). No matter how good the sampling frame, the techniques of sampling, and rates of response in surveys, without high-quality measures, these other matters are virtually irrelevant. Indeed, elsewhere I have argued that the various sources of survey error are nested, and measurement errors exist on the "margins of error"—that is, measurement issues act as a prism through which we view other aspects of survey data (Alwin, 2007, pp. 2–8).

Historically, improvements in reliability due to the formation of composites from similar survey questions, as reflected in ICR estimates, has served as a rationale for including multiple indicators in survey questionnaires. This is, however, often a justification for sloppy question design in surveys, in that analysts often believe that if they include a list of questions in batteries within sections of a questionnaire, that somehow the truth will emerge post-data collection, once one factor analyzes the results. The best counsel is probably to take a more deliberate approach, including questions that precisely tap the dimensions of interest, which in the end will result in greater thinking up front about what concepts are being measured. In addition, in many cases it is a sobering realization that improvements in composite reliability due to the addition of multiple indicators depend on far longer composites than the survey instrument can bear (see Alwin, 2007, pp. 308–315). Hence, improving survey measurement through the scrutiny of the nature and level of measurement errors is an important item on the collective agenda and the work of MTMM researchers will only serve to enhance the quality of survey measurement.

NOTES

1   This paper was prepared for presentation at the Workshop on Question Evaluation Methods, sponsored by the National Center for Health Statistics and the National Cancer Institute, Washington DC, October 21–23, 2009.

2   The author is the inaugural holder of the Tracy Winfree and Ted H. McCourtney Professorship in Sociology and Demography and Director of the Center on Population Health and Aging at The Pennsylvania State University, University Park, Pennsylvania, and Emeritus Senior Research Scientist at the Survey Research Center, Institute for Social Research, University of Michigan—Ann Arbor.

3   As will become clear in the course of this chapter, I make a distinction between multiple indicators and multiple measures, and it is important to emphasize in this context that the MTMM approach relies on the former rather than the latter.

4   I return to this distinction in Section 17.5 below.

5   This discussion relies on the presentation of similar material in Alwin (1974, 1989, 2007).

6   This includes all of the models discussed in this chapter. The CTST model is discussed in detail in Alwin (2007) and references cited there, and I will therefore not discuss it further here, except to illustrate its application to the approaches considered later in this chapter.

7   These are standard survey measures of cognitive function. In the present case, these measures are employed by the author in a study of cognitive aging that models their change across waves of the survey (Alwin, 2008; Alwin et al., 2008; see also Hofer and Alwin, 2008).

8   The survey question was worded as follows: "I'll read a set of 10 words and ask you to recall as many as you can. We have purposely made the list long so that it will be difficult for anyone to recall all the words—most people recall just a few. Please listen carefully as I read the set of words because I cannot repeat them. When I finish, I will ask you to recall aloud as many of the words as you can, in any order. Is this clear?" After reading the list, the interviewer asked, "Now please tell me the words you can recall." An example of the words on one of the lists is as follows: hotel, river, tree, skin, gold, market, paper, child, king, book.

9   The survey question was worded as follows: "A little while ago, I read you a list of words and you repeated the ones you could remember. Please tell me any of the words that you remember now."

10 Saris and his colleagues (e.g., Saris and van den Putte, 1988; Coenders et al., 1999; Saris and Gallhofer, 2007) have been critical of this approach. I do not focus on these criticisms here, as they have been dealt with elsewhere (Alwin, 2007).

11 In many panel studies the reinterview intervals are rarely distant enough to rule out memory as playing a large factor, and it is important to rule out memory as a source of consistency in reliability estimates. Our experience indicates (Alwin, 1989, 1991, 1992; Alwin and Krosnick, 1991) that panel studies with short reinterview intervals (1–4 months) have substantially higher levels of reliability than those with longer intervals, such as 2 years.

12 Andrews' original proposal to the National Science Foundation was written many years earlier (see Andrews, 1977). I served as a consultant on his project and for many years was able to witness firsthand the development and application of the ideas that culminated in Andrews' 1984 paper in *The Public Opinion Quarterly*.

REFERENCES

Alwin DF (1974). Approaches to the interpretation of relationships in the multitrait-multimethod matrix. In: Costner HL, editor. Sociological Methodology 1973–74. San Francisco, CA: Jossey-Bass; pp. 79–105.

Alwin DF (1989). Problems in the estimation and interpretation of the reliability of survey data. Quality and Quantity; 23:277–331.

Alwin DF (1991). Research on survey quality. Sociological Methods and Research; 20:3–29.

Alwin DF (1992). Information transmission in the survey interview: number of response categories and the reliability of attitude measurement. In: Marsden PV, editor. Sociological Methodology 1992. Washington D.C.: American Sociological Association; pp. 83–118.

Alwin DF (1997). Feeling thermometers vs. seven-point scales: Which are better? Sociological Methods and Research; 25:318–340.

Alwin DF (2007). Margins of Error: A Study of Reliability in Survey Measurement. Hoboken, NJ: John Wiley & Sons.

Alwin DF (2008). The Aging Mind in Social and Historical Context. Unpublished Book Manuscript. Center on Population Health and Aging. University Park, PA: Pennsylvania State University.

Alwin DF (2009). Assessing the validity and reliability of timeline and event history data. In: Belli RF, Stafford FP, Alwin DF, editors. Calendar and Time Diary Methods in Life Course Research. Thousand Oaks, CA: Sage.

Alwin DF, Jackson DJ (1979). Measurement models for response errors in surveys: issues and applications. In: Schuessler KF, editor. Sociological Methodology 1980. San Francisco, CA: Jossey-Bass; pp. 68–119.

Alwin DF, Krosnick JA (1991). The reliability of survey attitude measurement: the influence of question and respondent attributes. Sociological Methods and Research; 20:139–181.

Alwin DF, McCammon RJ, Wray LA, Rodgers WL (2008). Population processes and cognitive aging. In: Hofer M, Alwin DF, editors. Handbook of Cognitive Aging: Interdisciplinary Perspectives. Thousand Oaks, CA: Sage; pp. 69–89.

American Educational Research Association, the American Psychological Association and the National Council on Measurement in Education (1999). Standards for Educational and Psychological Testing. Washington DC: American Psychological Association.

Andrews FM (1977). Estimating the Validity and Error Components of Survey Measures (Revised Version). A Research Proposal to the National Science Foundation. Ann Arbor: Institute for Social Research, University of Michigan.

Andrews FM (1984). Construct validity and error components of survey measures: a structural modeling approach. Public Opinion Quarterly; 46:409–442. Reprinted in W.E. Saris W, van Meurs A. (1990). Evaluation of measurement instruments by meta-analysis of multitrait multimethod studies. Amsterdam: North-Holland.

Andrews FM, Herzog AR (1986). The quality of survey data as related to age of respondent. Journal of the American Statistical Association; 81:403–410.

Biemer PP, Groves RM, Lyberg LE, Mathiowetz NA, Sudman S, editors (1991). Measurement Errors in Surveys. New York: John Wiley and Sons.

Bound J, Brown C, Duncan GJ, Rodgers WL (1990). Measurement error in cross-sectional and longitudinal labor market surveys: validation study evidence. In: Hartog J, Ridder G, Theeuwes J, editors. Panel Data and Labor Market Studies. North Holland: Elsevier Science Publishers.

Browne MW (1984). The decomposition of multitrait-multimethod matrices. British Journal of Mathematical and Statistical Psychology; 37:1–21.

Campbell DT, Fiske DW (1959). Convergent and discriminant validation by the multitrait-multimethod matrix. Psychological Bulletin; 6:81–105.

Coenders G, Saris WE, Batista-Foguet JM, Andreenkova A (1999). Stability of three-wave simplex estimates of reliability. Structural Equation Models; 6:135–157.

Coleman JS (1964). Models of Change and Response Uncertainty. Englewood Cliffs, NJ: Prentice-Hall.

Coleman JS (1968). The mathematical study of change. In: Blalock HM, Blalock AB, editors. Methodology in Social Research. New York: McGraw-Hill; pp. 428–478.

Cronbach LJ (1951). Coefficient alpha and the internal structure of tests. Psychometrika; 16:297–334.

Cronbach LJ, Meehl PE (1955). Construct validity in psychological tests. Psychological Bulletin; 52:281–302.

Goldstein H (1995). Multilevel Statistical Models, 2nd ed. London: Arnold.

Greene VL, Carmines EG (1979). Assessing the reliability of linear composites. In: Schuessler KF, editor. Sociological Methodology 1980. San Francisco, CA: Jossey-Bass; pp. 160–175.

Groves RM (1989). Survey Errors and Survey Costs. New York: John Wiley and Sons.

Groves RM (1991). Measurement errors across disciplines. In: Biemer PP, Groves RM, Lyberg LE, Mathiowetz NE, Sudman S, editors. Measurement Errors in Surveys. New York: Wiley; pp. 1–25.

Heise DR (1969). Separating reliability and stability in test-retest correlation. American Sociological Review; 34:93–191.

Heise DR, Bohrnstedt GW (1970). Validity, invalidity, and reliability. In: Borgatta EF, Bohrnstedt GW, editors. Sociological Methodology 1970. San Francisco, CA: Jossey-Bass; pp. 104–129.

Hofer SM, Alwin DF, editors (2008). Handbook of Cognitive Aging: Interdisciplinary Perspectives. Thousand Oaks, CA: Sage.

Jöreskog KG (1970). Estimating and testing of simplex models. British Journal of Mathematical and Statistical Psychology; 23:121–145.

Jöreskog KG (1971). Statistical analysis of sets of congeneric tests. Psychometrika; 36:109–133.

Jöreskog KG (1974). Analyzing psychological data by structural analysis of covariance matrices. In: Krantz DH, Atkinson RC, Luce RD, Suppes P, editors. Contemporary Developments in Mathematical Psychology. Vol. II. San Francisco, CA: W.H. Freeman and Company.

Jöreskog KG (1978). Structural analysis of covariance and correlation matrices. Psychometrika, 43:443–477.

Juster FT, Suzman R (1995). An overview of the health and retirement study. Journal of Human Resources; 30:S7–56.

Krosnick JA (1999). Survey research. Annual Review of Psychology; 50:537–567.

Lord FM, Novick ML (1968). Statistical Theories of Mental Test Scores. Reading MA: Addison-Wesley.

Lyberg L, Biemer P, Collins M, de Leeuw E, Dippo C, Schwarz N, Trewin D (1997). Survey Measurement and Process Quality. New York: John Wiley and Sons.

Marquis KH (1978). Record Check Validity of Survey Responses: A Reassessment of Bias in Reports of Hospitalizations. Santa Monica, CA: The Rand Corporation.

Marquis MS, Marquis KH (1977). Survey Measurement Design and Evaluation Using Reliability Theory. Santa Monica, CA: The Rand Corporation.

Rodgers WL (1989). Measurement properties of health and income ratings in a panel study. Presented at the 42nd Annual Scientific Meeting of the Gerontological Society of America, Minneapolis, November.

Rodgers WL, Andrews FM, Herzog AR (1992). Quality of survey measures: a structural modeling approach. Journal of Official Statistics; 3:251–275.

Rodgers WL, Herzog AR, Andrews FM (1988). Interviewing older adults: validity of self-reports of satisfaction. Psychology and Aging; 3:264–272.

Saris WE, Andrews FM (1991). Evaluation of measurement instruments using a structural modeling approach. In: Biemer PP, Groves RM, Lyberg LE, Mathiowetz NE, Sudman S, editors. Measurement Errors in Surveys. New York: Wiley; pp. 575–597.

Saris WE, Gallhofer IN (2007). Design, Evaluation, and Analysis of Questionnaires for Survey Research. Hoboken, NJ: John Wiley & Sons.

Saris WE, van den Putte B (1988). True score of factor models: a secondary analysis of the ALLBUS test-retest data. Sociological Methods and Research; 17:123–157.

Saris WE, van Meurs A (1990). Evaluation of Measurement Instruments by Meta-Analysis of Multitrait Multimethod Studies. Amsterdam: North-Holland.

Schaeffer NC, Presser S (2003). The science of asking questions. Annual Review of Sociology; 29:65–88.

Scherpenzeel AC (1995). A question of quality: evaluating survey questions by multitrait-multimethod studies. PhD Thesis, University of Amsterdam.

Scherpenzeel AC, Saris WE (1997). The validity and reliability of survey questions: a meta-analysis of MTMM studies. Sociological Methods and Research; 25:341–383.

Schuman H, Presser S (1981). Questions and Answers: Experiments in Question Wording, Form and Context. New York: Academic Press.

Traugott M, Katosh JP (1979). Response validity in surveys of voting behavior. Public Opinion Quarterly; 43:359–377.

Traugott M, Katosh JP (1981). The consequences of validated and self-reported voting measures. Public Opinion Quarterly; 45:519–535.

van Meurs A, Saris WE (1990). Memory effects in MTMM studies. In: Saris WE, van Meurs A, editors. Evaluation of Measurement Instruments by Meta-Analysis of Multitrait Multimethod Matrices. Amsterdam: North-Holland; pp. 52–80.

Werts CE, Linn RL (1970). Path analysis: psychological examples. Psychological Bulletin; 74:194–212.

Wiley DE, Wiley JA (1970). The estimation of measurement error in panel data. American Sociological Review; 35:112–117.

18 Response to Alwin's Chapter: Evaluating the Reliability and Validity of Survey Interview Data Using the MTMM Approach

PETER PH. MOHLER

Universität Mannheim

18.1 INTRODUCTION

This chapter focuses on survey process quality with an emphasis on question quality. Using a multitrait-multimethod (MTMM) example from the European Social Survey (ESS) it complements Alwin (this volume) who concentrates on measurement issues. The thrust of this paper is a plea for sound premises/input into any statistical or cognitive assessment procedures in an attempt to avoid the "false premises-dubious conclusions" trap (Riemer, 1954; Clark, 1958).[1]

Typically, survey quality has been reported using single indicators. Most prominent are "response rate" or "sampling error." Response rate has been used as an indication of response bias which is a serious threat to survey quality (Groves and Couper, 1998). A closer look at response rate as a valid quality indicator revealed, however, that this indicator is fairly insensitive to response bias (Groves and Couper, 1998; Groves, 2006) Similarly, sampling error statements such as "+/−3 percent error" are insensitive to complex sampling designs, thus underestimating the true error rate due to sampling. More recently, complex statistical item quality assessments have been proposed that give also single indicators such as Saris' "Q" (Saris and Gallhofer, 2007). They take advantage of complex statistical estimation procedures using Structural Equation Modeling (SEM), Item Response Theory (IRT), MTMM, or Latent Class Models (see relevant chapters this volume). There is a general belief that one can "model away" almost all insufficiencies of survey measurement. However, statistical modeling is based on estimations, not on measurements (see Alwin, this volume). Moreover, modeling itself relies on robust and

Question Evaluation Methods: Contributing to the Science of Data Quality, First Edition.
Edited by Jennifer Madans, Kristen Miller, Aaron Maitland, Gordon Willis.
© 2011 John Wiley & Sons, Inc. Published 2011 by John Wiley & Sons, Inc.

valid assumptions (premises) about the modeling parameters, and modeling procedures can also be sensitive to really small changes in, often themselves estimated, input parameters. The impact of such nonrobust assumptions, biased input data, and volatile computational procedures is often ignored in statistical assessments. To get the most out of powerful modeling approaches, one needs in turn very clean and healthy input data and "true" premises or assumptions. Moreover, all point quality indicators mentioned here are generated ex post, that, is estimated and calculated after data have been collected and edited. Thus, they cannot help to improve the current survey but only call for the "paramedics," that is, statisticians, to tend the biases and errors or to give information for a future survey. Also, there is no real quality control of such bias and error indicators. For instance, response bias is estimated using control sheet data (contact data or other paradata—Stoop et al., 2010), or MTMM quality assessments are based on experimental data collected in a cross-section survey. There is, however, no quality report available indicating whether contact protocol data or experiments met specified quality standards in design, pretesting, and implementation.

In an up-to-date process quality paradigm, however, each production step serves as input into the next step and thus provides "true or false" premises for the next step. This is the turning point away from single indicator assessments to a complete survey production process (survey production cycle) assessment. This is done best by optimizing quality assurance and quality control measures at each stage of the survey production process. Faulty outcomes on one stage will be rejected as input into the next stage (Biemer and Lyberg, 2003). To show the differences between point quality indicators and process quality approaches, we will look firstly at a single indicator (point quality) on item quality using MTMM modeling ("Q," Saris and Gallhofer, 2007) that is estimated at the very end of the survey production cycle (i.e., analysis) and then look at previous stages such as question design, technical instrument implementation and data collection in an effort to identify the quality achieved at each stage, and finally to assess the validity of the modeling effort.

18.2   Q: A POINT QUALITY INDICATOR EXAMPLE

Point quality indicators serve to summarize the quality of a product in one single, final figure. Saris in following Andrews (1984) propagates "Q" as such a point quality measure for single items (Saris and Gallhofer, 2007; Oberski et al., 2010, p. 442). Q is based on statistical estimation techniques using a MTMM modeling approach (Saris and Gallhofer, 2007).

The thrust of Saris' argument presented in Table 18.1 is that using different response scales results in substantial variations of Q (i.e., measurement quality). Q is reported in Table 18.1 for two response scales: IS 11, an 11-point scale with labeled end points and answer options reflecting the question text (*is it good or bad for country?*), and 5 Lik, a 5-point Likert-type scale

TABLE 18.1. Q for an ESS Question Using Four Different Response Scales Sorted in Descending Order of Q for IS 11 (Saris, 2009)

Country	IS 11	5 Lik
Romania	0.88	0.29
Spain	0.83	0.46
Portugal	0.83	0.47
Latvia	0.81	0.24
Slovenia	0.81	0.37
GB	0.81	0.41
Ukraine	0.81	0.44
Austria	0.81	0.46
Cyprus	0.81	0.47
France	0.79	0.55
Ireland	0.77	0.37
Russia	0.77	0.42
Germany	0.77	0.43
Denmark	0.74	0.61
Netherlands	0.72	0.38
Belgium	0.72	0.51
Norway	0.72	0.67
Bulgaria	0.71	0.3
Switzerland	0.71	0.5
Finland	0.71	0.6
Poland	0.69	0.33
Slovakia	0.67	0.32
Estonia	0.55	0.41

The data used in Table 18.1 are from the 2006 European Social Survey (ESS, http://www.europeansocialsurvey.org). The coefficients were provided by Saris in a paper for the QEM 2009 conference in Washington.

(*agree-disagree*). IS-11pt was asked to all respondents, 5 Lik to a randomly assigned subsample.

Saris concludes in his Question Evaluation Method (QEM) 2009 paper "this table shows that the IS-11 point scale is *much* better than any of the other measures for all questions in all countries studied" (Saris, 2009, emphasis mine). We will come back to this statement. Before doing so, we will have a closer look at how Q is estimated.

18.2.1   Short Introduction into Estimating Q

Q is estimated using a specific MTMM model (Saris and Gallhofer, 2007), thus we will first have a look what MTMM is. It was developed as a method for estimating *construct validity* (Campbell and Fiske, 1959; Alwin, 2007, pp. 91–97; Chapter 17 this volume). It is about to decompose a given measurement (i.e., response to a survey item) into several components: random error, systematic error (bias), correlation with true value, and correlation of the true value with a construct that cannot be measured directly. To achieve this, several constructs and several measurement methods must be simultaneously examined in an effort to achieve enough information for estimating all necessary model parameters.

Construct validity can be defined as follows:

> Construct validity refers to the degree to which inferences can legitimately be made from the operationalizations in your study to the theoretical constructs on which those operationalizations were based . . . where external validity involves generalizing from your study context to other people, places or times, construct validity involves generalizing from your program or measures to the concept of your program or measures. (Trochim, 2006, last retrieved on August 27, 2010 at http://www.socialresearchmethods.net/kb/constval.php)

Note that constructs are "latent variables" with properties similar to nonlatent, directly measurable variables. Among such properties are distributional form (normal, skewed, etc.), variance, and correlations with other latent variables (constructs).

Campbell and Fiske's original MTMM (also called multi-trait multi-method matrix) requires latent constructs that are associated with each other as well dissociated with their counterparts (labeled "convergent" and "discriminant" by Campbell and Fiske, 1959). A full Campbell and Fiske MTMM model requires a minimum of three traits (latent constructs) measured using three different methods (Campbell and Fiske, 1959; Saris and Gallhofer, 2007, p. 209). Thus, all in all 45 correlations and variances are generated for a full MTMM model. That would not be feasible for general social surveys. Andrews and then Saris et al. finally brought the number of items and methods down to a bare minimum using split-ballot designs (still quite a burden for general surveys, but manageable—Saris and Gallhofer, 2007, p. 219; Oberski et al., 2010). That is an achievement in itself and should be highly appreciated.

18.2.2   MTMM Methods and Measurement

Search engines will provide an enormous amount of literature on MTMM analytical graphs representing specific models (search key "MTMM" and "analysis"). Not so much literature is available about designing and implementing MTMM survey instruments, that is why specific questions were used in MTMM experiments. Thus, one has to reconstruct core elements of MTMM studies via documentation of how the traits are measured, that is the survey instrument. It is especially interesting to learn what different "methods" are used. That is to understand why which method variation is thought to be helpful while being independent from other methods used (independence of measurement is required for MTMM models). Thus, we will to have a look at the full experimental instrument of the ESS 2006 of which two questions are mentioned in Table 18.1. The four items used in the 2006 MTMM experimental survey are given in Table 18.2 that consists of the question wording, the corresponding response scale (with instructions for the respondent), the intended response mode, and also information about the actual response mode used during data collection.

In the 2006 ESS MTMM experiment all respondents were first asked B38 as part of the core questionnaire, an attitude question using an 11-end point scale (11 IS). After the core questionnaire followed items on values followed by MTMM items HS4, HS16, and HS25 using one of three different response scales as well as change in wording, mode, and context. Thus, there are four "methods" in the nomenclature of MTMM used for estimating the model parameters. Each respondent had to answer only two of them. Saris showed in an earlier paper that such an approach is feasible (Saris and Gallhofer, 2007).

As one can see from Table 18.2 the differences in "independent method" can be classified as follows:

(a) Change of stimulus by changing the question text (e.g., "would you say" to "it is" or a change in the introduction of the item)

(b) Change of response scale (e.g., "bad-good" to "agree-disagree", or 11 points to 5 points

(c) Change of response mode (e.g., "face-to-face" [FtF] to "self-completion" (in some countries only)

(d) Change of all three components simultaneously.

We will discuss those differences in the light of survey design and quality later. Now, we will have a look at the core MTMM assumption about independence of methods (measurement) in cross-sectional surveys regarding the time lapse needed to assume independent measurements.

18.2.3 Independence of Measures in Cross-Sectional MTMM Experiments

A core assumption of MTMM is that each "method" must be independent from any other "method" used to measure a given construct: "Validation is typically convergent, a confirmation by independent measurement procedures. Independence of methods is a common denominator among the major types of validity (excepting content validity) . . . " (Campbell and Fiske, 1959, p. 81). To ask the same or similar questions in one cross-section survey thus requires an additional assumption about the independence of "methods," namely that responses given in a survey earlier do not influence responses given later on.

TABLE 18.2. Compilation of All Items Used in the 2006 ESS MTMM Experiment with Reference to Item B38

Question Text	Response Scale	Intended Mode	Actual Mode
B38 CARD 15 Would you say it is generally bad or good for (country's) economy that people come to live her from other countries? Pleas use this card.	*Showcard* Bad for the economy — Good for the economy 0 1 2 3 4 5 6 7 8 9 10	Face-to-face	Face-to-face
HS4 It is generally bad for (country's) economy that people to come to live here from other countries?	Now some questions about people from other countries coming to live in (country). Please read each question and tick the box on each line that shows how much you agree or disagree with each statement. Agree strongly (1) / Agree (2) / Neither agree nor disagree (3) / Disagree (4) / Disagree strongly (5)	Self-completion	See ESS documentation; mostly interviewer assisted
HS16 How much do you agree or disagree that it is generally bad for (country's) economy that people come to live here from other countries? Please tick one box.	Now some questions about people from other countries coming to live in (country). Disagree strongly (0) ... Agree strongly (10)	Self completion	See ESS documentation; mostly interviewer assisted
HS 25 It is generally bad for (country's) economy that people come to live here from other countries? Please tick one box.	Now some questions about people from other countries coming to live in (country). Please read each question and tick the box on each line that shows how much you agree or disagree with each of the following statements. Disagree strongly (01) ... Agree strongly (07)	Self completion	See ESS documentation; mostly interviewer assisted

There is a standard reference in the literature indicating that 20 minutes between two measurements guarantees measurement independence: "Van Meurs and Saris (1990) have demonstrated that after 20 minutes the memory effects are negligible. This time gap is enough to obtain independent measures in most circumstances" (Saris and Gallhofer, 2007, p. 220).

In contrast to this Duncan and Stenbeck (1988, p. 523) found: "in the course of a single interview . . . there is evidence that responses to one question may 'contaminate' responses to another or, alternatively, that several questions may be vulnerable to common but evidently transitory sources of contamination." In other words, they found interaction effects between early-on responses and those given at a later point in time in the interview. However, we are not stuck by these contradictory statements. A careful rereading of the original van Meurs and Saris paper reveals the following: They report on an experiment asking Dutch respondents after some time in the interview "*Can you remember exactly what you answered on the question about xyz. . . .?*" If respondents said they could not remember, they were asked to try to estimate their previous response (van Meurs and Saris, 1990, p. 138). The results show that a considerable number of respondents could correctly reproduce their previous answer (34% who were sure they could, 36% of those who were not sure that they could remember exactly). In a second attempt they also controlled for the time span between the first measurement and the recall attempt. The results are stunning. The number of correct response recall was about 45% for three attitudinal items and between 52% and 72% for three nonattitudinal items for time intervals greater than 20 minutes! For intervals less than 20 minutes they were even considerably higher. One should note that "correct" recall means that respondents arrived at the identical response (error free matching), while Duncan and Steinbeck reported statistical associations between measurements that included error and bias. Consequently, van Meurs and Saris are very cautious in their conclusions about MTMM experiments in a cross-sectional study such as the ESS. For instance, they recommend "a very long interview with many questions of the same kind concerning the same topic between repeated observations" (van Meurs and Saris, 1990, p. 145). In addition, they point to the fact that respondents who give extreme responses are "very capable" to make correct response recalls irrespective of the time interval. How these cautious recommendations and conclusions became a simple "20-minutes-rule" as stated by Saris and Gallhofer (2007) is not known.

What one can say now with confidence is that memory effects cannot be ruled out in cross-sectional surveys and thus, "all validity and reliability estimates will be subsequently over-estimated" (van Meurs and Saris, 1990, p. 146). From this follows that it is not correct to simply assume that 20 minutes between two measures guarantee independence of measures. One must instead demonstrate measurement independence in MTMM experiments. Otherwise, quality estimates could be overestimated which in turn means that the error components would be underestimated.

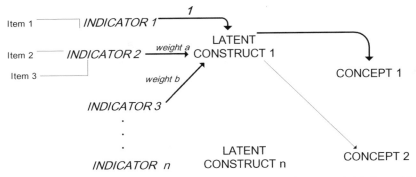

FIGURE 18.1. Simplified graphical representation of an MTMM model (adapted from Saris and Gallhofer, 2007, p. 311).

18.2.4 Weights in MTMM Models

Typically MTMM models are graphically represented. Figure 18.1 is simplified version of an example given by Saris and Gallhofer (2007, p. 311). We use here the terminology developed in comparative survey methodology instead of the indigenous MTMM labels (concepts, constructs, indicators, and items/questions; Harkness et al., 2010) where concepts and constructs are not directly measurable. Connecting lines indicate associations between elements in Figure 18.1 arrows indicate directed associations that are believed to represent causality. The directed (arrowed) associations between Indicator 1 or 2 and Latent Construct 1 are weighted using an "appropriate" weights a and b. What is meant by "appropriate" is often discussed in technical terms only ("experience shows," "A found that xy weights are appropriate"). Their substantive relevance seemingly plays no proper role in MTMM modeling considerations (see in contrast Campbell and Fiske, 1992). One also finds the weight "1" for the causal association between Indicator 1 and Construct 1, which "means that the observed and latent variable are seen as identical (no error)" (Saris and Gallhofer, 2007, p. 311). This is an interesting assumption, indeed, as there is no error-free measurement possible in empirical research. Moreover, latent variables (constructs) are clearly defined as not directly measureable. If one assumes that the observable indicator and the latent variable are identical, then one does not need to assume a latent dimension at all because the directly measured indicator explains 100% of the variance of the so-called latent construct. From this follows that Construct 1 should thus be eliminated from the model and replaced by the observable Indicator 1 or one should be more precise in what is meant to be latent and what is not.

We could go into more and more detail of implicit substantial or logical model implications, but we confine us here to conclude that model specifications such as the weights given here and elsewhere need both, a substantial and a logical assessment in addition to the usual technical, formal assessments undertaken. And it is obviously very helpful to read graphic representations of statistical models as meticulously as one is used to read mathematical formulas or tables.

18.2.5 Estimation of Q

Q is a composite score discussed in full by Saris and Gallhofer (2007, pp. 284–298, especially on pp. 296–297). The following citations from their publication is a condensation of their arguments:

"If the latent variable is called 'F' and the observed variable is called 'x' and the error variable 'e' it has been shown by several authors (Bollen, 1989) that

$$\text{Quality of } x = pFx2 = var(F)/var(x) = 1 - (var(e)/var(x))$$

Quality of $S = 1 - (var(eS)/var(S)) \ldots$ " (Saris and Gallhofer, 2007, p. 296).

They give an example of the computation of Q based on the estimation process:

$$Var(eS) = 0.312 \times 0.7 + 0.12 \times 0.42 + 0.812 \times 0.75 + 0.31 \times 0.1$$
$$\times 0.09 + 0.31 \times 0.81 \times 0.09 + 0.1 \times 0.81 \times 0.09 = 0.60.$$

The weights were estimated in such a way that the variance of the composite score is equal to 1. Hence, the quality of the composite score as an indicator for the concept "interest in political issues in the media" is Quality = $1 - (0.6/1) = 0.4$. It will be clear that a quality score of 0.4 is "not a very good result" (Saris and Gallhofer, 2007, p. 297).

One could, of course, argue about what is "high" or what is "low" after having made so many assumptions about weights and errors as well as the sensitivity of the model to outliers or misspecified weights.

Moreover, that chain of numbers must not be mistaken as hard rock digits. Quite to the contrary, they represent estimated outcomes from complex modeling. Many substantive decisions were made in establishing exactly those outcomes. One might be tempted to rush over some core statements such as "assuming the same method bias" or "regression weights give best quality." In addition, the literature on statistical modeling is full of caveats about the proper use of such approaches; sometimes they read like a list of side effects in medical package inserts (Campbell and Fiske, 1992; Tormarken and Waller, 2005). These caveats go way beyond the standard warnings about level of measurement or assumed independence of error terms. Thus, one should hold for a moment and consider the substantial correlates of model assumptions. If one does not feel happy with their implications, one might wish to reconsider either the model or the whole modeling approach.

Up to now we discussed some issues involved in understanding the figures in Table 18.1 representing Q. We identified Q as a point quality indicator that is estimated using complex modeling and measurement procedures from a published data set. In the next section we will introduce shortly the survey production process paradigm and examine more closely quality properties of the MTMM items presented above in the light of this approach.

18.3   PROCESS QUALITY VERSUS POINT QUALITY

Q is a point quality indicator that is not related to precision and accuracy of its input elements such as item formulation, data collection procedures, or comparability in the case of the ESS experiment. However, without knowledge about the quality of the production process of a survey resulting in data matrices, one cannot assess nor control the quality of point indicators such as Q.

In the last decade, survey methodology moved away from the point quality paradigm to a modern process quality paradigm (Biemer and Lyberg, 2003; Mohler et al., 2010, Pennell et al., 2010). The decline of point quality indicators was heralded for response rates by Groves and Couper (1998) who established the complex indicator "nonresponse bias"; for back-translation by Harkness (1999) pointing to the complex and iterative process to achieve quality translation; for random sampling by Groves and Heeringa (2006) who linked costs with controlling for survey errors in full probability sampling; and for survey error by Biemer and Lyberg (2003) introducing a quality process perspective for the first time in discussing total survey error. Similarly, item quality indications relying on an "it works" quality assumption were replaced by qualitative and quantitative item assessments such as cognitive interviewing and statistical models (CCSG, 2010—section on pretesting).[2]

However, one should also mention here that survey production process perspectives have a much longer tradition than the recent discussion suggests. For instance, Signal Systems were conceptualized long before the term paradata was coined (see Biemer and Lyberg, 2003), or the inadequacy of back-translation for comparative surveys has been ridiculed as a "translator test" in the 1960s (Scheuch, 1968/1993, p. 108). The new quality approaches benefit greatly from this earlier work.

In contrast to the old paradigm that tried to maximize a few often not related quality criteria, process quality control strives for a balanced quality throughout the entire survey production process. Basically, process quality systems require well-defined fail/pass assessments at each and every production stage in an attempt to optimize the amount of information obtained per monetary unit spent within the allotted time and meet the specified level of precision.[3] Thus, it is of no interest in the new paradigm to spent at the same time a huge part of the budget on hunting for the last respondent (Stoop, 2005) while interviewers are causing substantial selection bias, or implementing low-quality instruments while spending a fortune on complex data collection designs, or allotting much time and efforts on complex analyses while not controlling for data editing errors, and so on. In short, money spent on improving single indicators is often wasted due to major failures and faults at other production stages resulting in foul data.

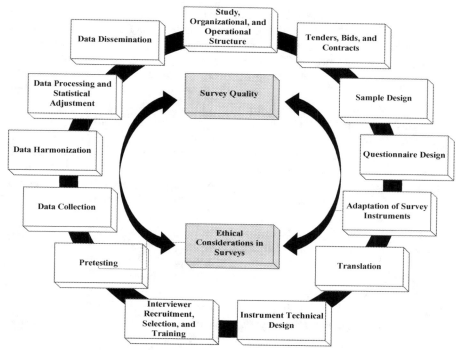

FIGURE 18.2. Survey production cycle (CCSG, 2010).

The most recent model of the survey production process (survey production cycle model—SPC) has been developed by the CCSG group (2010). It is given in Figure 18.2 below. Thirteen different production stages are identified plus two, Quality and Ethics, being integrated parts of all stages. The cyclical presentation indicates three important issues. The first concerns the aspiration of cumulative survey research, that is, earlier findings and strength of past experience are input into a new survey. Second, almost all stages of the SPC are iterative and recursive. For instance, Sample Design is a play between ideal design and the reality of available information about a sample frame as well as the balance between precision and costs. As, for instance in the case of questionnaire design, iteration between item formulating, testing, and reformulating should be now as standard as are iterations in multivariate statistics. And, third, iterations (cycles) occur between stages. For instance, sampling informs budgeting/contracting as the latter informs sampling.

The cycle indicates also a time sequence and hierarchy of stages—data collection requires all previous stages being successfully finished and having passed their predefined quality benchmarks.

Note that Q as discussed above has been estimated after the last production stage, data dissemination, and could thus only serve as input into further analyses or serve as information for the next survey. The quality of Q depends on the quality of all stages in the SPC.

On each stage, separately quality assurance and control processes have to be implemented and managed successfully. Figure 18.3 (shown later) depicts a prototypical question development process that sets out with defining the theoretical concept and cycles through until a given item meets all specified benchmarks. Note that it might be necessary to go back to a prior stage, if at the technical implementation stage or at the piloting quality issues arise that require a redesign of the question.

Quality benchmarks are defined cutoff procedures that allow one to differentiate between "acceptable" and "inacceptable" quality or "pass" and "fail." Benchmarking procedures can be statistical such as reliability or correlation or Statistical Process Control tools as well as nonstatistical such as expert rating. They are not naturally given but are carefully chosen and selected for specific purposes. At best, calibration studies bolster and verify benchmark indicators.

In the following paragraphs we will investigate three stages of the survey production cycle that are relevant for assessing the quality of the items used for the ESS 2006 MTMM experiment. These are Questionnaire Design, the stage where questions are developed either from scratch or adapted from previous surveys. Second, we will look at Instrument Technical Design. At that stage visuals and other formatting are decided. Finally, Data Collection is discussed, the stage where numerous factors play a role. For the discussion here, mode and implementation issues will be discussed.

As will become clear from our discussion, numerous quality issues arise already at these three stages of the survey production process. Many more can be found at other stages, such as translation and adaptation, sample design and its outcomes (where, for instance, crucial information for one country is still missing–Latvia is excluded from the integrated data set due to missing information), or pretesting where it would become clear that most items were only "technically" pretested.

All such quality issues could seriously affect the estimation of Q.

18.3.1 Questionnaire Design

Questionnaires are the heart of a social survey. If questions do not meet expected quality characteristics such as salience, reliability, or validity, all is lost. No statistical weight can really remedy foul questions. Other than many may like to believe, being able to ask questions in everyday life situations definitely does not qualify for profession survey questions design (Mohler, 2006). Moreover, questionnaire design is a team effort like translation (Harkness, 1999, 2010). Mohler identified more than 10 special qualifications needed for question design (2006, see also Noelle-Neumann and Petersen, 2005). They range from substantive and theoretical knowledge over text management or linguistics to pretesting specialists, etc. The question design process has many iterations and quality assessment steps as outlined in Figure 18.3. This production stage is also closely linked to Instrument Technical Design and Pretesting.

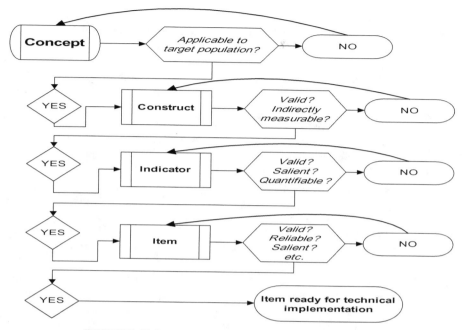

FIGURE 18.3. Prototypical item development process.

Without intensive, teamed up, and quality-controlled efforts, the outcomes of the questionnaire design stage will rely on hazardous or heroic assumptions that are all too often proven to be completely wrong in statistical analyses.

It is also important to accept that quality assurance on the questionnaire design stage involves items to be drafted according to established standards and best practice (Mohler, 2006). References and discussion on standards and best practice can be found in Groves et al., 2004/2009; Harkness et al., 2010; CCSG, 2010).

However, there seems to be a logical contradiction in many discussions on questionnaire design, as some say that no one knows how to "write" a good question while others refer to "best practice." True, up to the present time, it is (scientifically) unknown how to "write" a perfect or a good question. However, "writing" a survey question does not correctly identify the task at hand. Survey questions are part of the overall measurement instrument, the questionnaire. Scientific and technical instruments are designed and calibrated precisely and specifically for the targeted objects. In other words, one does not go into the woods, cuts of a twig one believes to be straight, and calls it a measurement instrument. Similarly, one cannot sit at a desk writing texts with question marks and call this a survey instrument. Instead a team of specialists will select appropriate questions or items from the universe of items using clearly defined methods (Mohler, 2006). Guidance is given by current standards and best practices that indicate what should be avoided (negation, double-barreled, context effects, etc.) in designing questions. Thus, they are no prescriptions or cookbooks for "good" questions, but there are well-established procedures governing the questionnaire design process. This process must be quality controlled using well-defined assessments and benchmarks (Presser et al., 2004). Whether a question is "good or bad" is, in the end, not the question. In quality terms, questions/items must optimally meet predefined measurement properties. If they fail to meet such properties, they should be changed or dropped. That is why today it is also standard to apply appropriate assessment techniques such as cognitive interviewing, group discussions, or quantitative experiments. All these belong to the realm of quality assurance, quality controls.

Earlier we discussed the problem of independence of measures as a necessary prerequisite for the validity of MTMM experiments. Our concern was that memory effects could bias estimates of reliability, validity, and finally Q. The designers of the 2006 ESS were well aware of this issue and also of the ethical problem that one should inform respondents about such experimental settings. Thus, they introduced the experimental part of the questionnaire as follows: "To help us improve our questions in the future, here are some final questions on a range of different topics which are similar to previous ones. *Please don't try to remember what you answered before* but treat them as if they were completely new questions" (see ESS documentation, emphasis mine). Asking someone to try not to remember might have counter effects such as stimulating efforts by respondents to remember. It would be worthwhile to study that problem. As it stands now, it looks like a paradoxical instruction that could be a further source of item bias.

Turning now to standard item design issues, we will have a closer look at the items in Table 18.2 above. The base item, Q1, is item B38 in the 2006 ESS. B38 is embedded into an item battery on attitudes toward foreigners/immigrants (quoted here from the ESS source questionnaire, which differs from language and country-specific versions not only due to language but also due to mode—see below technical instrument implementation and data collection) (Table 18.3):

TABLE 18.3. ESS 2006 Item B38 and Its Immediate Context

"Now some questions about people from other countries coming to live in (country)"

B35 CARD 14 Now, using this card, to what extent do you think (country) should allow people of the *same race or ethnic group* as most (country's) people to come and live here?

B36 STILL CARD 14 How about people of a *different* race or ethnic group from most (country) people?

B37 STILL CARD 14 How about people from the *poorer countries outside Europe?*

B38 CARD 15 Would you say it is generally bad or good for (country)'s economy that people come to live here from other countries? Please use this card.

Source: ESS Documentation

In the following paragraph we will discuss shortly some major characteristics of these items. First, some important characteristics of B38: the mode of data collection is personal interview (interviewer reads questions, respondent has show cards, interviewer and respondent setting next-to-next or FtF) not prescribed, in some countries Computer Assisted Self Interview, in some not, (see ESS documentation). The name of the country where the interview takes place is inserted accordingly. No other respondent instruction but "Please use this card" (Card 15 above), similar cards were used before. "Don't know" is a visible option for the interviewer only (in a next-to-next situation respondent might see the "don't know" option). And finally, labels of the response category on the show card do not use the word "generally." That is only part of the read-out question text.

From a qualitative point of view, there might be still some room for improvement, especially the burden for both interviewer and respondent about the usage of Card 15. However, let us assume here that Item B38 passes the quality control after the questionnaire design and is forwarded to technical instrument implementation.

The context for the three experimental MTMM differs greatly from B38. They are part of item batteries as shown for item HS4 below in Table 18.4.

The context of presentation for HS4 is quite different from ItemB38 which was part of a series, not a matrix-like item battery. That is a first change of "method" in terms of the MTMM design. The next is a dramatically different wording of the question. While B38 ask whether it is "good or bad" for the country, HS4 item asks for "bad" only. The attached agree-disagree answer scale camouflages the un-dimensionality of item HS4. In doing so, the designers assume that the underlying construct "effects of immigration" appears to have only one—negative dimensionality—or only the negative dimensionality is attempted to be measured by item HS4. One should note here, to say that one does not agree does not mean that you said something is good, you just disagree that it is bad. One could also assume that asking for good or bad does not imply that there has been a decision made that X is good or bad. Asking "is it bad?" implies that a decision has been made by the researcher that X is actually bad, but one could disagree with the researcher's decision.

In addition to the changed mode of presentation and substantial wording changes, the placement of HS4 is unfortunate in terms of sequence effect. The first three items (HS1, 2, and 3 are all formed that liberal respondents could agree while HS4 all of a sudden asks liberals to disagree. That is, if one agreed that more people should be allowed in my country, I could agree. Both, the content and the response scale go into the same direction. I one asks me "It is bad to let more people into my country" and I have liberal attitudes already expressed before, I have to stop and rethink because a positive answer would contradict my previous liberal statements.

According to experimental research, items containing negative components increase the respondent burden and definitely lowers item quality (Oberski et al., 2010). One should also note that changing the response direction is

TABLE 18.4. Item HS4 Context

HS4
Now some questions about people from other countries coming to live in (country). Please read each question and tick the box on each line that shows how much you agree or disagree with each statement.

	Strongly Agree	Agree	Neither Agree nor Disagree	Disagree	Disagree Strongly
HS1 (Country) should allow more people of the *same race or ethnic group* as most (country's) people come to live here.					
HS2 (Country) should allow more people of a *different* race or ethnic group from most (country's) people to come and live here.					
HS3 (Country) should allow more people from the *poorer countries outside Europe* to come and live here.					
HS4 It is generally bad for (country's) economy that people come to live here from other countries.					

usually used to identify response style behavior (Yang et al., 2010), but it is not understood to be good practice for substantive measurement.

Moreover, only in six countries HS4, HS16, and HS25 have been asked in the "standard" self-completion mode without don't know/can't choose option for the respondent (see ESS documentation). In all other countries asked in FtF mode with don't know as a response category visible for the interviewer only. In addition, in some countries using self-completion mode, interviewers could assist respondents, thus switching to FtF mode (see ESS documentation).

To sum it up, changing response direction in an item battery is a way to identify specific response styles (yea-saying). It should not be used for other purposes. Thus, item HS4 is defective and does not meet basic quality criteria. It should not have made it to the next stages in the survey production cycle. This assessment holds for the other two MTMM experimental items HS16 and HS25 given in Table 18.2.

18.4   INSTRUMENT TECHNICAL DESIGN

We leave adaptation and translation aside as our emphasis here is not on comparability of question texts, and have a closer look at the technical design of item B38 and HS4 (and implicitly also of HS16 and HS25) in different countries that might affect comparability. To answer B38 and HS4 (with the exempt of six countries using self-completion mode for HS4) respondents are asked to look at a show card for formulating their answer. The layout of the show cards differs from country to country. For but one example, look at the source questionnaire to be found on the ESS website (Card 15 for item B38) (Fig. 18.4).

Note that the both labels cover 0 and 2 as well as 9 and 10, respectively, in the source show card. However, in Romania and the UK, the right-hand label covers three response scale points (8–10) (see Figs. 18.5 and 18.6).

What one sees here are that the information given to respondents via show-card layout differs greatly between countries. In addition, looking at the show cards used for HS4, one can find some surprises. For instance, the German 5-point Likert scale is represented as a vertically ordered list of words. That is in line with the layout for the core questionnaire of the ESS 2006. But it is not in line with the source layout for the supplementary questionnaire that sets horizontal Likert scales as standard. So respondents in one (or more) countries respond to a vertical response scale, while others respond to a horizontal one.

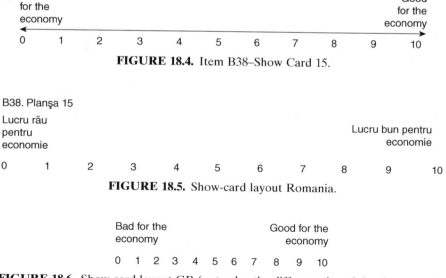

FIGURE 18.4. Item B38–Show Card 15.

FIGURE 18.5. Show-card layout Romania.

FIGURE 18.6. Show-card layout GB (note also the different size of the show card in Great Britain.

The reasons for such differences in layout are not documented, but the layout differences could be a source of bias.

18.4.1 Data Collection

Mode of data collection may have an impact on measurement characteristics (Harkness et al., 2010). The ESS uses FtF as a standard data collection mode with the exemption of the so called "supplementary questionnaire" containing the Schwartz's value scales and second MTMM measurements. Here countries can decide to use either FtF or a self-completion data collection mode (note that the layout is standardized for self-completion only). According to the ESS data documentation, six countries decided to use self-completion mode in 2006: Finland (98%), Hungary (75%), Ireland (84%), Latvia (78%),[4] the Netherlands (89%), Norway (100%), and Sweden (98%) (ESS, 2006, percentage in brackets represent self-completion mode in such countries, all other respondents were interviewed with interviewer assisting or in full FtF mode, that is, in Hungary about 25% of respondents).

The MTMM experiment in the 2006 ESS used a split-ballot design, that is, the three experimental items HS4, HS16, and HS25 were asked to separate subsamples of respondents. As mentioned above, using split-ballot designs for MTMM experiments is a major step forward in making such experiments feasible for general social surveys. But there is a price to be paid for this. Implementing and monitoring of random split-ballot designs is a heavy burden for data collection. From the outset it looks quite simple: Each selected respondent is randomly assigned to one of the subsamples. As one says in German after that assignment, "it gets downhill" because it is most unlikely that contacts, refusals, and responses are evenly distributed across the subsamples. Thus, much effort is put into supervising such complex data collection by field directors and supervisors. In addition, much more effort has to be put in the preparation of the interviewer material to ensure that the assigned experimental questionnaire is given to appropriate respondents. And finally, controlling that complex data collection process is near to a nightmare, and clients asking for such designs should listen very, very carefully to the comments of field directors.

With respect to the MTMM questions discussed here, one can summarize that there could be a mode effect between countries (FtF and self-completion), and in Hungary, Ireland, Latvia, and the Netherlands, an additional within-country mode effect due to the high proportions of respondents that have been assisted by interviewers in filling out the self-completion questionnaire.

18.5 REVISITING Q

Considering the complex design with several split versions, a change of mode in the middle of interview from FtF to paper and pencil (or not), interesting cognitive clues such as "do not try to remember," one could conclude that the complexity of the MTMM implementation without proper process quality monitoring leads to "incomparable" and dubious Q-indicators.

There is some indication for that in the data. For instance, a cross-tabulation of the two items in Spain (B38 and HS4) reveals that from the 123 respondents who answered firstly "bad for the country" (codes 0–3 on the 11-point scale) 17 (15%) seemingly "changed their mind" to "disagree/strongly disagree that it is bad for the economy." From those who first said "it's good for the economy" (codes 7–11) about 9% (22 of 258) reversed their opinion the second time. In the light of several panel studies investigating swapping from one substantial extreme to the other, these are rather implausible figures: van Deth (1983) reported about 2% change in extreme types of Inglehart's Materialism-Postmaterialism over a 6-year period, Mohler (1986) reported similar results from a three-wave test–retest panel (four weeks between each wave), and his results were independently confirmed by Jagodzinski (1986).

Thus, one can muse whether the two items did either not refer to the same latent construct or that a substantial part of respondents were just misled by the item and response scale formulation of HS4. It would be interesting to do some cognitive testing here.

Also of concern is the calibration of both the algorithms used and the substantive benchmarks. It would be extremely helpful to learn how the algorithms used for MTMM estimations are calibrated. Moreover, taking existing items of unknown measured quality (not estimated) as a benchmark for MTMM interpretations might be difficult to defend. Here too, more information beyond the Saris paper and the Saris and Gallhofer (2007) publication would be very helpful.

Finally, there is a real surprise in the data. Comparing the Q-value for the three experimental MTMM items (HS4, HS16, and HS24), HS4 has higher Q-values in all countries (Table 18.5). Why is that surprising? Because HS4 uses a standard fully labeled 5-point Likert response scale, while the other two use a 7- or 11-point response scale with only end points labeled (see Table 18.2). That a 5-point Likert response scale shows the best Q-values is counter to the common wisdom that 11-point scales are the best (see ESS documentation). Thus, one might muse how much quality is either in Q or our common wisdom.

18.5.1 Some Comments on Survey-Tech-Speak and Survey Theory

That all the many people involved in producing a social survey (usually several hundred people) understand what they are supposed to do and why it is necessary to target specific quality benchmarks requires a lingua franca that allows them to communicate. However, each group seemingly insists on its own lingo and tries to outperform the others in inventing new labels. For instance, harmonization is used in the archive and official statistics world, others speak of standardization, or methodologists speak of latent constructs while modelers

TABLE 18.5. Comparison of Q-Values for Three Response Scales

Country	5-Point Lik	11-Point Lik	7-Point Lik
Austria	0.46	0.32	0.32
Belgium	0.51	0.24	0.29
Bulgaria	0.30	0.13	0.22
Switzerland	0.50	0.20	0.49
Cyprus	0.47	0.53	0.36
Germany	0.43	0.32	0.38
Denmark	0.61	0.40	0.41
Estonia	0.41	0.17	0.22
Spain	0.46	0.24	0.21
Finland	0.60	0.38	0.37
France	0.55	0.31	0.25
UK	0.41	0.28	0.31
Ireland	0.37	0.02	0.16
Latvia	0.24	0.05	0.10
The Netherlands	0.38	0.23	0.29
Norway	0.67	0.09	0.36
Poland	0.33	0.10	0.19
Portugal	0.47	0.18	0.40
Romania	0.29	0.08	0.17
Russia	0.42	0.36	0.27
Slovenia	0.37	0.01	0.13
Slovakia	0.32	0.12	0.14
Ukraine	0.44	0.17	0.12

Source: Adapted from Saris, 2009.

use variable of interest instead, some pretesters use cognitive interviews, while some psychologists use in-depth interview, a new fashion is to use paradata for contact form data. It is obvious that such insular lingos hamper the progress of survey research. There are, however, attempts to create a lingua franca in being as consistent as possible in using a low level and clearly defined survey-tech-speak (Groves et al., 2004/2009, Harkness et al., 2003, 2010).

Survey methodology also lacks a common theoretical frame, a theoria franca that allows to discuss across different research areas how respondents perceive questions, how they are answered, how proper designs could be developed, or what best statistics to use in a given case. Surely, there are some question processing schemata or guidelines how to construct good questions (see Groves et al., 2004/2009; CCSG, 2010; Harkness et al., 2010). However, they do not explicitly take advantage of modern knowledge management, communication research, the revolution in cognition and survey research, not to speak of the neuro-sciences. On the other hand, statistical tools and methods are often not widely known or suffer from the hammer and nail syndrome: if you got a hammer, everything becomes a nail. In an essence, this is all about the compartmentalization of survey research into noncommunicative groups such as marketing or official statistics, or small groups such as cognitive testing or, here, MTMM.

Two examples about compartmentalization of survey theory might suffice here. The first is from Saris and Gallhofer who state "one item in a survey cannot present an attitude" (2007, p. 16). This statement contradicts successful psychological research into single-item scales (Just bing or google "single-item scale"). In the single-item-scale approach, one looks at general attitudes which are difficult to decompose into several items. Moreover, asking someone several times about "satisfaction" might, according to the cognitive literature, trigger a completely different communication process than the one intended by the researcher and thus miss the target of research.

The second is about experiments, small group research, and general surveys. Many of the effects predicted by cognitive survey researchers (Sudman et al., 1996, Schwarz et al., 2010) cannot or not easily be found in general household surveys. This has a number of reasons, such as less extreme formulations than the ones used in a psychological experiment, counter effects due to the heterogeneity of the population (recency and latency effects in student population are different from a general population), and a misunderstanding of what a psychological experiment is about. Until now, nonpsychologists organize their small group research mechanically along the lines of demographics (gender, education, age). It would be helpful for everyone to see that an experiment is not about representation of a population but checking whether a predicted effect occurs or not (homogeneous population in respect to the effect investigated).

In conclusion, we would like to reiterate the dependence of total survey quality indicators such as Q on high-quality input data, that is, optimal survey instruments, sampling designs, technical instrument implementation, adaptation, and translation, if needed, data collection, data editing, and documentation of all production stages. With high-quality surveys such as the ESS and its outstanding documentation, there is no excuse left for ignoring obvious quality issues at specific production stages. In other words, the days are over, where all data matrices could be taken at face value ("at night all cats are grey"). Instead, process quality assessments and demonstrations that assumptions for specific analyses are fulfilled is state-of-the-art today that sheds new light on how to improve survey quality. It would be embarrassing if one has to state 2012 what Campbell and Fiske said in 1992 about the substantial problems in linking psychology and modeling: "Our article had impact because it raised a problem: the links between psychological methods and psychological constructs. The problem is still with us 33 years and more than 2000 citations later" (Campbell and Fiske, 1992, p. 394).

NOTES

1   The syllogism "no valid conclusions from false premises" does not imply that true premises always lead to valid conclusions. Thus, true premises (input) are necessary

for valid conclusions (output) but are not sufficient. In more prosaic modern terms computational science indicated this problem using the term GIGO (garbage in-garbage out; Hand, 1993).

2   One should note that complex item quality assessments have been standard in psychological research for a long time; it is only recently that survey research redis-covered this strand of methodology.

3   This definition is adapted from Heeringa and O'Muircheartaigh (2010).

4   Note that Latvian data are, due to missing design weights, not part of the integrated ESS 2006, edition 3.6 data set used here.

REFERENCES

Alwin DF (2007). Margins of Error: A Study of Reliability in Survey Measurement. Hoboken, NJ: John Wiley & Sons.

Andrews FM (1984). Construct validity and error components of survey measures: a structural modeling approach. Public Opinion Quarterly; 46:409–442.

Biemer PP, Lyberg L (2003). Introduction into Survey Quality. Hoboken, NJ: J. Wiley & Sons.

Bollen KA (1989). Structural Equations with Latent Variables. New York: Wiley.

Campbell DT, Fiske DW (1959). Convergent and discriminant validation by the multitrait-multimethod matrices. Psychological Bulletin; 56:81–105.

Campbell DT, Fiske DW (1992). Citations do not solve problems. Psychological Bulletin; 112(3):393–395.

Clark GA (1958). Note on false premises and true conclusions. The Journal of Philosophy; 55(26):1148–1149.

Duncan OD, Stenbeck M (1988). No opinion or not sure? Public Opinion Quarterly; 52(4):513–525.

ESS—European Social Survey (2002/2010). Available online at http://www. europeansocialsurvey.org (accessed September 9, 2010).

Groves RM (2006). Nonresponse rates and nonresponse bias in household surveys. Public Opinion Quarterly; 70:646–675.

Groves RM, Couper MP (1998). Nonresponse in Household Interview Surveys. New York: J. Wiley & Sons.

Groves R, Heeringa S (2006). Responsive designs for household surveys. University of Michigan, Survey Methodology Program, Working Paper Series. Available online at http://www.isr.umich.edu/src/smp/Electronic%20Copies/127.pdf.

Groves R, Fowler FJ, Couper MP, Lepkowski JM, Singer E, Tourangeau R (2004/2009). Survey Methodology. Hoboken, NJ: John Wiley & Sons.

Hand DJ (1993) (in review). Journal of the Royal Statistical Society. Series C (Applied Statistics); 42(2):410.

Harkness JA (1999). In pursuit of quality: issues for cross-national survey research. International Journal of Social Research Methodology; Vol:2(2):125–140.

Harkness JA et al. (2003). Comparative Survey Methods. Hoboken, NJ: J. Wiley & Sons.

Harkness JA et al. (2010). Survey Methods in Multinational, Multiregional and Multicultural Contexts. Hoboken, NJ: J. Wiley & Sons.

Heeringa SG, O'Muircheartaigh C (2010). Sampling designs for cross-cultural and cross-national survey designs. In: Harkness JA, Braun M, Edwards B, Johnson TP, Lyberg L, Mohler P, Pennell B-E, Smith TW, editors. Multinational, Multicultural and Multiregional Survey Methods. Hoboken, NJ: John Wiley; pp. 251–268.

Jagodzinski W (1986). Black & White statt LISREL? Wie groß ist der Anteil von Zufallsantworten beim Postmaterialismus Index? ZA Informationen, Nr.; 19:30–51.

Mohler PPh (1986). Mustertreue Abbildung—Ein Weg zur Lösung des Stabilitäts-Fluktuationsproblems in Panelumfragen. ZUMA-Nachrichten, Nr.; 19:31–44.

Mohler PPh (2006). Sampling from a universe of items and the De-Machiavellization of questionnaire design. In: Mohler PPh, Braun M, editors. Beyond the Horizon of Measurement. ZUMA Nachrichten Spezial, Vol. 10. Mannheim: ZUMA; pp. 9–14.

Mohler PP et al. (2010). A survey process quality perspective on documentation. In: Harkness JA, Braun M, Edwards B, Johnson TP, Lyberg L, Mohler P, Pennell B-E, Smith TW, editors. Multinational, Multicultural and Multiregional Survey Methods. Hoboken, NJ: John Wiley; pp. 299–314.

Noelle-Neumann E, Petersen T (2005). Alle, nicht jeder—Einführung in die Methoden der Demoskopie. Berlin: Springer.

Oberski D, Saris WE, Hagenaars JA (2010). Categorization errors and differences in the quality of questions in comparative surveys. In: Harkness JA, Braun M, Edwards B, Johnson TP, Lyberg L, Mohler P, Pennell B-E, Smith TW, editors. Multinational, Multicultural and Multiregional Survey Methods. Hoboken, NJ: John Wiley; pp. 435–453.

Pennell B-E et al. (2010). Cross Cultural Survey Guidelines (CCSG). Ann Arbor, MI: ISR. Available online at http://ccsg.isr.umich.edu/ (accessed September 9, 2010).

Presser S, Rothgeb JM, Couper MP, Lessler JT, Martin E, Martin J, Singer E, editors (2004). Methods for Testing and Evaluating Survey Questionnaires. New York: Wiley.

Riemer S (1954). Premises in sociological inquiry. The American Journal of Sociology; 59(6):551–555.

Saris WE (2009). The MTMM approach to coping with measurement errors in survey research. Paper presented at the 2009 QEM Washington Workshop.

Saris WE, Gallhofer IN (2007). Design, Evaluation, and Analysis of Questionnaires for Survey Research. Hoboken, NJ: Wiley & Sons.

Scheuch EK (1968/1993). The cross-cultural use of sample surveys: problems of comparability. Historical Social Research; 18(2):104–138. First published in: Rokkan S, editor. (1968). Comparative Research across Cultures and Nations, Paris/The Hague: ISSC/Mouton, pp. 176–209.

Schwarz N, Oyserman D, Peytcheva E (2010). Cognition, communication and culture: implications for the survey response process. In: Harkness JA, Braun M, Edwards B, Johnson TP, Lyberg L, Mohler P, Pennell B-E, Smith TW, editors. Multinational, Multicultural and Multiregional Survey Methods. Hoboken, NJ: John Wiley; pp. 175–190.

Stoop IA (2005). The Hunt for the Last Respondent. Nonresponse in Sample Surveys. Den Haag: Social and Cultural Planning Office.

Stoop IA et al. (2010). Improving Survey Response. Hoboken, NJ: J. Wiley & Sons.

Sudman S, Bradburn NM, Schwarz N (1996). Thinking about Answers: The Application of Cognitive Processes to Survey Methodology. San Francisco, CA: Jossey-Bass.

Tormarken AJ, Waller NG (2005). Structural equation modeling: strengths, limitations, and misconceptions. Annual Review of Clinical Psychology; 1:31–65.

Trochim WM. The Research Methods Knowledge Base, 2nd Edition. Available online at http://www.socialresearchmethods.net/kb/.

van Deth J (1983). The persistence of materialist and post-materialist value orientations. European Journal of Political Research; 11:63–79.

van Meurs A, Saris WE (1990). Memory effects in MTMM studies. In: Saris WE, van Meurs A, editors. Evaluation of Measurement Instruments by Meta-Analysis of Multitrait Multimethod Matrices. Amsterdam: North-Holland; pp. 52–80.

Yang Y, Harkness JA, Chin T-Y (2010). Response styles and culture. In: Harkness JA, Braun M, Edwards B, Johnson TP, Lyberg L, Mohler P, Pennell B-E, Smith TW, editors. Multinational, Multicultural and Multiregional Survey Methods. Hoboken, NJ: John Wiley; pp. 203–223.

PART VII
Field-Based Data Methods

19 Using Field Tests to Evaluate Federal Statistical Survey Questionnaires

BRIAN A. HARRIS-KOJETIN
U.S. Office of Management and Budget

JAMES M. DAHLHAMER
National Center for Health Statistics

19.1 INTRODUCTION

In contrast to previous chapters of this volume that focus on specific evaluation methods or statistical models, field tests often incorporate multiple evaluation methods, have multiple goals, and involve the collaboration of questionnaire designers and other researchers, statisticians, survey managers, and personnel directly involved in data collection. By their nature, field tests, in contrast to laboratory or office-based methods, take place in real-world settings and are often designed to mimic the implementation of the main survey as closely as possible. Thus, field tests represent an integration of many of the topics covered in the Question Evaluation Method (QEM) workshop and this volume. For example, some previous chapters have covered methods either typically employed in field test settings, such as experiments and behavior coding, or focus on techniques that often depend largely upon data gathered in a large-scale field test, such as item response theory (IRT) modeling. Field test methods can also be used as part of ongoing data collection and may utilize paradata in order to identify potential question problems.

Field tests play an important role in a research program to develop new survey questions or to redesign an existing survey. In this chapter, we will briefly describe what field tests are, why they are conducted, and what methods

Question Evaluation Methods: Contributing to the Science of Data Quality, First Edition.
Edited by Jennifer Madans, Kristen Miller, Aaron Maitland, Gordon Willis.
© 2011 John Wiley & Sons, Inc. Published 2011 by John Wiley & Sons, Inc.

are often used in field tests to evaluate questions. Examples are provided to illustrate how U.S. federal statistical agencies conduct pretesting activities, the standards they follow when designing their surveys, as well as how field tests are used to reduce measurement error and improve the overall quality of survey data. Finally, the limitations of field testing are discussed and thoughts for future directions are posed, including the use of survey paradata for questionnaire evaluation.

19.2   WHAT ARE FIELD TESTS?

There is little consistency or agreement in terminology that is used to characterize different pretesting activities (Biemer and Lyberg, 2003). Indeed, a variety of activities, such as pilot surveys, feasibility tests, embedded experiments, and methodological studies can be considered field tests. A "conventional" pretest (e.g., Presser and Blair, 1994) typically involves conducting a small number of interviews that are completed by a handful of interviewers. The response distributions are then examined, and a debriefing of the interviewers is conducted (Presser et al., 2004). Field tests can range greatly in scope from a small number of interviews to large-scale tests that approximate a full ongoing production survey (Tucker et al., 1998). A large-scale test can serve as a scalable preparation, that is, a dress rehearsal, for the actual survey. For purposes of this chapter, field test studies will include any evaluative study involving preliminary data collection that uses procedures similar to those of the actual production survey.

With regard to question and questionnaire evaluation, the quantitative results from a field test can be analyzed to assess response distributions, the frequency of item missing data and reasons for missing information (e.g., "don't knows" vs. refusals), and determine whether the instrument's routing and skip patterns are being followed correctly. Depending on the size and representativeness of the sample, field tests may provide the data necessary for more sophisticated analyses of response distributions, including examinations of IRT parameters (see Reeve, this volume). These initial results may also inform the collapsing or the addition of new response categories.

Field tests, then, are not so much a method for performing survey questionnaire evaluations as they are mechanisms or settings that allow the utilization of other evaluation methods. Specifically, field tests offer the opportunity to employ other questionnaire evaluation methods, such as behavior coding (see Fowler, this volume), experiments and embedded experiments (see Krosnick, this volume; Tourangeau, 2004), interviewer debriefings, and respondent debriefings. In addition to analyzing the quantitative survey data, other supplementary data can be collected alongside the survey data that can indicate how survey questions are performing. Field tests typically include both field staff who will be involved in implementing the main survey as well as respondents who are part of the target population. Thus, field tests often incorporate some means of gathering feedback from the interviewers, respondents, or both on how the survey questionnaire worked.

With regard to respondents, there are a variety of techniques that survey researchers have used to obtain reactions to survey questions. Similar to cognitive interviews in the laboratory, respondent debriefings are typically conducted to determine whether concepts and questions are understood by respondents in the way in which the survey designers intended (DeMaio and Rothgeb, 1996). Although some field-based probes may be used for specific questions or as part of the interview in the field tests (Willis, 2005), respondent debriefings are usually conducted at the end of the interview so the content of the debriefing questions do not bias responses to later questions.

The most common approach to respondent debriefings is asking respondents directly how they interpret and define the terms in a question to see if their definition is consistent with that of the researchers or survey designers (DeMaio and Rothgeb, 1996; Martin, 2004). Other approaches have also been used, including asking respondents if they found certain questions sensitive, asking for their certainty or confidence in their responses, or probing to determine if the respondent failed to report something or reported incorrectly (see Martin, 2004). Another approach to gaining insight into respondents' understanding is to use vignettes, which are brief stories or scenarios that present hypothetical situations to which respondents are asked to react (Martin, 2004); their responses, it is argued, reveal how they interpret the concepts the researcher is attempting to measure.

Interviewer feedback, on the other hand, is typically collected after the completion of the entire field test project. Interviewers are asked to provide commentary on their experiences with the questionnaire, how respondents reacted to the survey, and aspects of survey operations (Biemer and Lyberg, 2003). Interviewer debriefings may be formal focus groups or informal discussions with interviewers, or may even include administering questionnaires to interviewers to gain more systematic feedback.

Although these techniques can provide useful insights into question problems as well as potential ways to fix those problems, they have significant weaknesses. Interviewers are not always accurate reporters of question problems. They may not know the underlying cause of the problem, they may report their own preferences for a question rather than the respondents, and some may change the wording of problem questions without realizing it (U.S. Census Bureau, 2010). Others have noted that the information collected in interviewer debriefings is often subjective, unsystematic, and shows a lack of agreement among interviewers (Oksenberg et al., 1991; Fowler, 2002). During a pretest or field test, interviewers may complete only a handful of interviews and recall the more extreme or atypical situations (Czaja, 1998). Although respondent debriefings and vignettes appear to be promising methods, they have not been rigorously evaluated (Martin, 2004). Interviewers may have difficulty in administering respondent debriefing questions, while researchers need to have a clear idea of the potential problems in order to create specific debriefing questions. And, unlike a laboratory setting, only a few debriefing questions can reasonably be administered (Fowler and Roman, 1992).

19.3   WHY SHOULD FIELD TESTS BE CONDUCTED?

As previously noted, it is best to think of a field test as a platform in which various QEMs are used. Multiple methods can provide information on different issues (Groves et al., 2009) and, in combination, can provide greater insight into questionnaire problems than a single-method-approach. These additional methods can also address some of the weaknesses of other pretesting methods and can provide a quantitative portrayal of how the questionnaire is working. For example, laboratory and office-based settings typically include only respondents who are paid and motivated to participate, and most likely do not represent the entire target population (Fowler and Roman, 1992). Field tests typically use representative samples, but even if a purposive sample is chosen, the respondents are in their natural setting, and so additional issues may arise that were not uncovered in the lab or office.

Additionally, some issues of concern can only be reasonably addressed by using field test methods. For example, Tourangeau (2004, p. 216) notes that "clearly, to the extent that an experiment is attempting to forecast the impact of a questionnaire redesign on an ongoing study, a realistic field experiment with a probability sample of respondents drawn from the same population (with the same sample design) as the ongoing survey is absolutely necessary." Other types of studies that seek to benchmark results or assess validity typically require field tests with large, representative samples or utilizing the production sample from the main survey.

Although this volume is on QEMs, field tests are often used for more than just evaluating questionnaires. Field tests provide useful data on how study procedures will work in practice, what response rates may be, and what non-response problems exist (Dillman et al., 2009). In some cases, these other concerns may be of greater concern to the survey managers than the questionnaire, so the field test may be more optimally designed to assess those issues rather than the questionnaire. As Groves et al. (2009, p. 253) observed, "a large survey is rarely fielded without a correspondingly large field pretest. The risk of a major operational failure is too great to go into the field without substantial pretesting."

19.4   THE FEDERAL CONTEXT: STANDARDS AND REVIEW PROCESS FOR FEDERAL SURVEYS

There appears to be no dearth of advice on the necessity of pretesting survey questionnaires. Field tests are often implicitly included or even explicitly referred to as a method for pretesting. For example, Presser et al. (2004) note that elementary textbooks and experienced researchers declare pretesting an essential step to evaluate, in advance, whether a questionnaire causes problems for interviewers or respondents. The American Association for Public Opinion Research (AAPOR) lists as one of its 12 "Best Practices" to "Pretest questionnaires and procedures to identify problems prior to the survey." AAPOR notes that "A pretest of the questionnaire and *field procedures* (emphasis added) is the only way of finding out if everything "works," especially if a survey employs new techniques or a new set of questions."[1] Similarly, Dillman's Tailored Design Method (Dillman et al., 2009) offers four guidelines on pretesting questionnaires, including Guideline 6.38: "Conduct a small pilot study with a subsample of the population in order to evaluate interconnections among the questions, the questionnaire, and the implementation procedures" (p. 228).

In the United States, the Paperwork Reduction Act (PRA) requires federal agencies to obtain approval from the Office of Management and Budget (OMB) prior to collecting information from 10 or more members of the public, whether from individuals, households, establishments, educational institutions, organizations, or other levels of government. All proposed information collections are subject to public comment as part of this process. This review is not limited to survey or to statistical agency data collections. Rather, the reviews extend to *all* collections of data, whether they originate for statistical, administrative, or regulatory uses. OMB's review and approval process provides the mechanism (1) to ensure that statistical methods are appropriate for intended uses, (2) to monitor use of classification standards, (3) to coordinate collections carried out by various departments, (4) to prevent duplicative requests, and (5) to reduce respondent burden. Surveys and other data collections are reviewed to ensure that they conform to the proper statistical methodology, standards, and practices. Furthermore, collections are approved for a maximum of three years and must be approved again if the agency plans to continue use.

19.4.1 OMB Standards

OMB is charged under the PRA with ensuring the quality, integrity, and accessibility of federal government statistical methodologies, activities, and products through the issuance of government-wide policies, guidelines, standards, and classifications that are developed in collaboration with the federal statistical agencies. OMB issues standards of various types, but of primary interest here are OMB's standards and guidelines for statistical surveys that are related to pretesting questionnaires. The standards and guidelines are intended to ensure that such surveys and studies are designed to produce reliable data as efficiently as possible and that methods are documented and results are presented in a manner that makes the data as accessible and useful as possible.

OMB has provided specific guidance to agencies on the use of pretesting and field testing activities. For example, in *Questions and Answers When*

Designing Surveys for Information Collections (U.S. Office of Management and Budget, 2006a), one of the questions asks "When should agencies conduct a pilot study, pretest, or field test?" One component of the answer specifically focuses on the use of field tests for questionnaire development or revision:

> Agencies may want to conduct pretests when developing new questionnaires to see how respondents actually answer questions and identify potential data quality problems, such as high item nonresponse rates. Agencies may also conduct pretests to gather data to refine questionnaire items and scales and assess reliability or validity. Sometimes agencies may also use a field test or experiment (a study to compare the effects of two or more procedures or questionnaires) when planning a change in methodology or questions in an ongoing survey. This enables comparisons and often provides quantifiable data to decide among the different methods or questions to use. An agency may further want to consider conducting a field test experiment on a representative sample to measure the effect of the change in methods or questions on resulting estimates.

In OMB's *Standards and Guidelines for Statistical Surveys* (U.S. Office of Management and Budget, 2006b), one of the standards focuses specifically on pretesting:

> **Standard 1.4:** Agencies must ensure that all components of a survey function as intended when implemented in the full-scale survey and that measurement error is controlled by conducting a pretest of the survey components or by having successfully fielded the survey components on a previous occasion.

Furthermore, the associated guidelines provide more specific information and best practices for implementing the standard. In this case, using a field test to develop or revise a survey questionnaire is just one of the uses that are noted:

> **Guideline 1.4.2:** Use field tests prior to implementation of the full-scale survey when some or all components of a survey system cannot be successfully demonstrated through previous work. The design of a field test should reflect realistic conditions, including those likely to pose difficulties for the survey. Elements to be tested include, for example, frame development, sample selection, questionnaire design, data collection, item feasibility, electronic data collection capabilities, edit specifications, data processing, estimation, file creation, and tabulations. A complete test of all components (sometimes referred to as a dress rehearsal) may be desirable for highly influential surveys.

19.4.2   Agency Standards for Conducting Field Tests

Because OMB standards and guidelines must cover a broad range of applications, agencies are encouraged to develop their own more specific standards for the statistical surveys and studies they conduct or sponsor. Several agencies have published standards for their statistical surveys, and several include standards or guidelines for field tests.[2] For example, the U.S. Census Bureau (2010) has a quality standard for "Developing Data Collection Instruments and Supporting Materials" that includes

> **Requirement A2-3:** Data collection instruments and supporting materials must be developed and tested in a manner that balances (within the constraints of budget, resources, and time) data quality and respondent burden.

> **Sub Requirement A2-3.3:** Data collection instruments and supporting materials must be pretested with respondents to identify problems (e.g., problems related to content, order/context effects, skip instructions, formatting, navigation, and edits) and then refined, prior to implementation based on the pretesting results.
> . . .
> 3. One or more of the following pretesting methods must be used:

> (a) Cognitive interviews.
> (b) Focus groups, but only if the focus group completes a self-administered instrument and discusses it afterwards.
> (c) Usability techniques, but only if they are focused on the respondent's understanding of the questionnaire.
> (d) Behavior coding of respondent/interviewer interactions.
> (e) Respondent debriefings in conjunction with a field test or actual data collection.
> (f) Split panel tests. (pp. 6–9)

This Census Bureau standard (2010) includes an appendix describing "prefield" and field techniques. Although the field techniques referred to above can be used to satisfy the Census pretesting requirement, the use of both field and pre-field techniques are recommended in the Census standards:

> Pretesting is typically more effective when multiple methods are used. Additional pretesting techniques should be carefully considered to provide a thorough evaluation and documentation of questionnaire problems and solutions. The relative effectiveness of the various techniques for evaluating survey questions depends on the pretest objectives, sample size, questionnaire design, and mode of data collection. The Census Bureau advocates that both pre-field and field techniques be undertaken, as time and funds permit. (p. 21)

While the Census Bureau standard is for pretesting the questionnaire, other agencies may take a broader focus by using field tests to evaluate multiple aspects of a survey design and methodology. For example, the U.S. Department of Education, National Center for Education Statistics (2002) standards and guidelines for pretesting includes the following:

> **Standard 2-4-2:** A second type of pretest is a field test. Components of a survey system that cannot be successfully demonstrated through previous work must be

field tested prior to implementation of the full-scale survey. The design of a field test must reflect realistic conditions, including those likely to pose difficulties for the survey. Documentation of the field test (e.g., materials for technical review panels, working papers, technical reports) must include the design of the field test; a description of the procedures followed; analysis of the extent to which the survey components met the pre-established criteria; discussion of other potential problems uncovered during the field test; and recommendations for changes in the design to solve the problems.

Guideline 2-4-2A: Elements to be tested and measured may include alternative approaches to accomplishing a particular task. Elements to be tested may include: frame development; sample selection; questionnaire design; data collection; response rates; data processing (e.g., entry, editing, imputation); estimation (e.g., weighting, variance computation); file creation; and tabulations.

19.5   SOME USES OF FIELD TESTS BY FEDERAL AGENCIES

To better illustrate the uses of field tests, some specific examples of how federal agencies have used field tests are presented. The primary goal is to offer a broad perspective highlighting the use of additional evaluation methods in field tests.

19.5.1   American Community Survey (ACS): Census Bureau

The ACS is the ongoing household survey that replaces the decennial census "long form." Nearly 3 million addresses (both housing units and group quarters) are included in the annual sample. Because it is part of the decennial census program, participation in the survey is mandatory, and all of the items on the survey are required by law, regulation, or are vital to administering federal programs. Due to the mandatory nature of the survey, it is important that the survey instrument minimizes respondent burden. In addition, it is important for the content of the survey to remain relatively stable over time in order to be able to provide the multiyear estimates needed for small geographies. Therefore, OMB and the Census Bureau developed a process for the development and testing of any new or revised items to be included on the ACS (see U.S. Census Bureau, 2009). The Interagency Committee for the ACS, co-chaired by the OMB and the Census Bureau and comprised of representatives from more than 30 federal agencies, provides oversight and coordination to this process.

First, agencies wishing to add or change existing items on the ACS must provide an initial justification and proposal to Census and OMB. If this initial proposal is approved, then the agency must form and lead a subcommittee to develop the alternative versions of question wording for testing. Other federal agencies are invited to participate and a Census Bureau staff person coordinates participation across Census Bureau divisions. These different question version(s) are then subjected to one or more rounds of cognitive testing. Items that have "passed" cognitive testing are then field tested in a split panel content test. For new content areas, the field test includes at least two versions of the question to compare with each other. For existing ACS items, one or more alternative versions are tested with the current version serving as the control for the proposed revised question(s). The field test is conducted on a large sample separate from the ongoing ACS, and the testing must be conducted in all three modes of data collection used in the ACS: mail, computer-assisted telephone interviewing (CATI), and computer-assisted personal interviewing (CAPI). The last ACS content test was conducted in fall 2010. The primary results are quantitative comparisons of different questions in terms of panel response rates, item response distributions, and item response rates. Additional qualitative and quantitative methods, such as interviewer debriefings and behavior coding are also often included.

In addition, the content tests have typically included a reinterview to assess reliability over time. The reinterview may also include specific probes to attempt to determine as closely as possible whether the respondent understood and answered the question correctly; this is then compared to the original answer provided. For example, in 2007 the ACS tested two different versions of a new Field of Degree question that was requested by the National Science Foundation's Division of Science Resources Statistics. The results from the content reinterview were vital to assessing which of the two versions were more reliable in capturing the field of degree. Data from the National Survey of College Graduates (NSCG), which asked more detailed questions, provided a benchmark to assess the two different questions. In this case, comparing distributions of content test data to the NSCG combined with the reinterview results showed that an open-ended question performed much better than a set of response categories.

19.5.2 Hospital Consumer Assessment of Health Plans Survey (HCAHPS): Centers for Medicare and Medicaid Services (CMS) and Agency for Healthcare Research and Quality (AHRQ)

AHRQ has developed a series of Consumer Assessment of Health Plan Surveys (CAHPS) instruments to measure consumer perceptions of their health plans, including communication with doctors, nurses, office staff, whether they were able to get appointments when needed, whether they were treated with respect, and others. The results from these survey ratings are available on the Medicare website for seniors and are routinely available for federal employees and for many private sector health plans as well.

For these surveys, AHRQ and its contractors have followed a similar process to develop the instruments through a rigorous series of steps including careful examination of the literature and existing instruments, rounds of iterative cognitive testing, and field tests of the instruments. One example of the field tests conducted on these surveys was the pilot study sponsored by CMS on the Hospital CAHPS or HCAHPS instrument in 2003 (CMS, 2003). This pilot study instrument contained 66 items and was administered to medical, surgical, and obstetric patients who had an overnight hospital stay and were discharged between December 2002 and January 2003 at 1 of 109 hospitals in Maryland, Arizona, or New York.

The focus of the pilot was to obtain sufficient data for psychometric analysis of CAHPS items. Exploratory factor analyses were conducted to guide refinements to the initially hypothesized structure, and the revised structure was evaluated using item-scale correlations, internal consistency reliability, and correlations with global ratings. The revised instrument included 32 questions tapping seven domains of care as well as several global items. Additional analyses were also conducted to identify variables for case-mix adjustment.

19.5.3 National Assessment of Education Progress (NAEP): National Center for Education Statistics (NCES)

The NAEP is a federal survey of student achievement at grades 4, 8, and 12 in subject areas such as reading, mathematics, writing, science, and others. The NAEP is the source of the Nation's report card, providing an evaluation of how well students and schools are performing. Under the No Child Left Behind legislation, every state participates in the reading and mathematics assessments at grades 4 and 8, and most states participate in the writing and science assessments. These state-level assessments require reporting results at the state as well as national level, resulting in very large sample sizes. NAEP is an adaptive test that is required by law to report results within 6 months of data collection, a very tight timeline.[3]

NCES employs several strategies to develop survey items, including small and large-scale pilot testing to test potential assessment items and select those that will be included on the final instruments. Pilot tests are conducted with nationally representative samples of students to gather information about performance across the whole spectrum of student achievement. In general, two items are tested for each one that will appear on the final instrument. In addition, NCES conducts "precalibration tests" of the final instruments (after pilot testing) to obtain IRT parameters in advance of the main assessment. Thus, the data from the main assessment can then be analyzed using the defined item parameters and produce scale scores much more quickly.

19.5.4 National Survey of Drug Use and Health (NSDUH): Substance Abuse and Mental Health Services Administration (SAMHSA)

The National Survey on Drug Use and Health (NSDUH) is an annual survey of the civilian, noninstitutionalized population age 12 or older that collects data on the use of alcohol, tobacco, and illicit substances to track the prevalence of substance use in the United States. The survey relies on self-reported information, but uses computer-assisted interviewing with sensitive questions administered via audio computer-assisted self-interviewing (ACASI) methods.[4]

In 2000 and 2001, SAMHSA conducted a field-based validity study on a sample of approximately 4000 respondents aged 12–25 using a revised version of the questionnaire. Some of the questionnaire changes included adding follow-up questions about drug use corresponding to shorter time periods in order to be comparable with the window of detection of most drugs in urine and hair (e.g., use of marijuana in the past 3 days, the past 7 days, and the past 180 days). A new module was added that re-asked many of the drug use questions with the shorter time periods. One-half of the respondents received these questions preceded by a detailed introduction that discussed the importance of the study and asked for honest responses (the appeal). The other half received the same questions, but was preceded by a very brief introduction (nonappeal). Other questionnaire changes included the deletion of several modules (e.g., social environment, health insurance, and other modules) in order to keep the total interview to about 1 hour in length. At the end of each interview, respondents were asked to provide a hair and urine specimen, for which SAMHSA was able to obtain high response rates. The specimens were mailed to a testing laboratory. For urine specimens, self-reported drug use (marijuana, cocaine, opiates, and amphetamines) and tobacco use was compared with the test results (there were problems encountered with the hair testing). Most youths (12–17) and young adults (18–25) reported their recent drug use accurately; however, there were some reporting differences in either direction—with some not reporting use and testing positive, and some reporting use and testing negative.

19.6 WHAT ARE THE WEAKNESSES OF FIELD TESTS?

There are several noteworthy drawbacks to conducting field tests for performing question evaluation. First, large-scale field tests and experiments can be difficult to implement. Unlike laboratories, where there is much greater control, there is much more "noise" and many more uncontrollable factors in a field setting (Tucker et al., 1998). Some implementation difficulties can also arise due to the differences between the research and production cultures (Dillman, 1996). Unlike laboratory experiments, the interviewers and researchers are not double-blind; everyone typically knows the different procedures or instruments (or knows of the existence of the other procedure or instrument even if they are only implementing one version) introducing potential Hawthorne effects. Additional problems may occur when different personnel implement one version of the instrument than the other, as the supervision, training, and experience of the different field staffs may affect the observed results (Tucker et al., 1998; Moore et al., 2004).

Additionally, field tests are typically not iterative (but see Esposito, 2004; Moore et al., 2004), so what is learned from the test may be quite limited and, like many of the other evaluation methods, field tests may identify question or questionnaire problems, but may not provide information on how to fix the problem or how much the problem actually affects data quality.

Perhaps the most significant drawback to conducting field tests, however, is the cost. Planning and then actually conducting a field test, along with analyzing the resulting data and incorporating those results into the main survey, is resource dependent. It takes months to prepare and complete and requires a staff with a wide range of expertise. In terms of funding, time, and staff, therefore, field tests can be prohibitively expensive. At the same time, many government surveys have been hard hit by budget cuts or stagnant budgets that have failed to keep up with the increasing costs of production work. In this type of budgetary context, surveys find themselves doing without field tests simply to maintain ongoing production. In order to perform evaluative research and, at the same time, maintain production, new and creative approaches must be discovered.

19.7   PARADATA ANALYSIS AS AN ALTERNATIVE OR COMPLEMENT TO FIELD TESTS

A relatively new approach that is gaining popularity is the use of survey paradata as a supplemental tool for both evaluating the usability of survey questionnaires and for identifying potentially problematic survey items. Here, we largely restrict our definition of paradata to the automatic by-product data of a computer-assisted field data collection. Examples include audit trails, trace files, and keystroke files that capture question-level date and time stamps and interviewer or respondent movement through the survey instrument (including backward movement). The use of function keys (e.g., pressing F1 to access a help screen for a question), occasions of programmed error messages, and the locations of interview breakoffs or terminations may also be captured in these files. Paradata of this sort have been used (1) to describe and classify response behavior, (2) to relate response behavior to data quality, and (3) to detect problematic survey items or features of survey design (Couper, 2000; Heerwegh, 2003).

Most surveys in the federal statistical system have moved to computer-assisted data collection and are likely collecting some form of paradata, regardless of whether or not it is being used for evaluative purposes. And while it can be collected during a pretest or large-scale field test, two clear advantages of paradata are that it is relatively cheap to capture, and it allows a main survey administration to be treated as a quasi-field test. In the next section, we provide examples of paradata analysis for evaluating the usability of computerized survey instruments and the performance of interviewers and respondents. Whether the focus is field tests prior to the main data collection, field tests as part of the main data collection, or simply the monitoring of an ongoing production instrument, paradata may provide an additional, relatively inexpensive tool for identifying problems, allowing for more cost-effective targeting of questionnaire design/redesign efforts.

19.7.1   Paradata and Questionnaire Evaluation: Examples from the Literature

19.7.1.1   *Survey Instrument Navigation and Function Key Use*   Some of the earliest examples of the use of paradata for usability evaluations of survey instruments and for identifying potentially problematic survey questions focused on the transition of the National Health Interview Survey (NHIS) from paper-and-pencil interviewing (PAPI) to CAPI in 1997 (Caspar and Couper, 1997; Couper et al., 1997a,b; Lepkowski et al., 1998). Trace file data (i.e., paradata) were used to explore interviewer navigation of the new CAPI instrument and the use of specific function keys (e.g., accessing item-specific help and recording item-level notes). Drawing on navigational data, Couper and Schlegel (1998) identified a subset of questions that were subject to a disproportionately high number of backups. Just over 40% of the visits to a question on health insurance coverage were the result of a back-up from a subsequent question. It was later revealed that the data entry format for the item was inconsistent with that used with other mark-all-that-apply items, generating considerable confusion for interviewers. The trace file data also revealed that help screens were rarely accessed in the NHIS CAPI instrument (Couper et al., 1997b; Couper and Schlegel, 1998). Nonetheless, five items, two of which focused on leisure-time physical activity, accounted for roughly two-thirds of the total help screen access, suggesting possible problems with respondent comprehension or interviewer difficulties with screen layout and design.

Lepkowski et al. (1998) provide a nice example of using paradata in a multiple method evaluation. Again, the focus was the 1997 NHIS CAPI instrument, and the evaluation utilized behavior coding, a laboratory-based usability evaluation, and the analysis of trace files collected with field interviews. The goals of the usability testing and trace file analysis were both overlapping with and distinct from the behavior coding. For example, the trace file analysis focused on the interaction of the interviewer with the CAPI instrument, especially the use of function keys, and provided an indication of the extent to which the interviewer–computer interaction was impeded by the design of the instrument. Behavior coding was performed on 154 interviews covering 542 unique questions or screens. Trace file analysis was performed on over 16,000 field interviews and covered 418 unique screens, while the usability evaluation was conducted on 38 laboratory interviews and covered 475 unique screens. Across the three methods, a total of 86 questions or screens were identified as problematic for either the interviewer–respondent or interviewer–computer interactions. Interestingly, only four of the 86 screens were identified by all three methods. An additional 19 screens were identified by two of three methods, while the remaining 63 screens were identified by only one method.

Behavior coding had the greatest overlap with the other two methods, but the trace file analysis and the usability evaluation tended to identify unique screen problems. The authors went on to note that the trace files provided the cheapest form of data collection, appeared to be a useful supplement to the others methods, and brought ". . . greater insight into the problems of the interviewer-computer interaction" (Lepkowski et al., 1998, p. 922).

Edwards et al. (2007) describe the use of paradata analysis, along with other methods, for evaluating two touch-screen ACASI questionnaires. Laboratory-based usability tests and rounds of behavior coding were performed, with the results used to revise screens prior to fielding of the final survey instrument. During the main data collection, paradata were collected on a random subset of roughly 3000 participants. Paradata-based variables included completion times, the number of times the audio for a question was repeated, and the rate at which online help was used. Of the more telling findings was the extensive use of help screens (used by 40.1% of participants) in the questionnaire covering health, lifestyle, and physical activity. This was considerably higher than use in the diet history questionnaire (7.8% of participants), prompting the investigators to revisit the health, lifestyle, and physical activity content.

Couper et al. (2009) demonstrate the use of paradata as an aid in evaluating the relative efficacy of an ACASI versus text-CASI component of the National Survey of Family Growth. Pretest evaluations of audio- versus text-CASI relied on substantive responses to questions, as well as observational data completed by interviewers, audit trails generated by the Blaise software, and respondent debriefings. Paradata analysis focused on the use of function keys (accessing help, replaying audio files, blanking the screen for privacy) and response times across the two modes. Consistent with findings from other evaluation methods, no significant differences were identified by mode.

Focusing on the ACASI component of the National Survey on Drug Use and Health (NSDUH), Penne et al. (2002) used navigational data (backups) captured in Blaise audit trails to explore the impacts of changed responses on estimates of drug use. All instances were identified where a respondent initially answered "yes" to ever using a drug or substance, advanced to one or more follow-up questions, and eventually returned to the initial gate question and changed the response to "no," "don't know," or "refused." Taking a "worst-case scenario" approach, all changes from "yes" to "no," "don't know," or "refused" were reclassified as "yes." For substances such as cigarettes and alcohol, slight increases in the number of "yes" responses had little to no effect. But for more rare substances (e.g., heroin and cocaine), small increases in "yes" responses had substantial impacts.

19.7.1.2 Item Times and Response Latencies Growing attention is given to item-level time stamps captured in audit trails and trace files for exploring response times or latencies in a survey setting, which can indicate the amount of information processing necessary to answer a question (Bassili and Fletcher,

1991; Heerwegh, 2003; Draisma and Dijkstra, 2004; Yan and Tourangeau, 2008). Bassili (1996) reported on studies exploring the associations between latencies and responses to various types of questions, including factual and attitudinal items. Of particular importance were the findings associated with a small set of value questions on merit and equality. While relatively brief in terms of words or characters, the questions elicited comparatively long response times. It was noted that both questions were poorly worded. One item, for example, contained a double negative and was vague in conveying the meaning of key concepts or phrases. Additional research by Bassili and Scott (1996) confirmed these findings, noting that double-barreled questions and questions with superfluous phrases took longer to answer.

Draisma and Dijkstra (2004) identified a link between response latency and response error in a survey of members of an environmental organization. Individual true scores were determined for several questions by linking respondent survey responses with membership records. Target questions included length of membership and amount of fees paid, along with questions that were deemed true for the entire sample (e.g., fake questions on nonexistent campaigns). Analysis revealed that response latencies were shortest for correct answers, followed by incorrect answers, and then nonsubstantive answers (e.g., "don't know"). Longer response times were associated with increased probabilities of response errors. The authors concluded by supporting Bassili's (1996) contention that response times or latencies may provide a powerful tool for screening questions during questionnaire development.

Item and screen level timers, along with rates of help screen access, were used in a field test of the 2004 National Study of Postsecondary Faculty (Heuer et al., 2007). Each question in the test instrument was entered into a spreadsheet, along with accompanying information on average administration time, rate of missing data, and the number of help text hits. The information was first used to guide interviewer debriefings, focusing on why certain items took longer to administer than expected, had high rates of help screen access, and/ or had high rates of item missing data. The debriefing data were then added to the spreadsheet. Using the combined information, questions were eliminated or rewritten, or definitions were added onscreen to clarify items for the main survey administration. The authors noted that the relatively inexpensive collection of paradata provided useful supplementary information for the evaluation of survey items.

More recently, Yan and Tourangeau (2008) explored the effects of both respondent and item characteristics on response times using data collected from four web surveys. In multivariate analyses, they found that age and education had significant effects on response times; respondents who completed high school answered more quickly than those who did not, and younger respondents answered more quickly than older respondents. At the item level, they found that the more clauses in the question and the more words per clause, the longer the response time. However, the authors noted that item characteristics such as fully labeled scales and the number of answer categories, both of which increased response times in their work, have been found to improve item reliability (Saris, 2005). Hence, increases in response times do not necessarily signal increased processing difficulties or reductions in data quality.

19.7.1.3 *Visual Layout and Question Format in Web Surveys* An increasing number of web-based survey studies have used paradata to explore the effects of visual design or question format on response times, the frequency and nature of changed responses, and respondent use of help or clarification features (Couper et al., 2004; Conrad et al., 2006; Stern, 2008). These studies consistently rely on experimental designs with key outcome measures constructed from paradata.[5] Heuer et al. (2007), for example, describe a split-ballot experiment implemented as part of a field test for the 2004/2006 Beginning Postsecondary Students Longitudinal Survey (BPS). Cases were randomly assigned to one of two coding systems for determining major field of study: text string entry to describe one's major field followed by a pair of drop-down boxes (with general and specific categories for coding the major), or text string entry to describe one's major field of study followed by a set of categories returned by a keyword search of the database (computer-assisted coding). Using item-level time information, manual coding with drop-down boxes took 0.9 minutes on average to complete compared to 0.4 minutes for the computer-assisted coding scheme. Separate analysis comparing the two experimental coding results to an expert coder's coding of the text string revealed that the computer-assisted coding approach produced more reliable results. Taken together, the results led to the incorporation of the computer-assisted coding scheme in the full-scale BPS study.

Stern (2008) experimentally varied formats for questions with long lists of response options in a web survey of Washington State University students. A question asking for the student's major was presented in four ways: an open-ended field for typing in the major, a drop-down menu, an alphabetical list of majors with radio buttons, and a grouped major list (by field) with radio buttons. Time data revealed that it took the least amount of time to complete the open-ended version, followed by identifying a major from an alphabetical list (+4 seconds on average), identifying a major grouped by field (+6 seconds on average), and the use of a drop-down menu (+10 seconds on average). Tourangeau et al. (2004) also demonstrated that respondents answered more quickly when response options followed a logical order from top to bottom than when options were not presented in a logical order.

19.7.1.4 *Computer-Assisted Recorded Interviewing (CARI)*[6] An additional form of paradata taking hold in computer-assisted interviewing (in-person or telephone) is CARI.[7] CARI can be used to digitally record the verbal exchanges between interviewers and respondents during field tests and production interviews. Recordings can begin or terminate at any predeter-

mined or randomly selected points in the interview (Biemer et al., 2000). A primary use of CARI has been to supplement traditional reinterview methods with a focus on interview falsification. However, CARI can also be used, among others, to evaluate interviewer adherence to interview procedures and guidelines, assess the extent to which specific question and screen designs hinder or facilitate the collection of accurate data, collect verbatim responses to open-ended questions, and aid in the identification of questionnaire problems (Biemer et al., 2000; Arceneaux, 2007; Dulaney and Hicks, 2010). Couper (2005) notes that the ease by which digital recordings can be implemented will likely lead to a revival and extension of techniques such as behavior coding and conversation analysis. Hicks et al. (2009) recently demonstrated a CARI-based system by which digital recordings of interview snippets are combined with the survey responses and limited audit trail data in a user-friendly interface for behavior coding. Coders can review a series of questions in one interview or one question across a series of interviews, and results can be summarized by item or interviewer.

A recent study used CARI digital recordings to assess interviewer coding errors in a CAPI survey (Mitchell et al., 2008). The analysis focused on four items that varied in the type of response (single response and multiple-response questions), number of response options (response lists of varying lengths), and overall question complexity (low, medium, high). Each question was asked in an open-ended format, and interviewers coded responses using lists of pre-coded response options (which were not read to respondents). Coders listened to the digital recordings and determined if coding errors were present. Overall, 85.5% of the questions were coded correctly. However, important differences emerged across question type. For example, questions with multiple response options produced more coding errors when the complexity (measured by the number of words or sentences elicited in responses) of the item increased. The authors noted that more cognitively complex items may be more prone to data entry error because respondents may not answer in a clear and logical fashion, making the interviewers' task more difficult. They also found that interviewers would opt to record more difficult responses as "other" in an attempt to maintain the pace of the interview (Mitchell et al., 2008).

19.7.2   Institutionalizing Paradata as a Tool to Identify Problems with Questions: The NHIS

Staff with the NHIS routinely use paradata, including audit trail data and interviewer observations, to explore various operational and data quality issues such as interviewer performance (Dahlhamer et al., 2010), the implications of participant reluctance for satisficing and measurement error (Dahlhamer et al., 2008), mode effects (Simile et al., 2007), and unit and subunit nonresponse and nonresponse bias (Dahlhamer et al., 2006; Bates et al., 2008; Dahlhamer and Simile, 2009). Question and section timing data are also central to annual quality assessments that inform decisions about the inclusion or exclusion of interview records from final NHIS data sets. In this context, NHIS staff is looking to make greater use of paradata as a means for identifying potentially problematic items and feeding that information into testing efforts. This includes more formal assessments of question functioning using timing data, as well as exploring other potential uses of audit trail data. For example, one approach could be to compute item- and section-level statistics on the recording of interviewer notes. These computations can be done at regular intervals (e.g., quarterly, annually) and provide a crude indication of whether certain questions or sections are the inordinate target of such notes. Past monitoring of similar statistics led to a review of interviewer notes associated with a handful of items in the NHIS health insurance section. Content analysis of the notes revealed several problems with an item on private health insurance premiums, including a reference period that forced difficult and unnecessary premium computations and data entry options inconsistent with the way some respondents reported. The notes for this and other items in the section suggested other potential problems such as routing errors in which persons without private health insurance coverage were being funneled through a set of follow-up questions on private health insurance plans. Taken as a whole, the content analysis prompted an expert review of the health insurance section, followed by a round of cognitive interviewing. This is an example of how paradata-based analysis was and can be useful for identifying questions or sections of the instrument in need of further investigation using more expensive and thorough testing methods (Couper, 2009).

Whether in the context of a field test or the monitoring of an ongoing production instrument, paradata provide an additional, relatively inexpensive tool for identifying problematic questions or design flaws. The data are available in real time and on all cases at varying levels of granularity. For example, paradata can be analyzed at the item or interviewer level and allow for various subgroup comparisons (Couper, 2009). And since paradata can be collected as part of the regular, production administration of a survey instrument, it does not suffer from issues of nonrepresentativeness, which may pose limitations on the results of some field- and laboratory-based evaluations.

As with any method, there are limitations to the use of paradata analysis for questionnaire evaluation. The data captured in trace files or audit trail are nonrectangular and must be summarized and transformed into a usable format for analysis (Mockovak and Powers, 2008; Couper, 2009). In addition, the data are voluminous, presenting storage issues (Bumpstead, 2002). But most importantly, it cannot be overemphasized that the data should be used in a supplementary fashion. Many examples cited in this chapter demonstrate the use of paradata in a supplementary or complementary role (see Lepkowski et al., 1998). As Stern (2008) notes, "the information paradata provide might gain additional value when it is integrated with information from substantive data as well as any cognitive interviews or field tests" (p. 395). The reason for this is straightforward. Paradata may suggest a problem but it cannot tell us *why* something happened (Couper, 2009). The data require interpretation to understand what the interviewer, respondent, or both were doing at any given time, and the reasons behind their actions (Couper et al., 1997b). Thus, an evaluation method such as cognitive interviewing would be necessary to understand why a problem is occurring and suggest ways to correct the problem (see Miller, this volume).

19.8 CONCLUSION

Although field tests may be conducted to provide information on other aspects of survey administration, they also have a great deal to offer in the evaluation of survey questions and questionnaires. Field tests involve more realistic production conditions and personnel and introduce many more variables than do laboratory methods that have typically been utilized prior to the field test. Exactly how the field test is designed and what specific methods are used depends upon the goals of the test and the resources available. Because they are resource-intensive, field tests are typically conducted after other developmental work, such as focus groups, cognitive interviews, and/or expert review have already been conducted. Multiple questionnaire evaluation methods can be and are often utilized as part of a field test, including respondent debriefings, interviewer debriefings, behavior coding, split-panel experiments, and analysis of item nonresponse, distributions, and psychometric properties. In practice, federal agencies utilize field tests for a variety of purposes, ranging from simple "feasibility" tests on the first several dozen cases to make sure the questionnaire instrument appears to be working correctly to large-scale parallel tests intended to provide data users with information on what effect changes in the survey instrument would have on key national estimates. Thus, despite their cost, field tests can be a key tool for evaluating survey questionnaires.

With growing use of technology in data collection, paradata analysis offers an important complement to other laboratory and field-based methods of questionnaire evaluation. Critically important in the evaluation process is the need to pair or match methods that can identify question problems (i.e., diagnostic utility) with those that can identify the source or nature of those problems (i.e., design utility), the latter being essential to corrective redesign. The relative low cost and ease of collecting paradata enable usability assessments and evaluations during field tests and main data collections, freeing up resources for more cost-effective targeting of design/redesign efforts. This is particularly useful for surveys where field tests are a luxury at best. Hence, we suggest that paradata analysis provides a relatively inexpensive means for survey organizations to identify question problems, lending itself as a particularly powerful complement to evaluation methods, such as cognitive interviewing, that have greater design utility.

NOTES

1   Available at http://www.aapor.org.
2   Note that some of these documents provided key background for OMB's recent revision and updating of its Standards and Guidelines for Statistical Surveys or were being developed within the agencies about the same time as the OMB standards.
3   For more information on NAEP see: http://nces.ed.gov/nationsreportcard/.
4   For more information on the NSDUH, see: http://oas.samhsa.gov/nsduhLatest.htm.
5   Using paradata one cannot only observe the outcome effect of an experimental manipulation but also its effect on the process that led to that particular outcome.
6   The term "computer audio-recorded interviewing" is also used in the literature.
7   Unlike much of the paradata discussed to this point, a CARI-based system would cost considerably more to develop and implement.

REFERENCES

Arceneaux TA (2007). Evaluating the computer audio-recorded interviewing (CARI) household wellness (HWS) field test. American Statistical Association 2007 Proceedings of the Section on Survey Research Methods. Washington, DC: American Statistical Association.

Bassili J (1996). The how and why of response latency measurement in telephone surveys. In: Schwarz NA, Sudman S, editors. Answering Questions. San Francisco: Jossey-Bass; pp. 319–346.

Bassili JN, Fletcher JF (1991). Response time measurement in survey research: a method for CATI and a new look at nonattitudes. Public Opinion Quarterly; 55:331–346.

Bassili JN, Scott BS (1996). Response latency and question problems. Public Opinion Quarterly; 60:390–399.

Bates N, Dahlhamer J, Singer E (2008). Privacy concerns, too busy, or just not interested: using doorstep concerns to predict survey nonresponse. Journal of Official Statistics; 24(4):591–612.

Biemer P, Herget D, Morton J, Willis G (2000). The feasibility of monitoring field interview performance using computer audio recorded interviewing (CARI). American Statistical Association 2000 Proceedings of the Section on Survey Research Methods. Washington, DC: American Statistical Association.

Biemer PP, Lyberg LE (2003). Introduction to Survey Quality. New York: Wiley.

Bumpstead R (2002). A practical application of audit trails. Office for National Statistics Survey Methodology Bulletin; 50:27–36.

Caspar RA, Couper MP (1997). Using keystroke files to assess respondent difficulties with an audio-CASI instrument. American Statistical Association 1997 Proceedings of the Section on Survey Research Methods. Washington, DC: American Statistical Association.

Centers for Medicare and Medicaid Services (CMS) (2003). CAHPS three-state pilot study analysis results. Baltimore, MD. Available online at http://www.cms.hhs.gov/HospitalQualityInits/downloads/Hospital3State_Pilot_Analysis_Final200512.pdf.

Conrad FG, Couper MP, Tourangeau R, Peytchev A (2006). Use and non-use of clarification features in web surveys. Journal of Official Statistics; 22:245–269.

Couper MP (2000). Usability evaluation of computer-assisted survey instruments. Social Science Computer Review; 18:384–396.

Couper MP (2005). Technology trends in survey data collection. Social Science Computer Review; 23:486–501.

Couper MP (2009). The role of paradata in measuring and reducing measurement error in surveys. Paper presented at the NCRM Network for Methodological Innovation 2009: The Use of Paradata in UK Social Surveys, August 24, London.

Couper MP, Schlegel J (1998). Evaluating the NHIS CAPI instrument using trace files. American Statistical Association 1998 Proceedings of the Section on Survey Research Methods. Washington, DC: American Statistical Association.

Couper MP, Hansen S-E, Sadosky SA (1997a). Evaluating interviewer use of CAPI technology. In: Lyberg LE, Biemer P, Collins M, de Leeuw ED, Dippos C, Schwarz N, Trewin D, editors. Survey Measurement and Process Quality. New York: Wiley; pp. 267–285.

Couper MP, Horm J, Schlegel J (1997b). Using trace files to evaluate the national health interview survey CAPI instrument. American Statistical Association 1997 Proceedings of the Section on Survey Research Methods. Washington, DC: American Statistical Association.

Couper MP, Tourangeau R, Conrad FG (2004). What they see is what we get: response options for web surveys. Social Science Computer Review; 22(1):111–127.

Couper MP, Tourangeau R, Marvin T (2009). Taking the audio out of audio-CASI. Public Opinion Quarterly; 73:281–303.

Czaja R (1998). Questionnaire pretesting comes of age. Marketing Bulletin; 9:52–66.

Dahlhamer JM, Simile CM (2009). Subunit nonresponse in the National Health Interview Survey (NHIS): an exploration using paradata. American Statistical Association 2009 Proceedings of the Section on Survey Research Methods. Washington, DC: American Statistical Association.

Dahlhamer JM, Stussman BJ, Simile CM, Taylor B (2006). Modeling survey contact in the National Health Interview Survey (NHIS). Proceedings of the 22nd Statistics Canada International Methodology Symposium.

Dahlhamer JM, Simile CM, Taylor B (2008). Do you really mean what you say? Doorstep concerns and data quality in the National Health Interview Survey (NHIS). American Statistical Association 2008 Proceedings of the Section on Survey Research Methods. Washington, DC: American Statistical Association; pp. 1484–1491.

Dahlhamer JM, Cynamon ML, Gentleman JF, Piani AL, Weiler M (2010). Minimizing survey error through interviewer training: new procedures applied to the national health interview survey. American Statistical Association 2010 Proceedings of the Section on Survey Research Methods. Washington, DC: American Statistical Association.

DeMaio TJ, Rothgeb JM (1996). Cognitive interviewing techniques: in the lab and in the field. In: Schwarz NA, Sudman S, editors. Answering Questions. San Francisco: Jossey-Bass.

Dillman DA (1996). Why innovation is difficult in government surveys. Journal of Official Statistics; 12:113–124.

Dillman DA, Smyth JD, Christian LM (2009). Internet, Mail, and Mixed-Mode Surveys: The Tailored Design Method. New York: Wiley.

Draisma S, Dijkstra W (2004). Response latency and (para) linguistic expressions as indicators of response error. In: Presser S, Rothgeb JM, Couper MP, Lessler JT, Martin E, Martin J, Singer E, editors. Methods for Testing and Evaluating Survey Questionnaires. New York: Wiley; pp. 131–148.

Dulaney R, Hicks W (2010). Computer assisted recorded interviewing (CARI): experience implementing the new Blaise capability. International Blaise Users Conference 2010 Conference Proceedings; pp. 234–246.

Edwards SL, Slattery ML, Murtaugh MA, Edwards RL, Bryner J, Pearson M, Rogers A, Edwards AM, Tom-Orme L (2007). Development and use of touch-screen audio computer-assisted self-interviewing in a study of American Indians. American Journal of Epidemiology; 165(11):1336–1342.

Esposito JL (2004). Iterative, multiple-method questionnaire evaluation research: a case study. Journal of Official Statistics; 20:143–183.

Fowler FJ (2002). Survey Research Methods. Newbury Park, CA: Sage Publications.

Fowler FJ, Roman AM (1992). A study of approaches to survey question evaluation. Unpublished manuscript, University of Massachusetts, Boston.

Groves RM, Fowler FJ, Couper MP, Lepkowski JM, Singer E, Tourangeau R (2009). Survey Methodology, 2nd ed. Hoboken, NJ: John Wiley & Sons.

Heerwegh D (2003). Explaining response latencies and changing answers using client side paradata from a web survey. Social Science Computer Review; 21:360–373.

Heuer R, Doherty J, Zwieg E (2007). Interview timing data: simple yet powerful survey instrument development tools. Paper presented at the 62nd Annual Conference of the American Association for Public Opinion Research, Anaheim, CA, May 17–20.

Hicks W, Edwards B, Tourangeau K, Branden L, Kistler D, McBride B, Harris-Kojetin L, Moss A (2009). A system approach for using CARI in pretesting, evaluation and training. Presented at the 2009 FedCASIC Workshop, Washington, DC, March.

Lepkowski JM, Couper MP, Hansen SE, Landers W, McGonagle KA, Schlegel J (1998). CAPI instrument evaluation: behavior coding, trace files, and usability methods. American Statistical Association 1998 Proceedings of the Section on Survey Research Methods. Washington, DC: American Statistical Association.

Martin E (2004). Vignettes and respondent debriefing for questionnaire design and evaluation. In: Presser S, Rothgeb JM, Couper MP, Lessler JT, Martin E, Martin J, Singer E, editors. Methods for Testing and Evaluating Survey Questionnaires. New York: Wiley.

Mitchell S, Strobl M, Fahrney K, Nguyen M, Bibb B, Thissen MR, Stephenson W (2008). Using computer audio-recorded interviewing to assess interviewer coding error. American Statistical Association 2008 Proceedings of the Section on Survey Research Methods. Washington, DC: American Statistical Association.

Mockovak WP, Powers R (2008). The use of paradata for evaluating interviewer training and performance. American Statistical Association 2008 Proceedings of the Section on Survey Research Methods. Washington, DC: American Statistical Association.

Moore J, Pascale J, Doyle P, Chan A, Griffiths JK (2004). Using field experiments to improve instrument design: the SIPP methods panel project. In: Presser S, Rothgeb JM, Couper MP, Lessler JT, Martin E, Martin J, Singer E, editors. Methods for Testing and Evaluating Survey Questionnaires. New York: Wiley.

Oksenberg L, Cannell C, Kalton G (1991). New strategies for pretesting survey questions. Journal of Official Statistics; 7(3):349–365.

Penne MA, Snodgrass J, Barker P (2002). Analyzing audit trails in the National Survey on Drug Use and Health (NSDUH): means for maintaining and improving data quality. Paper presented at the International Conference on Questionnaire Development, Evaluation, and Testing Methods (QDET), Charleston, South Carolina.

Presser S, Blair J (1994). Survey pretesting: do different methods produce different results? Marsden PV, editor. Sociological Methodology. 24:Washington, DC: American Sociological Association; pp. 73–104.

Presser S, Rothgeb JM, Couper MP, Lessler JT, Martin E, Martin J, Singer E, editors (2004). Methods for Testing and Evaluating Survey Questionnaires. New York: Wiley.

Saris W (2005). The structural equation modeling approach: the effects of survey characteristics on random and systematic errors in surveys. Paper presented at the Total Survey Error Workshop, Washington, DC.

Simile CM, Stussman B, Dahlhamer JM (2007). Exploring the impact of mode on key health estimates in the national health interview survey. Proceedings of the 23rd Statistics Canada International Methodology Symposium.

Stern MJ (2008). The use of client-side paradata in analyzing the effects of visual layout on changing responses in web surveys. Field Methods; 20:377–398.

Tourangeau R (2004). Experimental design considerations for testing and evaluating questionnaires. In: Presser S, Rothgeb JM, Couper MP, Lessler JT, Martin E, Martin J, Singer E, editors. Methods for Testing and Evaluating Survey Questionnaires. New York: Wiley.

Tourangeau R, Couper MP, Conrad FG (2004). Spacing, position, and order: interpretive heuristics for visual features of survey questions. Public Opinion Quarterly; 68: 368–393.

Tucker C, Bloxham J, Bowie C, Esposito J, Harris-Kojetin B, Kostanich D, Miller S, Polivka A, Robison E, Stump M (1998). Improving field tests. Unpublished report, U.S. Bureau of Labor Statistics, U.S. Bureau of the Census.

U.S. Census Bureau (2009). Design and Methodology: American Community Survey. Washington, DC: U.S. Government Printing Office. Available online at http://www.census.gov/acs/www/methodology/methodology_main/.

U.S. Census Bureau (2010). U.S. Census Bureau statistical quality standards. Washington, DC. Available online at http://www.census.gov/quality/standards/index.html.

U.S. Department of Education, National Center for Education Statistics (2002). Statistical standards. Washington, DC. Available online at http://nces.ed.gov/statprog/2002/stdtoc.asp.

U.S. Office of Management and Budget (2006a). Questions and answers when designing surveys for information collections. Washington, DC. Available online at http://www.whitehouse.gov/sites/default/files/omb/assets/omb/inforeg/pmc_survey_guidance_2006.pdf.

U. S. Office of Management and Budget (2006b). OMB Standards and guidelines for statistical surveys. Washington, DC. Available online at http://www.whitehouse.gov/sites/default/files/omb/assets/omb/inforeg/statpolicy/standards_stat_surveys.pdf.

Willis GB (2005). Cognitive Interviewing. Thousand Oaks, CA: Sage.

Yan T, Tourangeau R (2008). Fast times and easy questions: the effects of age, experience and question complexity on web survey response times. Applied Cognitive Psychology; 22:51–68.

INDEX

Question Evaluation Methods: Contributing to the Science of Data Quality, First Edition.
Edited by Jennifer Madans, Kristen Miller, Aaron Maitland, Gordon Willis.
© 2011 John Wiley & Sons, Inc. Published 2011 by John Wiley & Sons, Inc.

WILEY SERIES IN SURVEY METHODOLOGY
Established in Part by WALTER A. SHEWHART AND SAMUEL S. WILKS

Editors: *Mick P. Couper, Graham Kalton, J. N. K. Rao, Norbert Schwarz, Christopher Skinner*
Editor Emeritus: *Robert M. Groves*

The **Wiley Series in Survey Methodology** covers topics of current research and practical interests in survey methodology and sampling. While the emphasis is on application, theoretical discussion is encouraged when it supports a broader understanding of the subject matter.

The authors are leading academics and researchers in survey methodology and sampling. The readership includes professionals in, and students of, the fields of applied statistics, biostatistics, public policy, and government and corporate enterprises.

ALWIN · Margins of Error: A Study of Reliability in Survey Measurement

BETHLEHEM · Applied Survey Methods: A Statistical Perspective

BETHLEHEM, COBBEN, and SCHOUTEN · Handbook of Nonresponse in Household Surveys

BIEMER · Latent Class Analysis of Survey Error

*BIEMER, GROVES, LYBERG, MATHIOWETZ, and SUDMAN · Measurement Errors in Surveys

BIEMER and LYBERG · Introduction to Survey Quality

BIEMER · Latent Class Analysis of Survey Error

BRADBURN, SUDMAN, and WANSINK ·Asking Questions: The Definitive Guide to Questionnaire Design—For Market Research, Political Polls, and Social Health Questionnaires, *Revised Edition*

BRAVERMAN and SLATER · Advances in Survey Research: New Directions for Evaluation, No. 70

CHAMBERS and SKINNER (editors) · Analysis of Survey Data

COCHRAN · Sampling Techniques, *Third Edition*

CONRAD and SCHOBER · Envisioning the Survey Interview of the Future

COUPER, BAKER, BETHLEHEM, CLARK, MARTIN, NICHOLLS, and O'REILLY (editors) · Computer Assisted Survey Information Collection

COX, BINDER, CHINNAPPA, CHRISTIANSON, COLLEDGE, and KOTT (editors) · Business Survey Methods

*DEMING · Sample Design in Business Research

DILLMAN · Mail and Internet Surveys: The Tailored Design Method

FULLER · Sampling Statistics

GROVES and COUPER · Nonresponse in Household Interview Surveys

GROVES · Survey Errors and Survey Costs

GROVES, DILLMAN, ELTINGE, and LITTLE · Survey Nonresponse

GROVES, BIEMER, LYBERG, MASSEY, NICHOLLS, and WAKSBERG · Telephone Survey Methodology

GROVES, FOWLER, COUPER, LEPKOWSKI, SINGER, and TOURANGEAU · Survey Methodology, *Second Edition*

*HANSEN, HURWITZ, and MADOW · Sample Survey Methods and Theory, Volume 1: Methods and Applications

*HANSEN, HURWITZ, and MADOW · Sample Survey Methods and Theory, Volume II: Theory

HARKNESS, BRAUN, EDWARDS, JOHNSON, LYBERG, MOHLER, PENNELL, and SMITH (editors) · Survey Methods in Multinational, Multiregional, and Multicultural Contexts

*Now available in a lower priced paperback edition in the Wiley Classics Library.

HARKNESS, VAN DE VIJVER, and MOHLER (editors) · Cross-Cultural Survey Methods

KALTON and HEERINGA · Leslie Kish Selected Papers

KISH · Statistical Design for Research

*KISH · Survey Sampling

KORN and GRAUBARD · Analysis of Health Surveys

LEPKOWSKI, TUCKER, BRICK, DE LEEUW, JAPEC, LAVRAKAS, LINK, and SANGSTER (editors) · Advances in Telephone Survey Methodology

LESSLER and KALSBEEK · Nonsampling Error in Surveys

LEVY and LEMESHOW · Sampling of Populations: Methods and Applications, *Fourth Edition*

LUMLEY · Complex Surveys: A Guide to Analysis Using R

LYBERG, BIEMER, COLLINS, de LEEUW, DIPPO, SCHWARZ, TREWIN (editors) · Survey Measurement and Process Quality

MADANS, MILLER, and MAITLAND (editors) · Question Evaluation Methods: Contributing to the Science of Data Quality

MAYNARD, HOUTKOOP-STEENSTRA, SCHAEFFER, and VAN DER ZOUWEN · Standardization and Tacit Knowledge: Interaction and Practice in the Survey Interview

PORTER (editor) · Overcoming Survey Research Problems: New Directions for Institutional Research, No. 121

PRESSER, ROTHGEB, COUPER, LESSLER, MARTIN, MARTIN, and SINGER (editors) · Methods for Testing and Evaluating Survey Questionnaires

RAO · Small Area Estimation

REA and PARKER · Designing and Conducting Survey Research: A Comprehensive Guide, *Third Edition*

SARIS and GALLHOFER · Design, Evaluation, and Analysis of Questionnaires for Survey Research

SÄRNDAL and LUNDSTRÖM · Estimation in Surveys with Nonresponse

SCHWARZ and SUDMAN (editors) · Answering Questions: Methodology for Determining Cognitive and Communicative Processes in Survey Research

SIRKEN, HERRMANN, SCHECHTER, SCHWARZ, TANUR, and TOURANGEAU (editors) · Cognition and Survey Research

SUDMAN, BRADBURN, and SCHWARZ · Thinking about Answers: The Application of Cognitive Processes to Survey Methodology

UMBACH (editor) · Survey Research Emerging Issues: New Directions for Institutional Research No. 127

VALLIANT, DORFMAN, and ROYALL · Finite Population Sampling and Inference: A Prediction Approach